Water Resources Sustainability

ABOUT THE EDITOR-IN-CHIEF

Larry W. Mays, Ph.D., P.E., P.H., is Professor of Civil and Environmental Engineering at Arizona State University, and former chair of the department. He was formerly Director of the Center for Research in Water Resources at the University of Texas at Austin, where he also held an Engineering Foundation–endowed professorship. A registered professional engineer in seven states, and a registered professional hydrologist, he has served as a consultant to many organizations. Professor Mays is author of *Water Resources Engineering* (from John Wiley & Sons) and *Optimal Control of Hydrosystems* (from Marcel Dekker), and co-author of *Applied Hydrology* and *Hydrosystems Engineering and Management* (both from McGraw-Hill) and *Groundwater Hydrology* (from John Wiley & Sons). He was editor-in-chief of *Water Resources Handbook, Water Distribution Systems Handbook, Urban Water Supply Management Tools, Stormwater Collection Systems Design Handbook, Urban Water Supply Handbook, Urban Stormwater Management Tools, Hydraulic Design Handbook,* and *Water Supply Systems Security,* all from McGraw-Hill. In addition, he is editor-in-chief of *Reliability Analysis of Water Distribution Systems* and co-editor of *Computer Methods of Free Surface and Pressurized Flow.* He has over 90 refereed journal publications and chapters in other books and over 65 papers in the proceedings of national and international conferences. Among his honors is a distinguished alumnus award from the University of Illinois at Champaign-Urbana in 1999. Professor Mays lives in Mesa, Arizona, and Pagosa Springs, Colorado.

Water Resources Sustainability

Larry W. Mays, Ph.D., P.E., P.H.
Editor-in-Chief

Department of Civil and Environmental Engineering
Arizona State University
Tempe, Arizona

McGraw-Hill

New York   Chicago   San Francisco   Lisbon   London   Madrid   Mexico City
Milan   New Delhi   San Juan   Seoul   Singapore   Sydney   Toronto

WEF Press

Water Environment Federation   Alexandria, Virginia

The McGraw·Hill Companies

Cataloging-in-Publication Data is on file with the Library of Congress

Copyright © 2007 by The McGraw-Hill Companies, Inc. All rights reserved. Printed in the United States of America. Except as permitted under the United States Copyright Act of 1976, no part of this publication may be reproduced or distributed in any form or by any means, or stored in a data base or retrieval system, without the prior written permission of the publisher.

1 2 3 4 5 6 7 8 9 0 DOC/DOC 0 1 2 1 0 9 8 7 6

ISBN-13: 978-0-07-146230-3
ISBN-10: 0-07-146230-9

The sponsoring editor for this book was Larry S. Hager and the production supervisor was Pamela A. Pelton. It was set in Times by International Typesetting and Composition. The art director for the cover was Brian Boucher.

Printed and bound by RR Donnelley.

WEF and *WEFTEC* are trademarks of the Water Environment Federation.

McGraw-Hill books are available at special quantity discounts to use as premiums and sales promotions, or for use in corporate training programs. For more information, please write to the Director of Special Sales, McGraw-Hill Professional, Two Penn Plaza, New York, NY 10121-2298. Or contact your local bookstore.

This book is printed on acid-free paper.

Dedicated to Humanity and Human Welfare

Water Environment Federation

Improving Water Quality for 75 Years

Founded in 1928, the water Environment Federation (WEF) is a not-for-profit technical and educational organization with members from varied disciplines who work toward the WEF vision of preservation and enhancement of the global water environment. The WEF network includes water quality professionals from 79 Member Associations in over 30 countries.

For information on membership, publications, and conferences, contact

Water Environment Federation
601 Wythe Street
Alexandria, VA 22314-1994 USA
(703) 684-2400
http://www.wef.org

Contents

9 Water Sustainablity: The Potential Impact of Climate Change   181

10 Water Supply Security: an Introduction   193

11 Water Resources Sustainability Issues in South Korea   207

12 Water-Based Sustainable Integrated Regional Development   235

13 Community Management of Rural Water Systems in Ghana: Postconstruction Support and Water and Sanitation Committees in Brong Ahafo and Volta Regions 267

14 Is Privatization of Water Utilities Sustainable? Lessons from the European Experience 289

Contributors

Bernard Akanbang   *TREND, Kumasi, Ghana* (Chap. 13)

Keyvan Asghari   *Department of Civil Engineering, Isfahan University of Technology, Isfahan, Iran* (Chap. 5)

Alex Bakalian   *World Bank, Washington, DC* (Chap. 13)

A. Mark Bennett III   *Arkansas Natural Resources Commission, Little Rock, Arkansas* (Chap. 5)

John A. Dracup   *Professor of the Graduate School, Department of Civil & Environmental Engineering, University of California, Berkeley, California* (Chap. 9)

Jim Holway   *Associate Director, Global Institute of Sustainability, Professor of Practice, Civil, and Environmental Engineering, Arizona State University* (Chap. 4)

Katharine Jacobs   *Executive Director, Arizona Water Institute* (Chap. 4)

Hwan-Don Jun   *Department of Civil and Environmental Engineering, Korea University, Seoul, S. Korea* (Chap. 11)

Paul J. Killian   *STS Consultants Ltd. Green Bay, Wisconsin* (Chap. 5)

Joong-Hoon Kim   *Department of Civil and Environmental Engineering, Korea University, Seoul, S. Korea* (Chap. 11)

Sung Kim   *Sustainable Water Resources Research Center (SWRRC), Koyang, S. Korea* (Chap. 11)

Kristin Komives   *Institute of Social Studies, The Hague, Netherlands* (Chap. 13)

Christopher Lant   *Department of Geography and Environmental Resources, Southern Illinois University, Carbondale, Illinois* (Chap. 3)

Eugene Larbi   *TREND, Kumasi, Ghana* (Chap. 13)

Antonio Massarutto   *Dipartimento di scienze economiche, Università di Udine—IEFE, Università Bocconi, Milano, Via Tomadini 30/a, 33100 Udine, Italy* (Chap. 14)

Larry W. Mays   *Professor of Civil and Environmental Engineering, Arizona State University, Tempe, Arizona* (Chaps. 1, 2, 8, 10)

James McPhee   *Department of Civil and Environmental Engineering, University of California, Los Angeles, California, Departamento de Ingeniería Civil, Facultad de Ciencias Físicas y Matemáticas, Universidad de Chile, Santiago, Chile* (Chap. 6)

Ann W. Peralta   *Peralta and Associates, Inc. Hyde Park, Utah* (Chap. 5)

Richard C. Peralta   *Department of Biological and Irrigation Engineering, Utah State University, Logan, Utah* (Chap. 5)

Robert N. Shulstad   *College of Agricultural and Environmental Sciences, University of Georgia, Athens, Georgia* (Chap. 5)

Rich Thorsten   *University of North Carolina, Chapel Hill* (Chap. 13)

Benedict Tuffuor   *TREND, Kumasi, Ghana* (Chap. 13)

Yeou-Koung Tung   *Department of Civil Engineering, Hong Kong University of Science & Technology, Hong Kong, China* (Chap. 7)

Olcay Unver   *Kent State University, Ohio, Formerly President, GAP Regional Development Administration, Ankara, Turkey* (Chap. 12)

Wendy Wakeman   *World Bank, Washington, DC* (Chap. 13)

Dale Whittington   *University of North Carolina, Chapel Hill* (Chap. 13)

William W-G. Yeh   *Department of Civil and Environmental Engineering, University of California, Los Angeles, California* (Chap. 6)

Preface

Why develop a book on water sustainability? To start with, water has always been and continues to be one of the major challenges of civilization and could be the biggest problem of the 21st century. When you grew up I'm sure you loved water as all children do, but yet approximately one in five children on earth do not have access to safe water. The present day problem with lack of adequate water resources, particularly in the developing nations of the world, coupled with challenges such as global climate change make water sustainability a paramount problem society has to face.

The concept of sustainable resources has been around for a long time, as water resources engineers and managers have been taught the principles of sustained yield management, long before publications such as the *Limits to Growth* in 1974 and *Our Common Future* by Brundtland Commission in 1987. In June 1939, Dr. W. C. Lowdermilk presented in Jerusalem for the first time what has been called an "Eleventh Commandment," which is a definition for water resources sustainability. This commandment is given in Chapter 1 along with other detailed definitions of water resources sustainability.

Sustainability is both a vague and politicized term, which many in the world community have rallied around. Even within this book the term sustainability takes on different meanings. In Chapter 1 I define water resources sustainability as follows:

> *Water resources sustainability is the ability to use water in sufficient quantities and quality from the local to the global scale to meet the needs of humans and ecosystems for the present and the future to sustain life, and to protect humans from the damages brought about by natural and human-caused disasters that affect sustaining life.*

Many of the reports and presentations on water resources sustainability by the state water agencies and other water authorities in the U.S. basically talk about the great job they are doing in the name of water sustainability. However you might ask yourself, how sustainable is it to live in a world where approximately 1.1 billion people lack safe drinking water, approximately 2.6 billion people lack adequate sanitation, and between 2 million and 5 million people die annually from water-related diseases? It is this worldwide focus that we have attempted to address in this book. Because freshwater is one of the most critical natural resource issues facing humanity, water resources sustainability has become a very important topic in the recent past. The world's population is expanding rapidly, yet we have no more freshwater on earth than there was 2,000 years ago, when the population was less than 3% of the present.

Jared Diamond, in his book *Collapse: How Societies Choose to Fall or Succeed*, proposed a five-point framework for the collapse of societies, of which two factors are (1) the damage that people inadvertently inflict on their environment and (2) climate change. Both of these factors are very important to water resources sustainability. There is no doubt that many ancient societies did collapse at least partially as a result of water sustainability issues from the misuse and depletion of natural resources and/or climate change. This depletion of natural resources then led to other events that caused the eventual complete collapse.

In Chapter 2 examples of ancient civilizations that collapsed in the American southwest and Mesoamerica are used to discuss the relevance of the collapse of ancient civilizations upon modern societies, particularly in the southwestern U.S. where we continue to practice unsustainable water resources. Simply stated, the water resources practices and growth in the southwestern United States, as in many other parts of the world is unsustainable. We certainly are a much more advanced society than those of ancient societies, but will we be able to overcome and survive? The ancient ones have warned us. To quote Professor Fekri Hassan of the University of London, "The secret of the Egyptian civilization was that it never lost sight of the past."

Chapter 3 presents an ecological economics perspective to water resources sustainability. Chapter 4 discusses sustainability in Arizona linking climate, water management, and growth. Chapters 5 and 6 look at water resources sustainability through conjunctive use of groundwater and surface water. Uncertainties and risks in water resources projects are discussed in Chapter 7, followed by Chapters 8 and 9 on the impacts of global climate change on water resources sustainability. Chapter 10 is based upon water supply security issues resulting from 9/11 issues. Chapters 11 through 14 discuss water sustainability issues around the world including South Korea, Turkey, Ghana, and Europe. Issues include water resources planning and management, integrated regional development, community management of rural water systems in developing countries, and privatization of water utilities.

This year I finished my 30th year of teaching, first thirteen years at the University of Texas at Austin, then followed by seventeen years at Arizona State University. During that time I have felt that I have been teaching the concepts related to water resources sustainability, even though it has not been so labeled. Most of my research has been related to water resources systems analysis, developing methodologies that interface optimization and simulation for the planning, design, and management of water resources projects. To further my efforts in this area I decided to develop this book with the help of the authors, in response to the critical needs of not only water resources engineers, planners, and managers; but also politicians, lawyers, developers, and many others for an appreciation of the issues related to water resources sustainability, not only on a local or regional or national basis, but also on an international basis.

This book has been a part of my personal journey in life to learn as much as possible about water resources and to use this knowledge in my teaching, research, writing, and consulting. I sincerely hope that you will be able to use this book in your own personal journey of learning about water. What I have gained from my experiences in developing books cannot be expressed in words. I want to thank all the authors who contributed to this effort, as they are the ones that made *Water Resources Sustainability* possible. A book is a companion along the pathway of learning. Have a great journey!

Larry W. Mays

Mesa, Arizona, and Pagosa Springs, Colorado

Water Resources Sustainability

1 INTRODUCTION: WATER SUSTAINABILITY

Larry W. Mays
Professor of Civil and Environmental Engineering
Arizona State University
Tempe, Arizona

INTRODUCTION

History has been a continual ebb and flow of civilizations. Ancient civilizations faced many severe problems associated with water as a means for their existence. I attempt to examine similarities of water resources sustainability development issues between our present and past civilizations. In the title I use sustainability; however, *unsustainability* may be more appropriate because it better reflects what I am attempting to accomplish. The Brundtland Commission (Brundtland, 1987) presented the following definition: "Sustainable development is development that meets the needs of the present without compromising the ability of future generations to meet their own needs." Sustainable development has also been referred to as the process in which the economy, environment, and ecosystem of a region develop in harmony and in a way that will improve over time.

As pointed out in Chap. 3 "sustainability is both a vague and politicized term, yet it is precisely because the world community has rallied around sustainability and sustainable development as normative goals of ecological-economic performance that the stakes are high for defining the concept in a manner that is true to its spirit." How sustainable is it to live in a world where approximately 1.1 billion people lack safe drinking water, approximately 2.6 billion people lack adequate sanitation, and between 2 and 5 million people die annually from water-related diseases (Gleick, 2005)?

The *United Nations Children's Fund's* (UNICEF) report, "The State of the World's Children 2005: Children under Threat," provides an analysis of seven basic "deprivations" that children feel and that powerfully influence their futures. UNICEF concludes that more than half the children in the developing world are severely deprived of one or more of the necessities essential to childhood:

- 640 million children do not have adequate shelter
- 500 million children have no access to sanitation
- 400 million children do not have access to safe water
- 300 million children lack access to information
- 270 million children have no access to health-care services
- 140 million children have never been to school
- 90 million children are severely food deprived

There are approximately 2.2 billion children (with 1.9 billion living in the developing world and about 1 billion living in poverty) in the world, so approximately one in five children in the world do not have access to safe water.

Certainly, fresh water has emerged as one of the most critical natural resource issues facing humanity. The world's population is expanding rapidly, yet we have no more fresh water on earth than there was 2000 years ago, when the population was less than 3 percent of the current 6 billion. This is one of the most important topics facing society today and will continue to be so well into the future.

While we all may have heard of the crisis and tragedies occurring in Darfur from the news media, with world headlines in early 2003, we may not have been made aware of the extent to which the problems are related to the water resources. The crisis in Darfur can be traced back to the early 1970s, when desertification in the northern part of Sudan and drought in Darfur led to increased competition over water resources and fertile land. The quest for food and water led pastoral groups from semidesert areas in the north to move south, where tension soon arose with farmers. Over the past 30 years there have been increasing disputes over the region's dwindling resources.

Threats to drinking water systems during conflicts have plagued humans since the dawn of history. Water has always been a strategic objective in armed conflicts. Many historical conflicts have caused flooding by diversion or eliminated water supplies by building dams or other structures. In Mesopotamia a great system of canals were dug but had to be cleaned of silt. Stoppage of canals by silt depopulated villages and cities more effectively than invading armies. The Romans had a long history of developing alternate water supplies to Rome for the expanding water supply and security.

The Bible mentions conflicts over the Jordan River since inhabitants moved to the area in ancient times: "And Gideon sent messengers throughout all the hill country of E'phraim, saying, 'Come down against the Mid'ianites and seize the water against them as far as Beth-Bara'ch, and also the Jordan'." More recently, a primary cause of the 1967 Arab-Israeli war was the struggle for fresh water. The struggle for fresh water in the Middle East has also contributed to other military disputes in the region (see Gleick, 1993, 1994, 1998, 2000, and Lilach, 1997 for additional information).

The events of September 11, 2001 have significantly changed our approach to the management and protection of water supplies, with the threat of terrorist activities toward these systems. These present-day threats include cyber threats, physical threats, chemical threats, and biological threats. Chapter 10 includes discussions on water supply security issues.

Natural disasters such as drought and flooding have plagued civilizations throughout history and continue to do so today. We continue to build new infrastructure in urban flood plains and develop new areas without regard to the downstream effects. Houston, Texas is one of many examples, with approximately $5 billion of damage due to flooding during Tropical Storm Allison, in June 2001. A large portion of the flooding damaged infrastructure was built in the past three decades. Then there was Katrina!

WHAT IS WATER RESOURCES SUSTAINABILITY?

The concept of sustainable resources has been around for a long time, as water resources managers had been taught the principles of sustained yield management long before publications such as *Limits to Growth* (Meadows et al., 1974) and *Our Common Future* (Brundtland Commision, 1987). Loucks (2000) defined sustainable water resource systems as "water resource systems designed and managed to fully contribute to the objectives of society, now and in the future, while maintaining their ecological, environmental, and hydrological integrity." Most definitions of sustainable water resources are so broad that they defy any measurement or quantitative definition. Other attempts at defining water resources sustainability have been by ASCE (1998).

When invited to broadcast a talk on soil conservation in Jerusalem in June 1939, Dr. W. C. Lowdermilk (1953) gave for the first time what has been called an "Eleventh Commandment," which could also be a definition for water resources sustainability:

> Thou shalt inherit the Holy Earth as a faithful steward, conserving its resources and productivity from generation to generation. Thou shalt safeguard thy fields from soil erosion, thy living waters from drying up, thy forests from desolation, and protect thy hills from overgrazing by thy herds, that thy descendants may have abundance forever. If any shall fail in this stewardship of the land thy fruitful fields shall become sterile stony ground and wasting gullies, and thy descendants shall decrease and live in poverty or perish from off the face of the earth.

The definition of water resources sustainability used in this text is consistent with that of the Brundtland Commission report (Brundtland, 1987) and that of Dan Rothman:

> Water resources sustainability is the ability to provide and manage water quantity and quality so as to meet the present needs of humans and environmental ecosystems, while not impairing the needs of future generations to do the same.

Another definition of sustainability is:

> Water resources sustainability is the ability to use water in sufficient quantities and quality from the local to the global scale to meet the needs of humans and ecosystems for the present and the future to sustain life, and to protect humans from the damages brought about by natural and human-caused disasters that affect sustaining life.

Because water impacts so many aspects of our existence, whichever definition is used, there are many facets that must be considered. These are summarized as:

- Water resources sustainability includes the *availability of freshwater supplies* throughout periods of climatic change, extended droughts, population growth, and to leave the needed supplies for the future generations.
- Water resources sustainability includes having the *infrastructure* to provide water supply for human consumption and food security, and to provide protection from water excess such as floods and other natural disasters.
- Water resources sustainability includes having the *infrastructure* for clean water and for treating water after it has been used by humans before being returned to water bodies.
- Water sustainability must have adequate *institutions* to provide for both the water supply management and water excess management.
- Water sustainability can be defined on a local, regional, national, and international basis.

Sustainable water use has been defined by Gleick et al. (1995) as "the use of water that supports the ability of human society to endure and flourish into the indefinite future without undermining the integrity of the hydrological cycle or the ecological systems that depend on it." The following seven sustainability requirements were presented:

- A basic water requirement will be guaranteed to all humans to maintain human health.
- A basic water requirement will be guaranteed to restore and maintain the health of ecosystems.
- Water quality will be maintained to meet certain minimum standards. These standards will vary depending on location and how the water is to be used.
- Human actions will not impair the long-term renewability of freshwater stocks and flows.
- Data on water resources availability, use, and quality will be collected and made accessible to all parties.
- Institutional mechanisms will be set up to prevent and resolve conflicts over water.
- Water planning and decision making will be democratic, ensuring representation of all affected parties and fostering direct participation of affected interests.

FRAMEWORKS FOR CIVILIZATIONS TO COLLAPSE

Tainter (1990) outlines how civilizations historically tend to collapse precisely at the moment levels of increasing complexity are not sustainable. His eleven major themes of societal collapse are (a) depletion or cessation of vital resources; (b) establishment of a new resource base; (c) insurmountable catastrophe; (d) insufficient response to circumstances; (e) other complex societies; (f) invaders; (g) class conflict/societal contradiction/elite mismanagement or misbehavior; (h) social dysfunction; (i) chance concatenation of events; (j) mystical factors; and (k) economic factors.

There are many different theories as to why ancient civilizations failed (Linden, 2006). Gill (2000) found over a hundred different theories for the collapse of the Mayan civilization in researching material for his book, *The Great Maya Droughts: Water, Life, and Death.* Desertification, climate change (Alley, 2000), and deforestation (Perlin,1989) have been somewhat unappreciated factors, in addition to the factors given in Diamond (1997), *Guns, Germs, and Steel,* to the collapse of societies.

Tainter (1990) is skeptical that collapses of ancient civilizations were due to the depletion of environmental (natural) resources. He feels that complex societies were not likely to allow collapse through failure to manage their resources. In Chap. 2, I use the Teotihuacans, the Xochicalcoans, the Mayas, the Chacoans (Anasazi), and the Hohokams as examples of societies that have failed as a result of at least partial depletion and misuse of environmental (natural) resources, and climate change. The Mohenjo Daro in the Indus Valley (Pakistan) declined after 2000 BC, possibly due to climate change, river shifts, and water resources management problems. Droughts possibly caused or attributed to the collapse of the Akkafdran Empire in Mesopotamia around 2170 BC, the collapse of the Moche IV civilization on the Peruvian coast around AD 600, and the collapse of the Tiwanaka civilization in the Andes around AD 1100. In other words *the ancient ones have warned us.*

Diamond (2005) proposed a five-point framework for the collapse of societies:

- Damage that people inadvertently inflict on their environment
- Climate change
- Hostile neighbors
- Decreased support by friendly neighbors
- Society's responses to its problems

I refer to this framework throughout the discussions in Chap. 2. There is no doubt that the ancient societies in Mesoamerica and the southwestern United States discussed in Chap. 2 did collapse at least partially as a result of water sustainability issues from the misuse and depletion of natural resources and climate change. This depletion of natural resources then led to other events that caused the eventual collapse. Diamond (2005) points out that he does not know of any case in which a society's collapse can be attributed entirely to environmental damage as there are always other contributing factors.

MODERN DAY EXAMPLES OF UNSUSTAINABILITY

To talk about sustainability of water we must first talk about unsustainability and look at examples. One example of unsustainable development in our present time has been the region of the Aral Sea, resulting in many detrimental effects to the population of that region. The decision by the former Soviet Union that water from the rivers of the Aral Sea basin should be used for growing cotton instead of sustaining the fourth largest inland sea resulted in an environmental disaster rivaling Chernobyl (Postel, 1997).

Aral Sea

The Aral Sea is located in Central Asia between Uzbekistan and Kazakhstan (both countries were part of the former Soviet Union) as shown in Fig. 1.1. The Amu-Darya and the Syr-Darya (dar'ya means river in Turkic) flow into the Aral Sea with no outlet from the sea. Over more than 30 years, water has been diverted from the Amu-Darya and the Syr-Darya to irrigate millions of acres of land for cotton and rice production, which has resulted in a loss of more than 60 percent of the sea's water. The sea has shrunk from over 65,000 km^2 to less than half that size, exposing large areas of the lake bed. From 1973 to 1987 the Aral Sea dropped from fourth to sixth among the world's largest inland seas. The satellite photos in Fig. 1.2 show the Aral Sea in 1985 and 2003; and Fig. 1.3 illustrates the decrease in size of the Aral Sea from 1957 to 2000.

The lake's salt concentration increased from 10 percent to more than 23 percent, contributing to the devastation of a once thriving fishing industry. The local climate reportedly has shifted, with hotter, drier summers and colder, longer winters. With the decline in sea level, salty soil remained on the exposed lake bed. Dust storms have blown up to 75,000 tons of this exposed soil annually, dispersing its salt particles and pesticide residues. This air pollution has caused widespread nutritional and respiratory ailments, and crop yields have been diminished by the added salinity, even in some of the same fields irrigated with the diverted water. Additional reading material includes Ellis and Turnley (1990), Ferguson (2003), and Perata (1988, 1993).

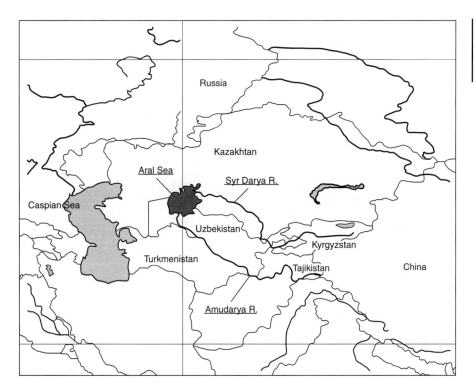

Figure 1.1
The Aral Sea basin. (*Courtesy of McKinney, 1996*)

The major consequences of the continuous desiccation of the Aral Sea since 1960 are summarized as follows:

- Climatic consequences such as mesoclimatic changes, increase of salt and dust storms, shortening of vegetation period
- Ecological/economic consequences including degeneration of the delta ecosystems, total collapse of fishing industry, decrease of productivity of agricultural fields

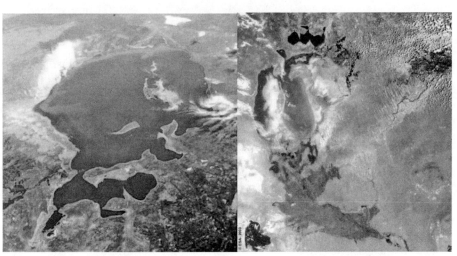

(1985)   (2003)

Figure 1.2
Comparison of Aral Sea 1985 and 2003. (*Courtesy of NASA*)

Figure 1.3

Aral Sea from 1957 to 2000 from the report Environment State of the Aral, developed by the International Fund for the Aral Sea (IFAS) and the UN Environment Programme (UNEP) under financial support of the Norway Trust Fund at the World Bank. Co-ordination from the side of IFAS was held by the Executive Committee of IFAS, and from the side of UNEP-UNEP/ GRID-Arendal. (*Courtesy of http://enrin.grida. no/aral/aralsea/ english/arseal arsea.htm#2*)

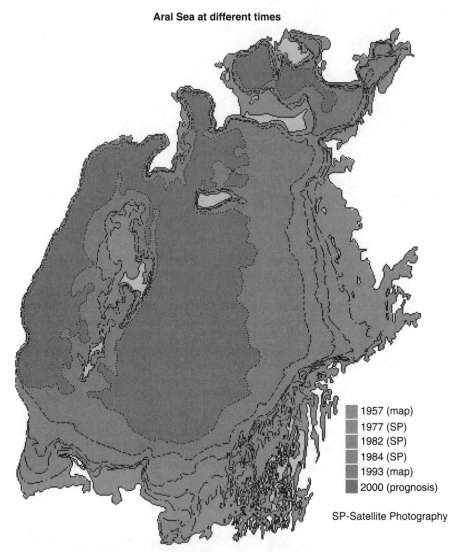

Aral Sea at different times

1957 (map)
1977 (SP)
1982 (SP)
1984 (SP)
1993 (map)
2000 (prognosis)

SP-Satellite Photography

- Health consequences such as increase of serious diseases, birth defects, and high infant mortality

People of the region did not make the decision to use the rivers of the Aral Sea basin but they have certainly suffered the consequences. As stated at the Conference of the Central Asian region ministers, "States of Central Asia: Environment Assessment," Aarhus, Denmark, 1998:.

The Aral crisis is the brightest example of the ecological problem with serious social and economic consequences, directly or indirectly connected with all the states of Central Asia. Critical situation caused by the Aral Sea drying off was the result of agrarian economy tendency on the basis of irrigated agriculture development and volume growth of irrevocable water consumption for irrigation.

Now let us look at present-day Mexico City (see Fig. 1.4), a very large urban center, as another example of water resources unsustainability. Mexico City is the cultural, economic, and industrial center for Mexico. This city is located in the southern part of the Basin of Mexico, which is an extensive, high mountain valley at approximately 2200 m above sea level and surrounded by mountains of volcanic origin with peak altitudes of over 5000 m above sea level.

Mexico City

Figure 1.4

Basin of Mexico and the Mexico City metropolitan area. (*Courtesy of NRC 1995*)

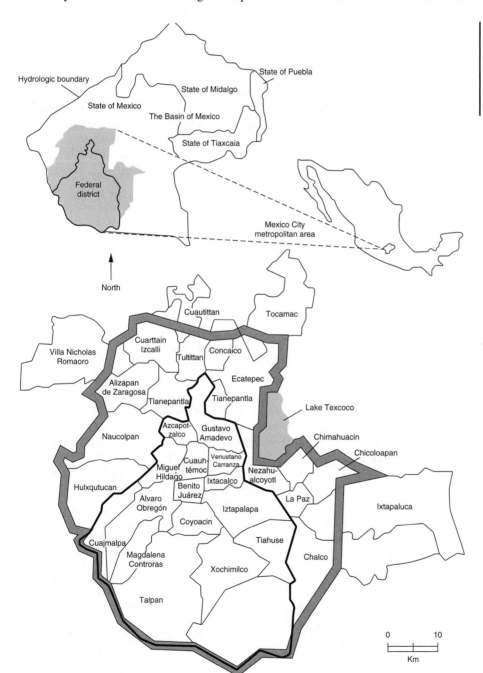

Beginning in the fourteenth century, the Aztecs made use of a system of aqueducts to convey spring water from the higher elevations in the southern portion of the Basin of Mexico to their city, Tenochtitlan. This ancient city was built on land reclaimed from the saline Lake Texcoco. The Spaniards defeated the Aztecs in 1520, after which they rebuilt the aqueducts and continued to use the spring water until the mid 1850s. Potable groundwater, under artesian conditions, was discovered in 1846. Over the next century, the increased groundwater extraction and the artificial diversions to drain the valley resulted in the drying up of many of the springs, the draining of lakes, loss of pressure in the aquifer with declining groundwater levels, and the consequent subsidence.

Because Mexico City is located on the valley floor, it has always been subject to flooding. Subsidence has worsened this problem by lowering the land surface of Mexico City below the level of Lake Texcoco, resulting in increased flooding. Drainage systems had to be dug deeper and Lake Texcoco had to be excavated. By 1950 dikes had to be built to confine stormwater flow, and pumping was required to lift drainage water under the city to the level of the drainage canals. By 1953 severe subsidence resulted in the closing of many wells that had to be replaced with new wells.

The *Mexico City metropolitan area* (MCMA) has become a magnet of growth, being the cultural, economic, and industrial center for Mexico, with an estimated population approaching 22 million people. A continual migration of people from rural areas to the city has occurred, with many of the people settling illegally in the urban fringe with the hope of eventually being provided public services. Providing water supply and wastewater services for Mexico City is a formidable challenge. Imagine that the city has the largest population in the world living in an enclosed basin with no natural outflow to the sea. The water supply situation has reached a crisis level, with the continued urban growth and poor system of financing by the government. The consequences include an inability to expand the water supply network to areas that are underserved or not served at all, repair leaks, and provide wastewater treatment. Mexico City cannot meet the water demands of its population.

Homero Aridjis, a successful novelist and journalist and one of Mexico's leading environmental figures, has stated "Mexico City, founded on water, may die of thirst." (Kunstler, 2001). The water problems in Mexico City are insurmountable. Aridjas has said that "the city is an urban disaster." James Kunstler in his book, *The City in Mind: Meditations on the Urban Condition*, describes Mexico City as a "hypertrophied metastasized organism destined to devour itself." In the preface to his book he referred to Mexico City as "a prototype of hypertrophic 'third world' urbanism, plagued by failed social contract, lawlessness, economic disorder, and a wrecked ecology."

The present-day water problems are summarized:

- Mexico City receives approximately 70 percent (55.5 m^3/s) of its water supply from the underlying aquifer system, with natural springs and runoff from the summer rains from the surrounding sierras and mountains supplying water to the aquifer. The other 30 percent (19 m^3/s) of the water supply comes from the Lerma River through a 15 km long aqueduct and the large-scale Cutzamala River project, which transfers water 120 km over a 1200 m elevation change. Skaggs et al. report 51 m^3/s are withdrawn from the aquifer and 23 m^3/s are imported from outside the city.

- An estimated 40 percent of the water supplied is lost through leaks in the aging municipal water supply system and through ineffective coordination between the various levels of government.

- Overexploitation of the aquifer has caused subsidence, which has been a problem since the early 1900s as a result of draining of the lake water (Lake Texcoco) from the basin floor. The subsidence has caused damage to the many parts of the infrastructure, including serious damage to the city's water supply and sewage system.

- Subsidence has also worsened the city's flooding problems as the city has sunk below the natural lake floor. The continuing subsidence has caused such a serious problem that now the pumping stations must run 24 hours a day year round to keep the summer rains from washing sewage and runoff back to the city (Morgan, 1996a).

- Only about 10 percent of Mexico City's wastewater is treated, with the remaining 90 percent being untreated, and diverted out of the Basin of Mexico through an extensive network of drainage. The primary destination of the untreated wastewater is the Mezquitl Valley (largest area in the world, irrigated with wastewater) in the state of Hidalgo, where it is eventually discharged into the Tula-Moctezuma-Panuco River system, which flows to the Gulf of Mexico. Wastewater is mixed with surface water from reservoirs for irrigation purposes.

The conclusion is obvious: the water resources situation in Mexico City is not sustainable by my definition or by any other definition. Additional sources of information include Mestre (1997), Paredes (1997), Perez de Leon and Biswas (1997), Tortajada and Biswas (2000), Skaggs et al. (2002), and *Mexico City: Opportunities and Challenges for Sustainable Management of Urban Water Resources* available at the Web site http://casestudies.lead.org/index.php?csp=15. We have very briefly examined two present day situations of water resources unsustainability, both resulting largely from unsustainable development.

THE PROBLEMS IN DEVELOPING NATIONS

The following press releases are only a couple of examples illustrating the threats associated with water sustainability in developing parts of the world such as Africa. The first is a UNICEF release (www.unicef.org):

Kenya: Worst drought in years threatens children

UNICEF Re-Issues Emergency Appeal for US$4 million

NAIROBI, 19 December 2005—UNICEF called urgent attention today to thousands of children in northern Kenya who face malnutrition due to deepening drought. The recent short rain season has been extremely poor in the northern and eastern pastoral districts. At a time of year when livestock should be healthy and feeding on new grass, carcasses are lying dead along the roadside. Many Government, UN and NGO experts meeting in Nairobi last week described the drought as the worst in years.

UNICEF Kenya Representative Heimo Laakkonen said in a statement today that rates of child malnutrition in districts like Wajir and Mandera may increase from the already alarming levels of almost 30 percent reported in assessments backed by the agency in October. "The dry weather is predicted to continue," said Laakkonen. "Given that situation can only get worse, it is imperative that all partners and the government act swiftly to protect the most vulnerable children and women." The World Food Programme has already more than doubled its estimate of the number of people needing food aid to about 2.5 million. It is estimated that about 560,000 people in 7 districts will require emergency supplies of water.

UNICEF has re-issued its October appeal which calls for US$4 million to assist more than 20,000 children estimated to be malnourished or at serious risk of malnutrition. The appeal also included programmes that aimed to keep children in school, ensure safe water supplies and provide emergency health care and protection.

Another news release of the Presbyterian News Service, August 22, 2005 is as follows:

Malawians face starvation, church leader says
Presbyterian cites drought, flooding as causes of famine in his homeland

by Toya Richards Hill

LOUISVILLE—If food isn't sent to the southern African nation of Malawi very soon, many people are sure to die, said the Rev. Winston Kawale, a leader of the Church of Central Africa Presbyterian (CCAP). . . .

"It's a pity, and I feel ashamed that we have hunger in Malawi," he said. "We have land, we have soil, we have water." The problem, he said, is a lack of expertise, especially in irrigation and water-resource management, which prevents year-round production. The country has no shortage of water—one of its greatest resources is Lake Malawi, the third-largest body of water in Africa, which covers almost one-fifth of the country. However, "At the moment, we only depend on the rains," Kawale said. With greater knowledge, he said, "hunger will no longer be there" and "poverty will be addressed."

These releases express the sadness associated with what is happening in Africa.

A Look at Sub-Saharan Africa

Sub-Saharan Africa has a history of poverty, war, and famine extending over millennia. What part has water sustainability played in the poverty, vulnerability, inequity, and threats to the social fabric in Sub-Saharan Africa? Only 45 percent of Sub-Saharan Africa's population has access to safe water (UNDP, 1994). As pointed out by Morgan (1996 b), the prospects of increased water scarcity are considerable, given the history of less precipitation in the past three decades and the reduction in volume of several water bodies, along with some evidence of worsening vegetation conditions or "desertification."

During the 1980s and the early 1990s, most of Sub-Saharan Africa witnessed serious economic decline or stagnation. The chief source of environmental degradation in Sub-Saharan Africa as a whole (including plant cover and species loss, destruction of fauna, climatic change, changes in water table levels and stream flow, and soil erosion) is deforestation, particularly when followed by overcultivation and overgrazing. The Sub-Saharan African economic and financial problems were made very much worse in the 1970s and in the 1980s by a combination of (Morgan, 1996 b):

1. An investment in growth and development that failed to earn the expected rewards
2. The international debt crisis, oil price hikes, and rising interest rates, plus the inadequacy of the aid programs that were meant to provide relief

3. Repeated drought, crop failure, and widespread famine

4. The failure of agricultural production to contribute significantly to growth and the increased dependence on imported food

5. Widespread warfare and civil unrest

6. The fact that *structural adjustment programs* (SAPs) have been only partially successful and that this success has been in terms of systems rather than people, who somehow have to survive in the hope that decisions linked to short- or medium-term deterioration based on theories that are unknown or unappreciated could solve their problems in the future

It is obvious that water sustainability issues have been a major factor in the poverty, vulnerability, inequity, and threatening of the social fabric of Sub-Saharan Africa.

FUTURE TARGETS FOR CURRENT GLOBAL WATER CHALLENGES

Current global water challenges and future targets are clearly stated in the "Millennium Development Declaration (2000)"—which includes the access to safe drinking water as one of its *millennium development goals*—and are strengthened and expanded in the plan of implementation of the world summit on sustainable development. The key role of sustainable water management for poverty eradication has been one of the key outcomes of the World Summit on Sustainable Development. The plan of implementation outlines several central statements related to freshwater and sanitation issues:

• Halve, by the year 2015, the proportion of people without access to safe drinking water

• Halve, by the year 2015, the proportion of people who do not have access to basic sanitation

• Combat desertification and mitigate the effects of drought and floods

• Develop integrated water resources management and water efficiency plans by 2005, with support to developing countries

• Support developing countries and countries with economies in transition in their efforts to monitor and assess the quantity and quality of water resources

• Promote effective coordination among the various international and intergovernmental bodies and processes working on water-related issues, both within the United Nations system and between the United Nations and international financial institutions

Global Climate Change

In December of 2005, I attended the International Water History Association meeting at the UNESCO headquarters in Paris, where I heard one of the most convincing talks by Robert Kandel (2005), Directeur de Recherche Emérite, CNRS—Laboratoire de Météorologie Dynamique—Ecole Polytechnique, France. The title of his talk was "Anthropogenic Global Warming and Water in the 21st Century." Dr. Kandel (also author of the book *Water from Heaven*) pointed out that for many observers, global warming of recent decades, although still modest, appears significantly stronger than climate variations of the previous thousand years, which is believed to result in part from anthropogenic intensification of the greenhouse effect, the trapping of thermal infrared radiation in the lower atmosphere. He emphasized that observers agree that anthropogenic alteration of atmosphere composition is dramatic in that the atmospheric concentration of carbon dioxide has increased by 30 percent since 1900, and methane has doubled. Climate modeling explains what is happening only when both natural forcing and anthropogenic forcing are included.

Because a major reduction of net carbon dioxide emissions is unlikely over the next few decades, a twenty-first century intensification of the greenhouse effect will occur, resulting in global warming. Dr. Kandel emphasized, "the coming rapid climate change will by necessity involve and depend on changes in the hydrological cycle, and some results suggest that it could be comparable to the strong climate changes of the past million years." He stated that, "what is new in the history of the planet is that this climate change will result from the activities of a species in part conscious of the consequences of its actions." Depending on the location of the acceleration and/or slowing of the hydrological cycle depending on location, the freshwater fluxes to the land biosphere can either increase or decrease.

We have tremendous advantages over past civilizations that were completely unaware of climate change. Figure 1.5, showing the decrease in the Arctic Sea ice boundary since 1979, gives us clear evidence of the oncoming climate change. Not only do we have the technologies to detect the oncoming of climate change, we are now able to date past climate events that the ancients faced. Now that we know the ancients have warned us, what will we do about this coming "strong warming"?

Can climate change spark international water wars? In recent years, we have seen water controversies among Israel, Jordan, and Palestine, between Turkey and Syria, between China and India, among Botswana, Angola, and Namibia, between Ethiopia and Egypt, and between Bangladesh and India.

Mega Water Projects: Can They Be a Partial Solution?

In Chap. 2 some of the ancient mega water projects are discussed, such as those developed by the Romans, the Mayans, and others. Presently, there are several mega water projects that are at different stages of development going on in the world. These include the southeast Anatolian project (GAP—*Guneydogy Anadolu Projesi*) in Turkey that involves the construction of 22 dams, 19 hydroelectric generation stations on the Tigris and Euphrates Rivers, and the irrigation of 1.7 million hectares of land; the great manmade river project in Libya, which is the construction of a massive conveyance system to transport over 6 million m^3 of water per day from aquifers in the southern parts of Libya to northern parts of Libya; the huge China south-to-north water project which includes three south-to-north canals, which will stretch across the eastern, middle, and western parts of China eventually linking the country's four major rivers—the Yangtze, Yellow, Huaihe, and the Haihe; and other smaller projects—the El Salaam project to divert Nile River water and irrigation return flows to the north Sinai and the Nile (Naga Hammadi) barrage that is 330 m long dam at Naga Hammadi in upper Egypt to divert water through a 1.1 km long canal. Are these projects the ultimate solution or are they simply bandages to the real problems associated with population growth and other societal concerns?

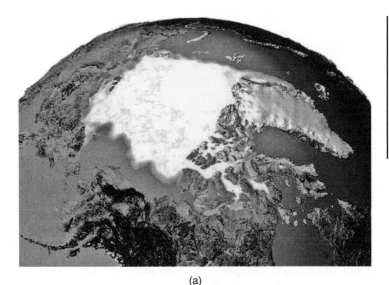

(a)

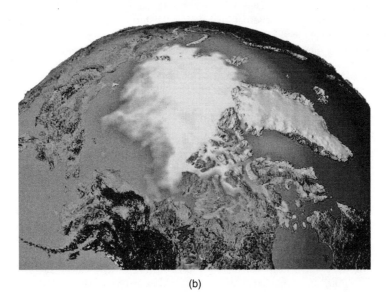

(b)

Figure 1.5
Decrease in the polar ice cap from 1979 to 2003 (a) Polar ice cap in 1979 (b) Polar ice cap in 2003. (*Courtesy of NASA*)

REFERENCES

Alley, R. B., *The Two-Mile Time Machine: Ice Covers, Abrupt Climate Change, and Our Future*, Princeton University Press, Princeton, NI, 2000.

ASCE, Task Committee on Sustainability Criteria, *Sustainability Criteria for Water Resources Systems*, American Society of Civil Engineers, Reston, VA, 1998.

Brundtland G.(ed), *Our Common Future: The World Commission on Environment and Development,* Oxford University Press, Oxford, 1987.

Diamond, J., *Guns, Germs, and Steel: The Fates of Human Societies*, Norton, New York, 1997.

Diamond, J., *Collapse: How Societies Choose to Fall or Succeed*, Viking, New York, 2005.

Ellis, W. S., and D. C. Turnley, "A Soviet Sea Lies Dying," *National Geographic Magazine*, Vol. 177, No. 2, pp. 73–93, February, 1990.

Ferguson, R. W., *The Devil and the Disappearing Sea: Murder & Mayhem Amid the Aral Sea Disaster*, Raincoast Books, Vancouver, 2003.

Frederiksen, H. D., "Water Crisis in Developing World: Misconceptions about Solutions," *Journal of Water Resources Planning and Management*, ASCE, Vol. 122, No. 2, pp. 79–87, March/April 1996.

Gill, R. B., *The Great Maya Droughts: Water, Life, and Death*, University of New Mexico Press, Albuquerque, 2000.

Gleick, P. H., "Water and Conflict," *International Security*, Vol. 18, No. 1, pp. 79–11, 1993.

Gleick, P. H., "Water, War, and Peace in the Middle East," *Environment*, Vol. 36, No. 3, Hedref Publishers, Washington, DC, p.6. 1994.

Gleick, P. H., "Water and Conflict," in P. H. Gleick (ed.) *The World's Water 1998–1999*, Island Press, Washington, DC, pp. 105–135, 1998.

Gleick, P. H., "Water Conflict Chronology," available at http://www.worldwater.org/conflict.htm, 2000.

Gleick, P. H., *The World's Water, The Biennial Report on Freshwater Resources, 2005*, Island Press, Washington, 2005.

Kandel, R., "Anthropogenic Global Warming and Water in the 21st Century," abstract of presentation at the International Water History Association meeting held at UNESCO Headquarters, Paris France, December 3, 2005.

Kunstler, J. H., *The City in Mind: Meditations on the Urban Condition*, The Free Press, Simon & Schuster, New York, 2001.

Lilach, G., "Jordan River Dispute," ICE Case Studies, http://gurukul.ucc.american.edu/ted/ice/jordan.htm, 1997.

Linden, E., *The Winds of Change: Climate, Weather, and the Destruction of Civilizations*, Simon & Schuster, New York, 2006.

Loucks, D. P., "Sustainable Water Resources Management," *Water International*, Vol. 25, No. 1, pp. 3–10, March, 2000.

Lowdermilk, C., *Conquest of the Land Through Seven Thousand Years*, Soil Conservation Service, U.S. Department of Agriculture, Washington, DC, pp. 14–25, 1953.

McKinney, D., "Sustainable Water Management in the Aral Sea Basin," *Water Resources Update*, Universities Council on Water Resources, Issue No. 102, Winter, 1996.

Meadows, D. H., D. L, Meadows, J. Randers, and W. W. Behrens, III, *Limits to Growth*, Report for the Club of Rome's Project on the Predicament of Mankind, 2d ed., Universe Books, New York, 1974.

Mestre, J. E., "Integrated Approach to River Basin Management: Lerma-Chapala Case Study-Attributions and Experiences in Water Management in Mexico," *Water International*, Vol. 22, No. 3, 1997.

Mexico City: Opportunities and Challenges for Sustainable Management of Urban Water Resources, available at http://casestudies.lead.org/index.php?csp=15.

Morgan, A., "Mexico City: A Megacity with Big Problems," Department of Earth Sciences, University of Waterloo Web site, 1996 a.

Morgan, W. B., "Poverty, Vulnerability, and Rural Development," in G. Benneh, W. B. Morgan, and J. Uitto (eds.), *Sustaining the Future:Economic, Social, and Environmental Change in Sub-Saharan Africa*, United Nations University Press, Tokyo, 1996 b.

National Research Council (NRC), *Mexico City's Water Supply: Improving the Outlook for Sustainability*, National Academy Press, Washington, D.C., 1995.

Paredes, A. J., "Water Management in Mexico: A Framework," *Water International*, Vol. 22 No. 3, 1997.

Perera, J., "Where Glasnost Meets the Greens," *New Scientist*, New Scientist Publications, London, England, Vol. 120, No. 1633, pp. 25–26, October 8, 1988.

Perera, J, "A Sea Turns to Dust," *New Scientist*, New Scientist Publications, London, England, Vol. 140, No. 1896, October 23, pp. 24–27, 1993.

Perez de Leon, M. F. N., and A. K. Biswas, "Water, Wastewater, and Environmental Security Problems: A Case Study of Mexico City and the Metzquital Valley," *Water International*, Vol. 22, No. 3, 1997.

Perlin, J., *A Forest Journey: The Role of Wood in the Development of Civilization*, Norton, New York, 1989.

Postel, S., *Last Oasis: Facing Water Scarcity*, Norton, New York, 1997.

Skaggs, R. L., L. W. Vail, and S. Shankle, "Adaptive Management for Water Supply Planning: Sustaining Mexico City's Water Supply," in L. W. Mays (ed.) *Urban Water Supply Handbook,* McGraw-Hill, New York, 2002.

Tainter, J. A., *The Collapse of Complex Societies*, Cambridge, New York, 1990.

Tortajada, C., and A. K. Biswas, "Environmental Management of Water Resources in Mexico," *Water International*, Vol. 25, No. 1, 2000.

UNDP (United Nations Development Programme), Human Development Report 1994, Oxford University Press, Oxford and New York, 1994.

2

WATER SUSTAINABILITY: PARALLELS OF PAST CIVILIZATIONS AND THE PRESENT

Larry W. Mays
Professor of Civil and Environmental Engineering
Arizona State University
Tempe, Arizona

INTRODUCTION

In this chapter, I discuss ancient civilizations that collapsed in the American southwest and Mesoamerica to further illustrate the concept of sustainability. Then I briefly discuss the relevance to modern societies of the collapse of ancient civilizations, particularly in the southwestern United States. Throughout history humans have struggled to find a lasting adjustment to the land and the natural resources; unfortunately though, civilizations have failed (collapsed) for various reasons; among those is the unsustainability of water resources.

The earliest known civilizations arose in the Tigris and Euphrates Valleys of the Fertile Crescent and in the Nile Valley of Egypt, where centralized political organizations or bureaucracies existed. The fertile alluvial plains of Mesopotamia and the valley of the Nile are the location where at least 7000 years ago, agriculture had its beginnings and where the first successful efforts to control the flow of water were made. Since these beginnings many civilizations made advances in the use of water. Water control systems appeared in many areas; including the Indus Valley, the Minoan civilization (Knossos on Crete), the Greeks, Mesoamericans, the Romans, the Nabataeans in present day Jordan, and many others. These ancient civilizations faced problems associated with sustainable development, particularly water-related issues. The challenges of the sustainability of water resources included many of the challenges that we face today, such as water supply security, climate variations, population growth, scarce natural resources, food security, and natural disasters such as floods and droughts.

Many civilizations, great centers of power and culture, were built in locations that could not support the populations that developed. Throughout history, arid lands seem to have produced more people than they could sustain. I will attempt to make comparisons between civilizations of the past and our present world in the context of water resources sustainability.

Water has always been a very important factor in the development and survival of societies. The collapse of societies for environmental and/or other reasons often masquerades as military defeats. The fall of the Western Roman Empire has been debated as a collapse, possibly masqueraded as barbarian invasions, with AD 476 as the year when the last emperor of the West was deposed. Was the fall a result of the barbarians becoming more numerous or better organized, having better weapons, or did the barbarians profit from climate change of the Central Asian steppes? Or were the barbarians unchanged and Rome became weakened by a combination of economic, political, environmental, and other problems?

ANCIENT MEGAWATER PROJECTS

Various ancient civilizations developed water projects that would have been considered as megawater projects during their times, and even today. Examples included the Romans throughout many parts of Europe, the Middle East, and northern Africa, the Hohokams and the Anasazi in the southwestern United States, the Egyptians, the Mesopotamians, and many others. I will use examples of Roman projects to illustrate ancient megawater projects.

The Romans built magnificent structures for water supply, including the following four that I will briefly describe: the water system of aqueducts and dams to Merida, Spain (see Fig. 2.1*a*); the system of aqueducts with various structures to Lyon, France (ancient Lugdunum) (see Fig. 2.2*a*); the system of aqueducts to Rome (see Fig. 2.3*a*); and the aqueduct of Nimes (ancient Nemausus) (see Fig. 2.4*a*).

Romans

In 25 BC, Emerita Augusta (Merida, Spain) became a colony and a century later the Romans had built a water supply system including three aqueducts (Fig. 2.1), two of which were supplied by dams (the Cornalvo dam and the Proserpina dam). The three aqueducts were the Cornalvo aqueduct (enters on the east side of Merida), the Proserpina aqueduct (enters on the northeast side of town), and the Las Thomas aqueduct (from springs on the north and northeast side of Merida).

The Cornalvo aqueduct was built first and was about 17 km long. Cornalvo dam (Fig. 2.1*b*) is an earthen dam approximately 194 m long, 20 m high, and has a 8 m dam crest width. A few remains of the aqueduct are visible as shown in Fig. 2.1*c* near the present day "bull ring."

The Proserpina dam (see Fig. 2.1*d*) is an earthen dam, 427 m long, 12 m high, located north of Merida, and supplied water to the 10 km long Los Milagros aqueduct. This aqueduct entered the town on the north side with an aqueduct bridge over the Rio Albarregas (see Fig. 2.1*e*), also referred to as the Los Milagros (the miracles) by the Spanish, with a maximum height of 30 m.

The Las Thomas aqueduct included an aqueduct bridge 1600 m long (across the Rio Albarregas), of which only three pillars (16 m high) remain (see Fig. 2.1*f*). Materials from this aqueduct bridge were used by the Arabs in the sixteenth century to construct

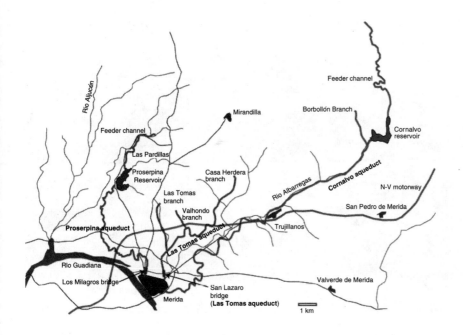

Figure 2.1 (a)

Map showing the three Roman aqueducts in Merida, Spain. (*Courtesy of C.W. Passchier of Mainz, Germany*)

Figure 2.1 (b)

Cornalvo dam near Merida, Spain. (*Photo copyright by L. W. Mays*)

the San Lazaro aqueduct bridge. Figure 2.1*g* shows the aqueduct near the Roman theatre and amphitheatre.

Four aqueducts were used to supply water to the ancient city of Lugdunum (Lyon, France). They were the Mont d'Or, the Yzeron, the Brevenne, and the Gier. Figure 2.2*a* shows the route of the aqueduct of the River Gier, which was the longest and the highest

Figure 2.1 (c)

Cornalvo aqueduct supplying water from Cornalvo dam to Merida. Only a few remnants of this aqueduct are still visible. (*Photo copyright by L. W. Mays*)

Figure 2.1 (d)
Proserpina dam near Merida, Spain. (*Photos copyright by L. W. Mays*)

of the four aqueducts. Approximately half of the aqueduct was subterranean with at least nine tunnels and four siphons. Figures 2*b* and 2*c* are photos of the aqueduct near Chaponost. This aqueduct had four siphons and over 80 manholes.

The aqueduct system in Rome evolved over a 500-year time period, with the first aqueduct, the Aqua Appia, being constructed around 313 BC. This system is illustrated in Fig. 2.3*a*, with 11 aqueducts that eventually supplied water to Rome from mostly springs, and two

Figure 2.1 (e)
Los Milagros
aqueduct bridge
across the Rio
Albarregas in
Merida, Spain.
(*Photo copyright
by L. W. Mays*)

Figure 2.1 (f)
Las Thomas
aqueduct bridge
across the Rio
Albarregas and
located near the
hippodrome. Only
the three pillars
remain of this
aqueduct bridge.
(*Photo copyright
by L. W. Mays*)

Figure 2.1 (g)
Las Thomas aque-
duct in Merida near
the Roman theater
and amphitheater.
Shown on the right
side of the aque-
duct is a lion head
of stone which was
used as a gutter
spout. There are
also remains of a
castellum nearby.
(*Photo copyright by
L. W. Mays*)

were supplied from the Anio River and one from Lake Alsietinus. All the major eastern aqueducts entered Rome at the Porta Maggiore (see Figs. 2.3*b* and 2.3*d*).

The aqueduct of Nemausus (built around 20 BC) conveyed water approximately 50 km from Uzes to the *castellum* in Nimes (see Fig. 2.4*a*). From an engineering point of view this was a remarkable construction project in that the elevation difference over the length of the aqueduct was only 17 m, making the slope only 0.0008.5 m/m, with the profile shown in Fig. 2.4*b*. The Pont du Gard, shown in Fig. 2.4*c*, is one of the most spectacular aqueduct bridges ever built and is the most photographed in the world. Figures 2.4*d* and 2.4*e* illustrate the *castellum divisorium* at Nimes.

I use the Teotihuacans, the Xochicalcoans, the Mayas, the Chacoans (Anasazi), and the Hohokams to illustrate that the ancient ones have warned us.

Many Mesoamerican civilizations developed and failed for various reasons. The period or era from about AD 150 to AD 900 (called the *classic*) was the most remarkable in the development of Mesoamerica (Coe, 1994). Figure 2.5 shows Mexico during the classic period. During the classic period the people of Mexico and the Maya area built civilizations comparable with advanced civilizations in other parts of the world. In Mesoamerica

Mesoamerica

Figure 2.2 (a)
Route of the aqueduct on the River Gier to Lugdunum (Lyon, France). (*Courtesy of Les Aqueducs Romains de Lyon, L'Araire, Le Borg, 69510 Messimy en Lyonnais*)

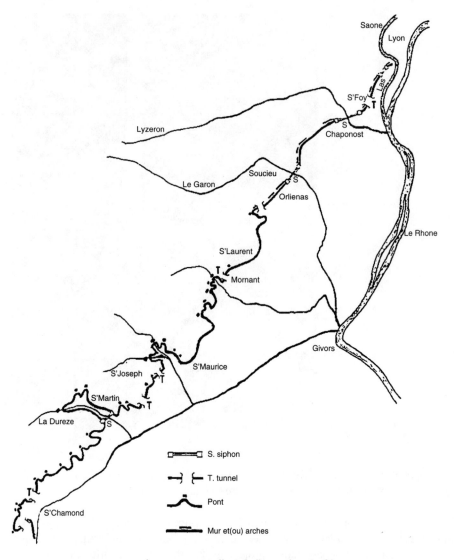

Les ouvrages d'art de l'aqueduc du Gier

the ancient urban civilizations developed in arid highlands where irrigation (hydraulic) agriculture allowed high population densities. In the tropical lowlands, however, there was a dependence on slash-and-burn (*milpa*) agriculture, which kept the bulk of the population scattered in small hamlets. Sanders and Price (1968) suggest that the nonurban lowland civilization resulted from responses to pressures set up by the hydraulic, urban civilization. Teotihuacan (city of the Gods) in Mexico is the earliest example of highland urbanism.

Figure 2.2 (b) and (C)

Aqueduct of the River Gier near Chaponost, France. (*Photo copyright by L. W. Mays*)

Teotihuacan was a very impressive civilization, which evolved about 25 mi north of Mexico City around the same time as Rome. Prior to 300 BC, Teotihuacan valley had a small population spread over the valley and was the dominant urban center in Mesoamerica throughout the classic period. By AD 100, Teotihuacan covered an area of 12 km², which has been linked to the development of so-called hydraulic agriculture (Haviland, 1970).

Teotihuacans

Figure 2.3 (a)

Aqueducts in ancient Rome. (I) Termini of the major aqueducts (*from Evans, H.B., Water distribution in Ancient Rome: the Evidence of Frontinus, University of Michigan Press, Ann Arbor, 1999*)

(II) The area of the Spes Vetus showing the courses of the major aqueducts entering the city above ground (*from R.Lancianni, Forma Urbis Romae, as presented in Evans, 1994*)

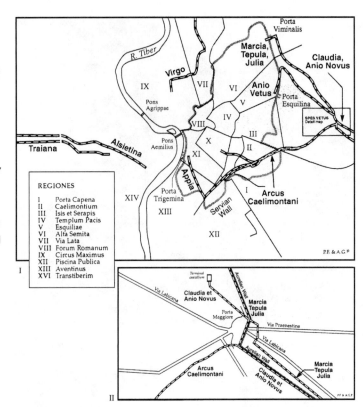

REGIONES

I	Porta Capena
II	Caelimontium
III	Isis et Serapis
IV	Templum Pacis
V	Esquiliae
VI	Alta Semita
VII	Via Lata
VIII	Forum Romanum
IX	Circus Maximus
XII	Piscina Publica
XIII	Aventinus
XVI	Transtiberim

Figure 2.3 (b)

View of Porta Maggiore (double-arched gate) on the Aurelian wall where all the eastern aqueducts entered Rome (*Photo copyright by L. W. Mays*)

Figure 2.3 (c)
Aqueducts Claudia (top) and the Anio Novus (bottom) above the Porta Maggiore. (*Photo copyright by L. W. Mays*)

Figure 2.3 (d)
Three aqueducts Julia (top), Tepula (center), and Marcia (lower). (*Photo copyright by L. W. Mays*)

Figure 2.4 (a)

Route of the
aqueduct of
Nimes, France.
(*From Hauck,
1988*)

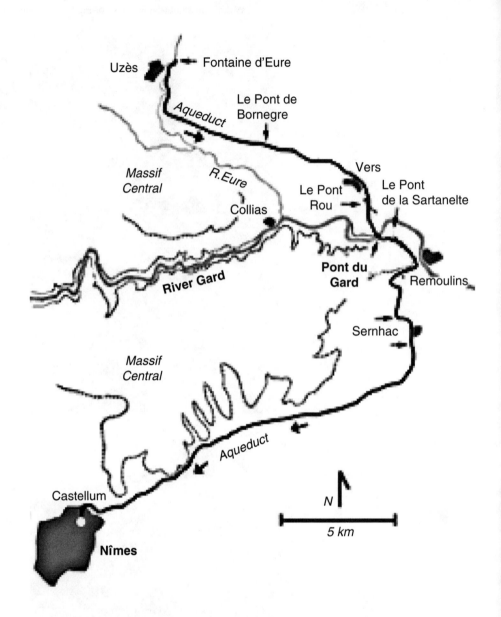

As the urban area expanded in size, there was an increased socioeconomic diversity, and an expanding political influence. At its height, around 600 AD, Teotihuacan was fully urban with a population of approximately 85,000 people and covering an area of 19 km^2 (Haviland, 1970). Others have estimated that the maximum population was approximately 125,000 during the Xolalpan phase (Millon, 1993). Teotihuacan was the largest urban center of the time in Mesoamerica.

Around 300 BC, the use of canals for irrigation rapidly spread throughout the central highland basin, the location of Teotihuacan (Doolittle, 1990). South of Teotihuacan

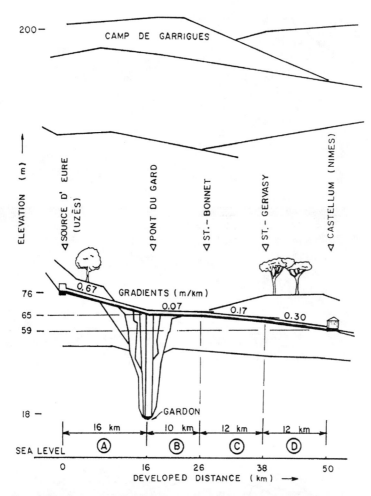

Figure 2.4 (b)
Profile of the aqueduct of Nimes. (*Courtesy of Hauck, G., The Aqueduct of Nemausus, McFarland, Jefferson, NC, 1988*)

Figure 2.4 (c)
Pont du Gard aqueduct bridge showing the three levels. (*Photo copyright by L. W. Mays*)

Figure 2.4 (d)

Views of the castellum divisorium at Nmes. Water enters the circular basin through the rectangular aqueduct opening. Refer to Fig. 2.4e for the plan view with dimensions. (*Photos copyright by L. W. Mays*)

Figure 2.4 (e)

Plan view of the Nimes castellum divisorium showing the 10 outlets and the 3 drains in the floor of the 1.5 m diameter basin. (*Courtesy of J. P. Adam, La Construction Romaine, Paris (1984) as drawn in Hodge, 2002*)

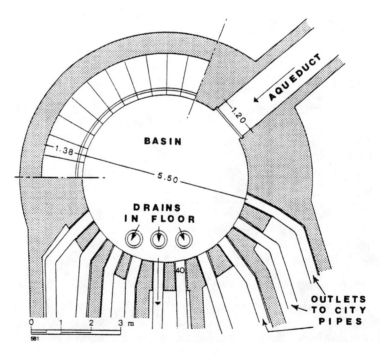

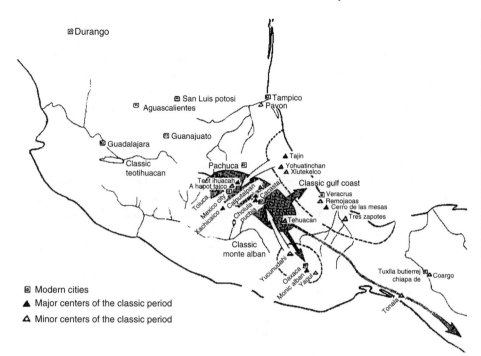

Figure 2.5
Classic period in Mexico showing the area (shaded) covered by the Teotihuacan civilization and its extensions in Mexico. (*Courtesy of Coe, M. M., Mexico: From the Olmecs to the Aztecs, 4th ed., Thames and Hudson, New York, 1994*)

near Amanalco, Texcoco east of the Basin of Mexico, irrigation would have consisted of diverting water from shallow spring-fed streams into simple irrigation canals and then onto fields only a few meters away. Flood water systems also were used. Northeast of Teotihuacan, south of Otumba, a series of ancient irrigation canals (dating between 300 BC and 100 BC) were excavated. Other canals in the same area date to AD 900 to AD 1600. Evidence of canals that were built between AD 200 and AD 800 near Teotihuacan also exists. One of these was the first confirmed relocation of a natural stream. The reader can refer to Doolittle (1990) for further information on these canals.

The city of Teotihuacan was abandoned mysteriously around AD 600 to AD 700. During this time the collapse of civilized life occurred in most of central Mexico. One possible cause was the erosion and desiccation of the region resulting from the destruction of the surrounding forests that were used for the burning of the lime that went into the building of Teotihuacan. The increasing aridity of the climate in Mexico may have been a related factor. The entire edifice of the Teotihuacan state may have perished from the loss of agriculture. Even though the city had no outer defensive walls, Millon (1993) believes that it was not an open city easy for hostile outsiders to attack. The collapse of Teotihuacan opened civilized Mexico to nomadic tribes from the north. Human malnourishment has been indicated from skeletal remains.

After the disintegration of Teotihuacan's empire in the seventh century AD, foreigners from the Gulf Coast lowlands and the Yucatan Peninsula appeared in central Mexico. Cacaxtla and Xochicalco, both of Mayan influence, are two regional centers that became

Xochicalco

important with the disappearance of Teotihuacan. Xochicalco (in the place of the house of flowers), was located on hill top approximately 38 km from Cuernavaca, Mexico, and became one of the great Mesoamerican cities in the late classic period (AD 650 to AD 900). Figures 2.6*a* and 2.6*b* are photographs of the city, which was well thought out with terraces, streets, and plazas among the buildings. Despite the very Mayan influences, the predominant style and architecture is that of Teotihuacan.

There were no rivers or streams or wells to obtain water. Water was collected in the large plaza area and conveyed into cisterns such as the one shown in Fig. 2.6*c*. From the cisterns water was conveyed to other areas of the city using pipe as shown in Fig. 2.6*d*. The collapse/abandonment of Xochicalco, most likely, resulted from drought, warfare, and internal political struggles. The reliance upon collecting rainwater for water supply is very vulnerable and unsustainable through periods of low rainfall and even more important to drought conditions.

Crisis overtook all the classic civilizations of Mesoamerica (including the Mayans), forcing the abandonment of most of the cities. Some anthropologists believe the crisis may have been a lessening of the food supply caused by a drying out of the land and a

Figure 2.6

Photos of Xochicalco (a) View showing Xochicalco on hilltop (b) View from Xochicalco (c) Cistern (d) Water pipes. (*Photos copyright by L. W. Mays*)

(a)

(b)

(c)

(d)

loss of water sources to the area. Speculation is that this might have been caused by a combination of a climatic shift toward aridness that appears to have happened all over Mexico during the classic period and the deforestation of the valley. Originally there were cedar, cypress, pine, and oak forests; today there are cactus, yucca, agave, and California pepper trees. Such a change in vegetation indicates a significant climate shift.

The Maya

The ancient Maya lived in a vast area covering parts of present-day Guatemala, Mexico, Belize, and the western areas of Honduras and El Salvador as illustrated in Fig. 2.7. Mayans settled in the last millennium BC and their civilization flourished until around AD 870. The environment that the Mayans lived in was less fragile than that of the semiarid lands where the Anasazi and Hohokam lived.

Figure 2.7

Maya sites during the classic period. (*Courtesy of Coe, M. D., The Maya, 5th ed., Thames and Hudson, New York, 1993*)

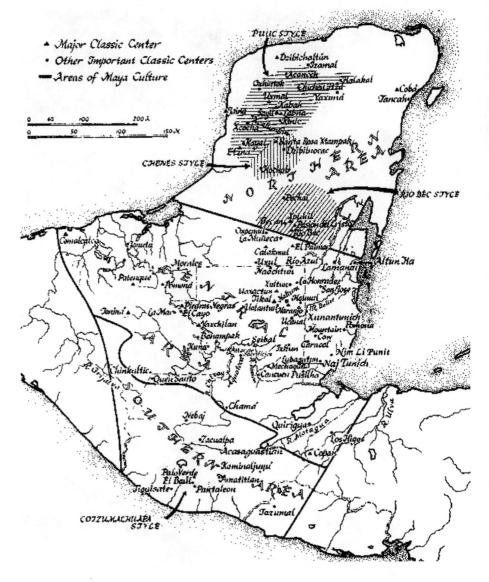

Tikal was one of the largest lowland Maya centers, located some 300 km north of present day Guatemala City. The city was located in a rain forest setting with a present-day average annual rainfall of 135 cm. The urbanization of Tikal was not because of irrigated (hydraulic) agriculture. A number of artificial reservoirs were built in Tikal, which became more and more important as the population increased. With the continued growth, the pressure for land and food increased to such an extent that the population growth stopped or reached a point of unsustainability.

The Mayas settled in the lowlands of the Yucatan Peninsula and the neighboring coastal regions (see Fig. 2.7). The large aquifer under this area is in an extensive, porous lime-stone layer (Karst terrain), which allows tropical rainfall to percolate down to the aquifer.

Because of this and the fact that few rivers or streams exist in the area, surface water is scarce. One important water supply source for the Maya, particularly in the north, was the underground caves (see Fig. 2.8) called *cenotes* (se-NO-tes), which also had religious significance (portals to the underworld where they journeyed after death to meet the gods and ancestors). In Yucatan there are over 2200 identified and mapped cenotes.

Figure 2.8
Sacred cenote at Chichen Itza (which means mouth of the well of the Itzas). The word *cenote* is derived from *tz'onot,* the Maya term for the natural sinkholes. (a) This cenote, which measures about 50 m from north to south and 60 m from east to west, was used for sacrifices of young men and women, warriors, and even children to keep alive the prophecy that all would live again. Shown at the left and in (b) is the remains of a building once used as a steam bath, or temezcal, to purify those who were to be sacrificed. Those sacrificed were tossed from a platform that jutted out over the edge of the cenote. (*Photos copyright by L. W. Mays*)

In the south the depths to the water table was too great for cenotes. Natural surface depressions were lined to reduce seepage losses and were used as reservoirs. Another source was water that collected when soil was removed for house construction in depressions called *aguados*. The Maya also constructed cisterns called *chultans* in limestone under buildings and ceremonial plazas. Drainage systems were developed from buildings and courtyards to divert surface runoff into the chultans. In the lowlands the Maya typically used one or more of these methods for obtaining and storing water supplies (Matheny, 1983).

Rainfall varies significantly from the north (18 in./year) to the south (100 in./year) of the Yucatan Peninsula. The soils are also deeper in the southern part, resulting in more productive agriculture; the area consequently supported more people. Rainfall was very unpredictable, resulting in droughts that destroyed crops. Ironically though, the water problems were more severe in the wetter southern part. Ground elevations increased from the north to the south, causing the depths down to the water table to be greater in the south.

Centuries before the Spanish arrived, the collapse of many other great Mayan cities occurred within a fairly short time period. Several reasons have emerged as to why these cities collapsed, including overpopulation and the consequent exhaustion of land resources possibly coupled with a prolonged drought. A drought from AD 125 until AD 250 caused the preclassic collapse at El Mirador and other locations. A drought around AD 600 caused a decline at Tikal and other locations. Around AD 760, a drought started that resulted in the Mayan classic collapse in different locations from AD 760 to AD 910.

The soil of the rain forest is actually poor in nutrients so that crops could be grown for only two or three years, then to go fallow for up to 18 years. This required ever-increasing destruction of the rain forest (and animal habitat) to feed a growing population. Other secondary reasons for the collapse include increased warfare, a bloated ruling class requiring more and more support from the working classes, increased sacrifices extending to the lower classes, and possible epidemics. The Maya collapsed as a result of four of the five factors in Diamond's (2005) framework. Trade or cessation of trade with friendly societies was not a factor for the Maya. Water resources sustainability was certainly a factor in the collapse of the Maya.

American Southwest

Three major cultures—the Anasazi, the Hohokam, and the Mogollon—inhabited the American southwest during the late precontact period (see Fig. 2.9). The concept of prehistoric regional systems has been used to describe these cultures (Crown and Judge, 1991). The Hohokam and Chaco regional systems have received particular attention as two of the most important. The extent of the Hohokam regional system has been defined by ball courts and material culture, and the Chaco regional system has been defined by roads and other architectural criteria. Each of these occupied a distinctive ecological niche within the southwestern environment, and as a consequence, their infrastructures significantly differed. The American southwest is a difficult and fragile environment consisting of arid and semiarid lands, with minimal water resources.

The Hohokam (300 BC to AD 1450)
Hohokam, translated as "the people who vanished," is the name given to their prehistoric predecessors by the present-day Pima Indians. The Hohokam built a complex irrigation

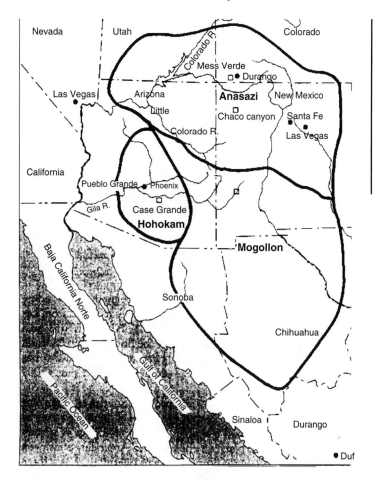

Figure 2.9
Three cultures in the American southwest. (*Courtesy of Thomas, D. H., Exploring Ancient Native America: An Archaeological Guide, Macmillan, New York, 1994*)

system in the desert lowlands of the Salt-Gila River Basin, Arizona, with no technology other than stone tools, sharpened sticks, and carrying baskets. They built more than 300 mi of major canals and over 700 mi of distribution canals in the Salt River Valley of present-day Phoenix, Arizona (see Fig. 2.10). The Hohokam civilization started in the Valley somewhere between 300 BC and AD 1 (see Crown and Judge, 1991) and extended to AD 1450 (Lister and Lister, 1983). Comparing Figs. 2.10 and 2.11, the similarities of the layout of the present-day canal systems with that of the Hohokams can be seen. A schematic representation of the major components of a Hohokam irrigation system is shown in Fig. 2.12.

In AD 899 a flood caused decentralization and widespread population movement of the Hohokams from the Salt-Gila River Basin to areas where they had to rely upon dry farming. The dry farming provided a more secure subsistence base. Eventual collapse of the Hohokam regional system resulted from a combination of several factors. These included flooding in the 1080s, hydrologic degradation in the early 1100s, and larger communities forcibly recruiting labor or levying tribute from surrounding populations (Crown and Judge, 1991). In 1358, a major flood ultimately destroyed the canal networks, resulting in the depopulation of the Hohokam area. Culturally drained, the Hohokam faced obliteration

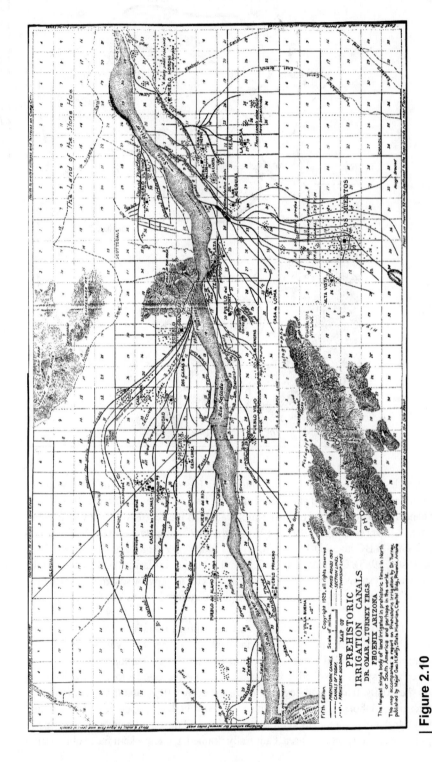

Figure 2.10
Hohokam canal system in Salt River Valley. (*Courtesy of Turney, O., "Prehistoric Irrigation in Arizona," Arizona Historical Review, Vol. 2, No. 5, Phoenix, 1929*)

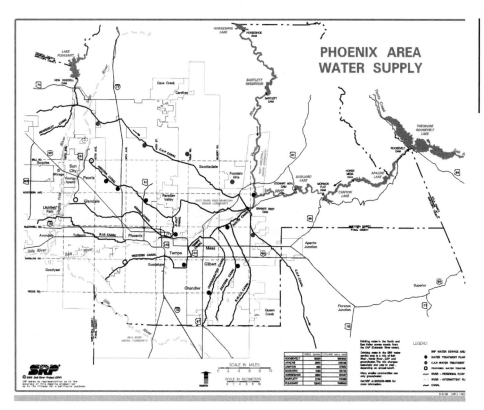

Figure 2.11

Present day canal system in Salt River Valley. (*Courtesy of Salt River Project*)

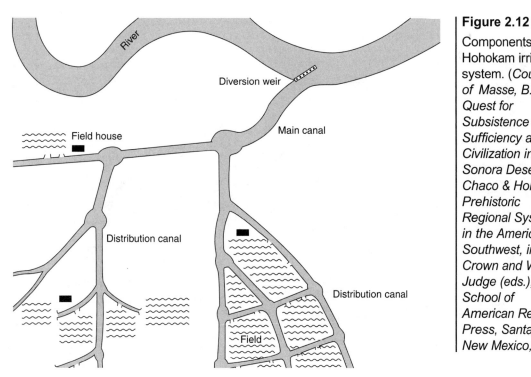

Figure 2.12

Components of Hohokam irrigation system. (*Courtesy of Masse, B., The Quest for Subsistence Sufficiency and Civilization in the Sonora Desert, in Chaco & Hohokam: Prehistoric Regional Systems in the American Southwest, in P.L. Crown and W.J. Judge (eds.), School of American Research Press, Santa Fe, New Mexico, 1991*) in about 1450. Parts of the irrigation system had been in service for almost 1500 years, which may have fallen into disrepair, canals silted in need of extensive maintenance, and problems with salt. See Haury (1978), Masse (1981), and Woodbury (1960) for further information.

The Chaco Anasazi (AD 600 to AD 1200)

In the high deserts of the Colorado Plateau (see Fig. 2.13), the Anasazi (a Dine' (Navajo) word meaning "enemy ancestors"), also called the "ancient ones," had their homeland. When the first people arrived in Chaco Canyon, there were abundant trees, a high groundwater table, and level floodplains without arroyos. This was most likely an ideal environment (conditions) for agriculture in this area. Chaco is beautiful, with four distant mountain ranges: the San Juan Mountains to the north, the Jemez Mountains to the east, the Chuska Mountains to the west, and the Zuni Mountains to the south.

The first Anasazi settlers, also called "basket makers," arrived in Mesa Verde (southwestern Colorado) around AD 600. They entered the early Pueblo phase (AD 700 to AD 900), which was the time they transitioned from pit houses to surface dwellings, evidenced by their dramatic adobe dwellings, or pueblos. Chaco Canyon was the center of Anasazi civilization, with many large pueblos probably serving as administrative and ceremonial centers for a widespread population of the Chaco regional system. Also of particular note is the extensive road system, built by a people who did not rely on either wheeled vehicles

Figure 2.13

Anasazi region showing Chaco Canyon. (*Courtesy of Lekson, S. H., T. C. Windes, J. R. Stein, and W. J. Judge, "The Chaco Canyon Community," Scientific American, Vol. 259, No. 1, pp. 100–109, July 1988)*

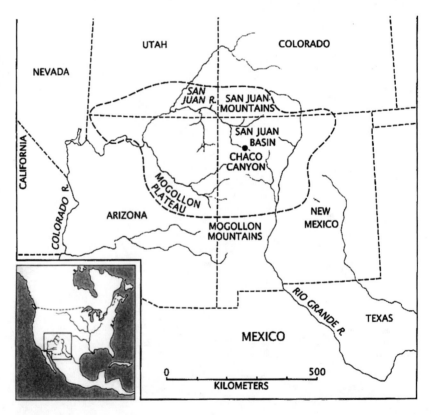

or draft animals. The longest and best-defined roads (constructed between AD 1075 and AD 1140) extended over 50 mi in length. The rise and fall of the Chacoan civilization was from AD 600 to AD 1200, with the peak decade being AD 1110 to AD 1120.

Chaco Canyon is situated in the San Juan Basin in northwestern New Mexico as shown in Fig. 2.13. The basin has limited surface water, most of which is discharged from ephemeral washes and arroyos. Figure 2.14 illustrates the method of collecting and diverting runoff throughout Chaco Canyon. The water, collected from the side canyon that drained from the top of the upper mesa, was diverted into canals by either an earthen or a masonry dam near the mouth of the side canyon. These canals averaged 4.5 m in width and 1.4 m in depth (Vivian, 1990); some were lined with stone slabs and others were bordered by masonry walls. The canals ended at a masonry head gate, where water was then diverted to the fields in small ditches or to overflow ponds and small reservoirs.

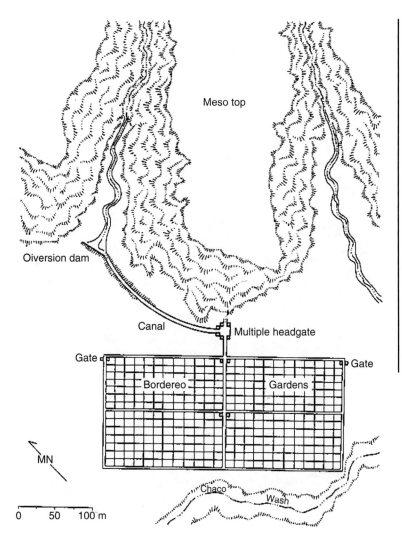

Figure 2.14
Water-control system in Chaco Canyon. (*Courtesy of Vivian, R. G., "Conservation and Diversion: Water Control Systems in the Anasazi Southwest," in T. Downing and M. Gibson (eds.), Irrigation Impact on Society, Anthropological papers of the University of Arizona, No. 25, pp. 95–112, University of Arizona, Tucson, 1974)*

The diversion of water into the canals combined with the clearing of vegetation resulted in the eroding (cutting) of deep arroyos to depths below the fields being irrigated. By AD 1000 the forests of pinyon and juniper trees had been deforested completely to build roofs, and even today the area remains deforested as shown in Fig. 2.15.

Between AD 1125 and 1180, very little rain fell in the region. After 1180, rainfall briefly returned to normal. Another drought occurred from 1270 to 1274, followed by a period of normal rainfall. In 1275, yet another drought began which lasted 14 years.

Of the five-factor framework for social collapse suggested by Diamond (2005), the only factor that did not play a role in the collapse of the Anasazi was hostile neighbors. Water sustainability was affected by the deforestation, the erosion (cutting) of the arroyos from the diversion of water resulting in lowering the groundwater levels and the supply source to the irrigated fields, and finally, the repeated periods of drought caused the final collapse.

Figure 2.15

Chaco Canyon. (*Photos copyright by L. W. Mays*)

THE BEGINNING OF THE WATERING OF THE WEST

The Colorado River blocked the exploration of sections in the west for several centuries. After the end of the Mexican War in 1848, the United States acquired Arizona, New Mexico, and California. In 1857, Lt. J.C. Ives explored the Colorado River from the Gulf of California to over 400 mi upstream near the present day Hoover Dam. Twelve years later Major John Wesley Powell, a one-armed veteran of the Civil War, explored the Colorado River from the Green River in Wyoming to the Virgin River in Nevada. His party was the first known to have traveled through the Grand Canyon and lived to tell about it. Powell was founder of the *United States Geological Survey* (USGS) and is considered as "the father of reclamation" in the United States (Espeland, 1998).

On the basis of Powell's studies and explorations, he created a comprehensive plan for developing the West, most of which was ignored. He was the first to argue the idea that large-scale irrigation was necessary to settle the West and that government, not private industry, would need to develop irrigation on the scale needed to sustain agriculture in the West. He recognized that the resources, technology, and coordination required were far beyond the means of individuals or private industry (Espeland, 1998). His *Report on the Lands of the Arid Region,* in 1878, was the first important stimulus to the national irrigation movement.

Approximately one-third of the United States, including most of the West, requires irrigation to sustain tilled agriculture. During the 1870s and early 1880s many private irrigation companies were created to meet the demand for irrigation, relying on eastern capital to make fast money. Most of the companies went bankrupt within 10 years, causing the irrigation boom to bust. After years of drought (1888–1897), farms failed, people left, and some began pressuring the federal government to invest in irrigation in the West.

President Theodore Roosevelt, being a strong backer of the federal development of irrigation and reclamation, maneuvered the Reclamation Act of 1902, creating the Reclamation Service or a new branch of the USGS. The Reclamation Service was moved from the USGS in 1907 to the Department of the Interior and renamed the Bureau of Reclamation in 1923. It is interesting to note that the Reclamation Act was conceived and sold as a regional home-building program, a political strategy to appease legislators who were concerned that subsidized water for large farms would cause unfair competition for eastern farmers.

Once the Reclamation Service was created, it was flooded with project requests. The Salt River Project in Arizona was one of the first projects authorized in 1903, illustrating what became a prominent pattern in the *United States Bureau of Reclamation* (USBR) development.

Water Law and Policy in the Southwest

Water management decisions are most often underlain by water laws. In the United States, water law has two basic functions:

1. The creation of supplemental private property rights in scarce resources
2. The imposition of public interest limitation on private use

For our purposes, water law is divided into surface water law and groundwater law. Surface water law is further categorized into riparian law and appropriation law. Riparian law is based on the riparian doctrine, which states that the right to use water is considered real property, but the water itself is not property of the landowner (Wehmhoefer, 1989). Appropriation law states that the allocation of water rests on the proposition that the beneficial use of water is the basis, measure, and limit of the appropriative right, the first in time is prior in right. In the western United States, surface water policy generally follows this doctrine of "first in time, first in right." In order to appropriate water, the user need only demonstrate availability of water in the source of supply, show an intent to put the water to beneficial use, and give priority to more senior permit holders during times of shortage. Beneficial use of water under the law includes domestic consumption, livestock watering, irrigation, mining, power generation, municipal use, and others. The states of Arizona and New Mexico follow the appropriation law of surface water, and in California and Texas both the appropriation doctrine and the riparian doctrine coexist.

Groundwater allocation is handled quite differently and is typically divided into common law or statutory law. Common law doctrines include the overlaying rights doctrines of absolute ownership, reasonable use, and correlative rights. These doctrines give equal rights to all landowners overlying an aquifer. Arizona, California, and Texas have adopted these principles for groundwater allocation.

The above surface and groundwater laws serve as the basis for individual state water policies. The burden of developing water policies lies upon each state. This is often achieved by the state proposing a water project and securing federal funds for the construction. It is also up to the states to agree on apportionment in interstate waters; if the states cannot agree, then the courts will intervene and settle the dispute by decree. The federal government only gets involved in such disputes where federal lands and Indian reservations are concerned.

Arizona

It is no secret that throughout Arizona's history, water policy has been directed at supporting the unconstrained growth of its population and major revenue-producing activities. Starting with mining, ranching, and farming, with the gradual shift to municipal and industrial uses, the water policy of the state has been directed at obtaining imported supplies. This has been an effort to augment what has appeared to be an insufficient and indigenous resource. Waterstone (1992) points out that the "state's water policies have led to the protracted exercise to capture and secure the *Central Arizona Project* (CAP), the ongoing infatuation with weather and water shed manipulation, the current experimentation with groundwater recharge and effluent use, and the recent spate of purchases of remote water farms." In Arizona, the state's water policy and management focused more on surface water than groundwater prior to 1980, when the Groundwater Management Code was developed; thereafter, the emphasis has been on groundwater. In regards to surface water, Arizona law defines surface water as "the waters of all sources, flowing in streams, canyons, ravines or other natural channels, or in definite underground channels, whether perennial or intermittent, flood, waste, or surplus water, and of lakes, ponds and springs on the surface." These surface waters are subject to the "doctrine of prior appropriation"

(ADWR, 1998). In Arizona, surface water rights are obtained by filing an application with the Department of Water Resources for a permit to appropriate surface water. Once the permit is issued and the water is actually put to beneficial use, proof of that use is made to the department and a certificate of water right is issued to the applicant. Once a certificate is issued, the use of the water is subject to all prior appropriations.

Because water law in the state of Arizona has changed substantially over the years, Arizona is now conducting a general adjudication of water rights in certain parts of the state. Adjudications are court determinations of the status of all state law rights to surface water and all claims based upon federal law within the river systems. These adjudications will provide a comprehensive way to identify and rank the rights to the use of water in some areas. The adjudications will also quantify the water rights of the federal government and the Indian reservations within Arizona.

In Arizona, groundwater problems arise from the overdrafting of water from the aquifers. Groundwater overdrafts cause many problems, such as increased well pumping costs and water quality issues. In areas of severe groundwater depletion, the earth's surface may also subside, causing cracks or fissures that can damage roads or building foundations. In order to manage groundwater pumping in Arizona, the Arizona groundwater management code was developed in 1980 as state legislation. The Arizona groundwater management code was named as one of the nation's 10 most innovative programs in state and local government by the Ford Foundation in 1986. This achievement came from the cooperation of Arizonans working together and compromising when necessary in order to protect the future of the state's water supply.

The Groundwater Management Code has three primary goals (ADWR, 1998):

1. Control the severe overdraft currently occurring in many parts of the state
2. Provide a means to allocate the state's limited groundwater resources to most effectively meet the changing needs of the state
3. Augment Arizona's groundwater through water supply development

In order to achieve these goals, the code set up a comprehensive management environment and established the Arizona Department of Water Resources.

The code outlines three levels of water management. Each level is based on different groundwater conditions. The lowest level applies statewide, and includes general groundwater provisions. The next level applies to *irrigation nonexpansion areas* (INAs), and the highest level applies to *active management areas* (AMAs) where groundwater depletion is the highest. The boundaries that divide the INAs and AMAs are determined by groundwater basins and not by political jurisdiction. The main purpose of groundwater management is to determine who may pump groundwater and how much may be pumped. This includes identifying existing water rights and providing new ways for nonirrigation water users to initiate new withdrawals. In an AMA or INA new irrigation users are not allowed. Even with the original publicity and enthusiasm, many people now feel that the efforts under the groundwater management code have been very costly with very little savings in water, making the success questionable.

Colorado River Basin and the Central Arizona Project (CAP)

Of the many river basins in the southwest, the Colorado River Basin has been the center of many controversies. The Colorado River Basin is divided into two sections, the upper and lower basins. The upper Colorado River Basin consists of the states of Arizona, Colorado, New Mexico, Utah, and Wyoming; the lower Colorado River Basin consists of Arizona, California, New Mexico, and Utah. Due to the doctrine of prior appropriation, the states in the upper Colorado River Basin became worried that the rapidly developing California would obtain a large portion of the appropriated water, leaving them with a shortage in the future. As an attempt to settle the issues, the upper basin states agreed to support California on the Hoover Dam proposal that it needed to obtain Colorado River water for its growing development. In return, the states requested a guaranteed amount of water from the river for their own future development. This agreement between the states resulted in the Colorado River Compact in 1922, which Arizona did not ratify until 1944. Table 2.1 lists the U.S. Federal laws of the Colorado River.

Under the Colorado River Compact, it was agreed that the upper Colorado River Basin would receive 7.5 maf, and the lower Colorado River Basin would receive 7.5 maf. It was also agreed that the lower basin would have the right to increase its beneficial consumptive use by 1 maf annually. All of the states supported the compact except Arizona, which opposed the compact and refused to sign it. The dispute over the water continued as the Boulder Canyon Project Act was passed. The Boulder Canyon Project Act was passed on December 21, 1928 by Congress, which authorized the construction of Boulder Dam (now Hoover Dam). However, the one stipulation was California must agree to limit its use of Colorado River water to an amount of 4.4 maf. Arizona and California fought over both the Colorado River Compact and the Boulder Canyon Act. Arizona was against the act and

Table 2.1

Federal Laws of the Colorado River

Year	Action
1922	**Colorado River Compact** apportions 7.5 MAF to lower basin states of California, Arizona, and Nevada
1928	**Boulder Canyon Project Act** authorizes Hoover Dam and All American Canal. Apportions lower Colorado River water, CA-4.4 MAF; AZ-2.8 MAF; NV-0.3 MAF
1945	**Mexican Water Treaty** apportions 1.5 MAF to Mexico
1948	**Upper Colorado River Basin Compact.** Arizona is apportioned 50,000 AF of water for territory in upper Colorado River Basin drainage
1964	***Arizona vs. California.*** U.S. Supreme Court Decree. Ratification of 1928 apportionment of the Colorado River water supply
1968	**Colorado River Basin Project Act.** Authorizes construction of the Central Arizona Project. Sets forth law governing the distribution and use of the CAP water
1974	**Colorado River Basin Salinity Control Act.** Authorizes works to control salinity of Colorado River water below Imperial Dam as part of Mexican Treaty obligation.

Source: Hermes and Mays, 2002.

did not want California to have any of their water. In order to help in settling the dispute, the U.S. Congress made it clear to Arizona that until they could settle the dispute of water allocation in the lower basin, the state would not receive any support for their water canal system, the CAP, which would later become a controversy in itself. Arizona finally agreed to share its water with California in order to receive funding for the CAP. As a result of the case *Arizona v. California*, which took place in 1964, the Supreme Court decreed that California would receive 4.4 maf of Colorado River water, Arizona would receive 2.8 maf, and Nevada would receive 300,000 maf.

CAP and the Users   The CAP was the largest, most expensive, and most politically volatile water-development project in the U.S. history; it was also the most ambitious basin project that the Bureau of Reclamation attempted (Espeland, 1998). Even early on in 1947, the strategy of CAP supporters was to paint CAP as a "rescue" operation. This was the project necessary to replace the "exhausted" groundwater supply in order to save the local economy. By 1963, the CAP was still justified as a "rescue" project; a doubling of the population over the previous 10 years supposedly made the project even more urgent. Economic development was assumed to be driven by agricultural development. The thought was that without more irrigated farmland, urban growth (which reduces irrigated farmland) would be stymied. How did the population grow so fast despite the previous prediction that water supply would limit economic growth?

In 1968, Congress authorized the construction of the CAP under the Colorado River Basin Project Act. The main purpose for the authorization was to assist Arizona in reducing its water deficiencies. By 1971, the first *environmental impact statement* (EIS) on the CAP was written and then finalized in 1972. The 1976 EIS was devoted solely to the Orme Dam, to become the beginning of a series of EISs in the major features of the CAP. In 1971, the *Central Arizona Water Conservation District* (CAWCD) was created to provide a means for Arizona to repay the federal government for the reimbursable costs of construction and to manage and operate the CAP once complete. The construction began in 1973 at Lake Havasu and was completed in 1993. The entire cost of the project was more than $4 billion. Under the Colorado River Basin Project Act, the CAP would be the first to take shortages in the lower Colorado River Basin.

The CAP is a 336 mi long system of aqueducts, tunnels, pumping plants, and pipelines. As shown in Fig. 2.16a, the CAP carries water from the Colorado River at Lake Havasu, through Phoenix, to the San Xavier Indian reservation southwest of Tucson. The main purpose of the CAP was to help Arizona conserve its groundwater supplies by importing surface water from the Colorado River.

The users of the CAP water fall into three categories. The first category is municipal and industrial. These customers include cities and water ulities which are responsible for treating drinking water and delivering it to residences, commercial buildings, and industries. The next water use category is agricultural. These agricultural users are primarily irrigation districts. The last category is the Indian community. These communities receive water from the CAP under contracts with the federal government. Agriculture has been the main water user in the past; however, due to the increasing development of Arizona, cities will soon become the largest customer for the CAP. The three priorities

Figure 2.16 (a)

The Central Arizona project. (*Courtesy of CAP*)

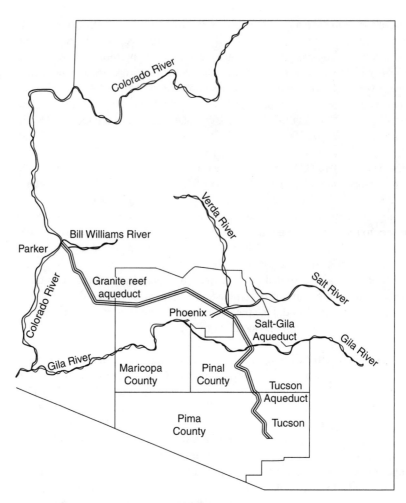

for water are (1) the municipal and industry of the Indians (2) agriculture (3) miscellaneous. Under shortages the order of issuing water would be the miscellaneous, agriculture, and then municipal and industry of the Indians.

One of the main criticisms of the CAP is the cost of the water and how the revenue is obtained. The price of the CAP water is determined annually by the CAWCD board of directors, and is based on projections of energy and operation, maintenance, and replacement costs. The payment shares for the municipal and industrial category, as well as the Indian agriculture, are based on their full annual CAP entitlement. The non-Indian agriculture user has the "take or pay" payment option. "Take or pay" means that the charge for the water is based on the amount available for delivery, not what is requested. The users essentially must pay for the water even if they do not use it all. This type of payment scenario was based on the assumption that non-Indian agriculture subcontractors would seek to purchase the remaining CAP water entitlement. Non-Indian agriculture obtains irrigation water from other less-expensive sources such as groundwater. This chain of events becomes very importatnt to the future of the CAP, because if Arizona does not use its full entitlement of Colorado River water it could

Figure 2.16 (b)

Central Arizona project (CAP) aqueduct through a residential are in Scottsdale, Arizona. (*Courtesy of CAP*)

possibly lose it. However, the protection of Arizona's Colorado River water entitlement is protected by law but can be changed by Congress.

THE PAST AND THE FUTURE

What relevance does the collapse of ancient civilizations have upon modern societies? When I look at the rapid development and population increases in the southwestern United States, a region with limited water resources, I continue to hope that our society will awaken. So many areas are being developed without regard to the future availability of water. We have developed the southwest with limited knowledge (short records) of historical rainfall and streamflow data. In many cases *paper water* (water created on paper that really does not exist) is being used to justify these projects. In recent decades we have not been exposed to the repeated extremely severe droughts that historically have occurred. Neither have we been faced with the realties of what a global climate change might bring rather rapidly. Simply stated, we have developed water policies under the assumptions that the past decades are typical of the future—and they may not be.

One might argue that if the ancient societies had our present-day technologies, they would not have failed. However in my opinion, presently we have the technologies to prevent future collapse of areas such as the southwestern United States, but there is still a good chance we will fail. I don't think that even newer (undeveloped) technologies are the answer for our present-day problems. The technologies exist to have prevented

Relevance of the Ancients

many of the problems associated with the water sustainability of Mexico City. The technologies have been available to have prevented much of the flood damage in New Orleans resulting from Hurricane Katrina. What we need is for society to have the political and institutional will to fund and to apply the available technology/solutions.

Quoting Falkenmach and Lindh (1993), "Water's fundamental importance in sustaining life and a culture makes any threat to an area's water supply a threat to its economic life as well."

The Unsustainable American Southwest

The American southwest is an excellent example of a highly developed region that is heading toward an unsustainable water base for the future. I have chosen to discuss the American southwest for several reasons. First it is where I live and it is a region I deeply love. I share my time between the "Valley of the Sun" in Arizona (the desert) and Pagosa Springs, Colorado (the mountains), two vastly different areas. Phoenix, Arizona is in a region with a semiarid climate that has attracted a large and rapidly growing population. More importantly this is an area that is growing rapidly and so far, very successfully, but may have severe problems in the future if we do not establish limits on its growth. Unfortunately the people of this region have relied on expensive supply-side projects such as the CAP (Fig. 2.16) to import water from the Colorado River.

Through the importation of water to southern California cities via aqueducts from northern California and the Colorado River, we have diverted nearly the entire flow of the Colorado River to supply water to Southern California and Arizona to irrigate crops and lawns in the desert and to fill swimming pools. Because of these diversions, today the Colorado River delta in northern Mexico is "a desiccated place of mud-cracked earth, salt flats, and murky pools," as noted by Postel (1997). The Coca Indians, "the people of the river," who have fished and farmed in the region for hundreds of years, are now a culture at the risk of extinction. After a canoeing trip of the Colorado River delta in 1922, Aldo Leopold, in his book *A Sand County Almanac*, referred to the delta as "a milk and honey wilderness."

Many barriers exist to the efficient management of the region's water resources, including a legal system from the gold and silver mining in the nineteenth century, "first in time, first in right" or "use it or lose it." Many areas rely on water pumped from groundwater aquifers that have been overdrafted for years and others are being allowed to develop without adequate water supplies for the future. Many areas have severe irrigation problems that are similar to what many ancient civilizations faced, with the irrigated soil becoming increasingly salty. California's San Joaquin Valley is an example where without irrigation, abundant crop yields are impossible. With irrigation, the land will very likely become impossible to farm. Modern methods do not seem to be helping the San Joaquin Valley avoid this fate. Farmers have tried to cleanse the salts from the soil by flushing it with water and draining it into the sea.

Simply stated, the water situation in the southwestern United States, as in many other parts of the world, is unsustainable as it is presently used and operated. Diamond (2005) states, "Just think today of the dry U.S. West and its urban and rural policies that profligate water use, after drawn up in wet decades on the tacit assumption that they were typical." We certainly are a much more advanced society than those of ancient societies, but will be able to overcome the obstacles to survival before us? Remember, the Ancients have warned us!

REFERENCES

Arizona Department of Water Resources (ADWR), "Overview of Arizona's Groundwater Management Code," available at Web site: www.adwr.state.az.us/Azwaterinfo/, 1998.

Brundtland, G. (ed.), Our Common Future: The World Commission on Environment and Development, Oxford University Press, Oxford, 1987.

Coe, M. D., *The Maya*, 5th ed., Thames and Hudson, New York, 1993.

Coe, M. D., *Mexico: From the Olmecs to the Aztecs*, 4th ed., Thames and Hudson, New York, 1989.

Crown, P. L., and W. J. Judge (eds.), *Chaco & Hohokam: Prehistoric Regional Systems in the American Southwest*, School of American Research Press, Santa Fe, New Mexico, 1991.

Diamond, J., Collapse: How Societies Choose to Fall or Succeed, Viking, New York, 2005.

Doolittle, W. E., *Canal Irrigation in Prehistoric Mexico: The Sequence of Technological Change*, University of Texas Press, Austin, TX, 1990.

Espeland, W. N., *The Struggle for Water: Politics, Rationality, and Identity in the American Southwest,* The University of Chicago Press, Chicago, IL, 1998.

Falkenmach, M., and G. Lindh, "Water and Economic Development," in P. H. Gleick (ed.), *Water in Crisis: A Guide to the World's Freshwater Resources*, Oxford University Press, New York, 1993.

Haviland, W. A., "Tikal, Guatemala and Mesoamerican Urbanism," *World Archaeology*, Vol. 2, No. 2, pp.186–198. October 1970.

Hauck, G., *The Aqueduct of Nemausus*, McFarland & Company, Inc. Jefferson, NC, 1988.

Haury, E. W., *The Hohokam: Desert Farmers and Craftsmen*, The University of Arizona Press, Tucson, 1978.

Hermes, V., and L. W. Mays, "Regional Water System Development and Management in the U.S. Southwest," in E. Cabrera, R. Cobacho, and J. Lund (eds.) *Regional Water System Management: Water Conservation, Water Supply, and System Integration*, A. A. Balkema Publishers, Lisse, The Netherlands, 2002.

Hodge, A. T., *Roman Aqueducts and Water Supply*, 2nd ed., Gerald Duckworth, London, 2002.

Lekson, S. H., T. C. Windes, J. R. Stein, and W. J. Judge, "The Chaco Canyon Community," *Scientific American*, Vol. 259, No. 1, pp. 100–109, July 1988.

Lister, R. H., and F. C. Lister, *Those Who Came Before*, Southwestern Parks and Monuments Association, The University of Arizona Press, Tuscon, AZ, 1983.

Masse, B.,"Prehistoric Irrigation Systems in the Salt River Valley, Arizona," *Science*, Vol. 214, No. 3, pp. 408–415, 1981.

Masse, B., "The Quest for Subsistence Sufficiency and Civilization in the Sonora Desert," in P. L. Crown and W. J. Judge (eds.), *Chaco & Hohokam: Prehistoric Regional Systems in the American Southwest*, School of American Research Press, Santa Fe, New Mexico, 1991.

Mestre, J. E., "Integrated Approach to River Basin Management: Lerma-Chapala Case Study-Attributions and Experiences in Water Management in Mexico," *Water International*, Vol. 22, No. 3, 1997.

Millon, R., "The Place Where Time Began: An Archaeologist's Interpretation Of What Happened in Teotihuacan History, in K. Berrin and E. Pasztory (eds.), *Teotihuacan: Art from the City of the Gods*, The Fine Arts Museums of San Francisco, San Francisco, 1993.

Perez de Leon, M. F. N., and A. K. Biswas, "Water, Wastewater, and Environmental Security Problems: A Case Study of Mexico City and the Metzquital Valley," *Water International*, Vol. 22, No. 3, 1997.

Postel, S., *Last Oasis: Facing Water Security*, Norton, New York, 1997.

Sanders, W. T., and B. J. Price, *Mesoamerica: The Evolution of a Civilization*, Random House, Incorporated, New York, 1968.

Thomas, D. H., *Exploring Ancient Native America: An Archaeological Guide*, Macmillan, New York, 1994.

Tortajada, C., and A. K. Biswas, "Environmental Management of Water Resources in Mexico," *Water International*, Vol. 25, No. 1, 2000.

Turney, O., "Prehistoric Irrigation in Arizona," *Arizona Historical Review*, Vol. 2, No. 5, Phoenix, 1929.

Vivian, R. G., "Conservation and Diversion: Water Control Systems in the Anasazi Southwest," in T. Downing and M. Gibson (eds.), *Irrigation Impact on Society,* Anthropological papers of the University of Arizona, No. 25, pp. 95–112, University of Arizona, Tucson, 1974.

Vivian, R. G., *The Chacoan Prehistory of the San Juan Basin*, Academic Press, San Diego, CA, 1990.

Waterstone, M., "Of Dogs and Tails: Water Policy and Social Policy in Arizona," *Water Resources Bulletin,* American Water Resources Association, Vol. 28, No. 3, pp. 479–486, 1992.

Wehmhoefer, R. A., *Water and the Future of the Southwest*, in Z. A. Smith (ed.), University of New Mexico Press, Albuquerque, New Mexico, 1989.

Woodbury, R. B., "The Hohokam Canals at Pueblo Grande, Arizona," *American Antiquity*, Vol. 26, No. 2, pp. 267–270, October 1960.

3 WATER RESOURCES SUSTAINABILITY: AN ECOLOGICAL-ECONOMICS PERSPECTIVE

Christopher Lant
Department of Geography and Environmental Resources,
Southern Illinois University
Carbondale, Illinois

INTRODUCTION

"Sustainability" is both a vague and politicized term, yet it is precisely because the world community has rallied around sustainability and sustainable development as normative goals of ecological-economic performance that the stakes are high for defining the concept in a manner that is true to its spirit. To do so, one must counteract definitions that either suit particular interests or are so broad and vague that most of what people do for self-interested reasons fits within them. Like other fields, water resources has struggled to bring the concept of sustainability to bear in the realm of practice. For example, what allocation of water in the Klamath River basin best achieves sustainability? Are plans to pipeline fossil ground water from the Ogallala of North Texas to the growing cities of Dallas and San Antonio consistent with sustainability? Is it sustainable to forgo renewable hydroelectric power in hopes that it will prevent the extinction of a strain of chinook or coho salmon? Is the recent completion of the Three Gorges Dam project on the Yangtze River an example of sustainable development? How sustainable is it to live in a world where over 1 billion people lack access to safe drinking water and 2 billion lack access to the basic benefits of the sanitation revolution (DeVilliers, 2000)?

Ecological economics helps us make more sustainable water resources decisions in three important ways. First, it provides a needed *theoretical* revision to neoclassical economic analysis. Second, this theoretical perspective points us toward better *methodologies for measuring* the value of water in competing uses. Third, it helps us identify a program of *institutional reform* that has the best chance of delivering more sustainable water resources management practices. Building a functional and operational definition of sustainability is the challenge. To bring life to the concept and goals of sustainability, it must guide us toward the best answers to these questions.

AN ECOLOGICAL-ECONOMICS VIEW OF SUSTAINABILITY

An ecological-economics view of sustainability is inevitably based on systems thinking (e.g., Capra, 2002; Costanza et al., 1993; and Costanza, 2001). Figure 3.1 presents a systems conceptualization of sustainable development where natural, human, intellectual, and manufactured capital are transformed continuously, one into another, by the processes of the market economy. The system is driven by low-entropy solar energy and evolves through the process of interactions among its interdependent components (after Capra, 1996), releasing high entropy heat as waste. A component of this system is the market economy as analyzed by neoclassical economics, where land, labor, and capital are obtained as factors to produce goods and services for consumption and investment that are measured as economic output or gross product.

In contrast to neoclassical economics, however, ecological economics views production and consumption of marketable goods and services as only an important part of a larger process. Neoclassical economics views manufactured capital (i.e., infrastructures of various kinds) as essential to economic production. Recent literature emphasizes the

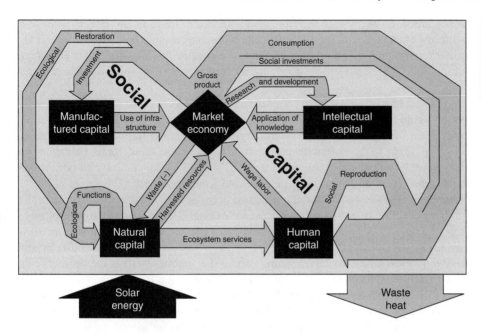

Figure 3.1
A systems conceptualization of sustainable development.

critical importance of intellectual capital as the driving force of the information revolution. Social scientists have extended the analysis to include human and social capital, and ecological economics has extended it further to include natural capital. Human capital is the set of attributes (e.g., knowledge, skills, attitudes, mental and physical health, and so forth) that determine individuals' capacity to contribute to society. While definitions of social capital vary, it is usefully conceived as the set of historically developed institutions and social networks that structure the productive and reproductive process as a whole. Natural capital is both the standing stock of natural resources that await future use and the characteristics of ecosystems that maintain ecological and environmental processes such as biological productivity and diversity and biogeochemical cycling. When these are beneficial to humans, they are ecosystem services.

Economic production is absolutely dependent in the medium- to-long term on each of these forms of capital: natural, human, manufactured, intellectual, and social. Consequently, increasing economic output in the short term by diminishing one or more of these capital stocks is "unsustainable," unless that capital stock is in long-term surplus supply. For instance, the eighteenth- and nineteenth-century American pioneers found a frontier enormously rich in natural capital stocks of forests, fertile soils, fish and wildlife populations, useful mineral deposits, unpolluted waters, well-functioning ecosystems, and so forth, but impoverished in human-derived forms of capital suitable to their purposes (as opposed to Native American forms and purposes). For them, liquidating this natural capital in order to transform it into manufactured and human capital increased the value of the overall capital stock available to frontier society and was necessary for the development process to be sustaining. But times have changed. Natural capital, historically taken for granted as a free good or accounted for only when it is used as an industrial raw material, is more and more often one of the limiting factors in the development of the system as a whole, in the same sense that phosphorus is often the limiting factor in algal growth in aquatic ecosystems.

A sustainable economy must therefore limit withdrawals from and produce investments in all forms of capital, working synergistically with noneconomic processes of natural and social reproduction, to ensure that no form of capital is diminished in order to increase short-term output of marketable goods and services. That is sustainable development.

If various forms of capital were completely substitutable, the "weak" sustainability criterion would be satisfactory. As long as we maintain the aggregate capital stock, shortages in one form of capital (e.g., natural capital) could be substituted for by investments in other forms (e.g., manufactured capital). But because these forms of capital are incompletely substitutable in practice, the "strong" sustainability criterion should hold—each form of capital must be protected from degradation (Pearce et al., 1992; Tietenberg, 2003). Manufactured capital can occasionally substitute for natural capital in the production of ecosystem services; for example, levees and flood control reservoirs contain flood waters formerly held by wetlands and organic matter in soils. Wastewater treatment plants accelerate the rate at which aerobic bacteria oxidize organic matter. In most cases, however, natural capital (i.e., nature itself) is the most efficient and effective producer of ecosystem services. Even in the example given, levees and reservoirs can provide flood control services in lieu of wetlands, but they do not provide equitable services in terms of wildlife habitat or biogeochemical cycling services such as *denitrification*. This illustrates the need to maintain natural capital as the best means to generate multiple ecosystem services in most instances. Achieving a better understanding of ecosystem services, the ecosystem functions that maintain them, and the ways in which they contribute to human capital is, consequently, a key research agenda as identified by the *National Science Foundation* (NSF Advisory Committee for Environmental Research and Education, 2003).

Sustainable development as an evolving political program focuses on the reform of social capital (i.e., institutional rules and cultures) such that the economic production process sustains stocks of human, intellectual, manufactured, and natural capital. Natural capital is of special concern because it provides not only essential resources for future economic use, but also essential ecosystem services such as nutrient cycling, waste treatment, disturbance regulation, atmospheric gas exchange, soil formation and binding, and habitat for the tremendous, but diminishing, diversity of life on this planet. The ecosystem functions that generate these ecosystem services are the biogeochemical processes that make some parts of the Earth, and no other place that we know of, habitable.

An ecological-economics approach to sustainability provides a valuable critique of neoclassical macroeconomics in a number of ways. First, it points out that economic growth can occur in positive, neutral, or negative ways with respect to sustainability. Sustainable economic growth occurs when new applications of knowledge (intellectual capital) allow a society to increase the efficiency with which various forms of capital are utilized to produce goods and services. This occurs through new or improved technologies, better systems of social organization, better means of making the experience of work an investment in rather than a withdrawal from human capital, or more efficient transformation of natural capital into products. Hawken et al. (1999) provides convincing evidence that the modern western industrial system, especially that of North America and Australia, is very efficient at utilizing labor and manufactured capital, but it is not an efficient transformer of natural capital into economic value. Huge improvements can thus be made with current technologies such as hybrid cars, wind turbines, passive solar designs, and drip irrigation. The developed Japanese, western European, and Israeli economies provide working examples of some of these improvements. Because this type of economic growth reinforces each of the system components in the long term, it is the core of sustainable development.

Neutral economic growth occurs when important social processes (e.g., cooking meals, raising children, growing food) or natural processes (e.g., maintenance of soil fertility) that have heretofore occurred in the nonmarket spheres of ecological or social reproduction are incorporated within the market economy, increasing measured economic output without necessarily improving the effectiveness of the larger ecological-economic system. Examples can be found in the rapid development of the low-wage service economy in Western societies and the transformation of subsistence to commercial farming in developing countries.

Unsustainable economic growth occurs when an increase in the output of market goods and services comes at the expense of reductions in the value of natural capital (e.g., pollution, use of renewable resources beyond sustainable yield), human capital (e.g., labor exploitation), intellectual capital (e.g., reduced investment in education and research), and/or manufactured capital (e.g., severe depreciation of urban water supply infrastructures) that exceed the value of the additional goods and services produced. When this occurs, the processes of social or ecological reproduction are disrupted, undermining the entire system's ability to recreate itself in the long term. Repetto (1992), for example, has documented how relatively high rates of economic growth in Costa Rica and Indonesia are the consequence of the liquidation of natural capital stored as forests, soil, wildlife, and watershed protection. Bartelmus (1994) offers a modification of national income accounts to take natural capital and ecosystem services into account when measuring economic (i.e., ecological-economic) performance.

Water plays at least three critical, but distinct, roles in the ecological-economic process diagrammed in Fig. 3.1. First, water is a raw material, a factor of production, of a number of marketable commodities, some of which are themselves factors of production of other final goods. Electricity, transportation, crops, livestock, industrial goods of various kinds, and residential and commercial landscapes each generate a derived demand for water. Second, because of its contribution to human health, treated potable water for domestic use is enormously valuable in producing human capital, whether it is delivered as a commodity by a private-sector firm, as a public service by a government-owned utility, or by some other institutional arrangement. Third, water in oceans, estuaries, rivers, lakes, wetlands, soil, and other components of the hydrologic cycle is a, if not the, critical factor of production of ecosystem services. In fact, one could argue that without water no ecosystem services could be generated; this is why the exploration of Mars by remote-sensing devices focused on the search for water. Wetlands are the most illustrative example of water-defined environments that produce multiple ecosystem services such as flood control, water purification, wildlife habitat, carbon sequestration, nitrogen cycle regulation, and sediment control to name just a few (Mitsch and Gosselink, 1993). Watersheds serve as particularly useful geographic and planning units for the management of natural capital (e.g., Gottfried, 1992; Lant et al., 2005; Ruhl et al., 2003) because they are spatially arranged in a hierarchy of scales, are usually easily definable geographically, and generate ecosystem services such as nutrient cycling, biodiversity, and control of the hydrologic cycle in a semiclosed system.

The contribution of water to sustainable development in these various uses must be evaluated if we are to understand the "highest and best use" of water, keeping in mind that it is the *marginal* value of water (or marginal value of changes in water quality) that needs to be compared among various uses. Here "marginal" refers to incremental changes from a base condition, or the rate of change in total costs and benefits. For example, the summer 2002 low-flow tragedy on the Klamath River, where 33,000 spawning salmon died, and on the Rio Grande, where the silvery minnow nearly lost its fight against extinction, are cases where the *marginal ecological opportunity costs* of reduced flows in those rivers under conditions of drought exceeded the *marginal economic benefits* of the agricultural products made possible by the water allocated to irrigation from those rivers. Results from a number of studies, for example, indicate the low marginal value of water applied inefficiently to crops that are in surplus supply or are used as livestock feed rather than for direct human consumption (e.g., Zilberman, 2002). In contrast, most ecosystem services provided by water are public goods and are thus subject to all the problems of market failure where private property rights to flows of value are not well established or where it is difficult or impossible to exclude benefits from those not paying for their production (e.g., Randall, 1983). The key, then, is to redesign policies and institutions so that local water managers making short-term decisions about water allocation, water quality, and the physical condition of aquatic ecosystems account for the costs and benefits of their actions on natural capital and ecosystem service flows. For example, had a system of leasing water rights similar to that applied in California in the 1990s drought been in place, the high ecological costs incurred in the Klamath and Rio Grande rivers might have been avoided. In other situations, changes may be required in the evaluation of the costs and benefits of water resources engineering projects, water prices, property rights and access rules to water, or the roles of various levels of government, nongovernment organizations (NGOs), and private sector firms.

MEASURING THE ECOLOGICAL-ECONOMIC VALUE OF WATER

Ecological economics improves our ability to *measure* the relative value of water in competing uses. In a widely read and controversial paper, Costanza et al. (1997) estimated that the annual value of ecosystem services is $16 to $54 trillion with a mean value of $33 trillion, slightly exceeding the annual output of goods and services in the world economy of $31 trillion. Of course this estimate is inaccurate, but it shows that "utility," viewed as contributions to human capital, is derived from ecosystem services in addition to utility derived from marketable goods and services (along with other sources of utility such as social reproduction). Moreover, ecosystems that generate the greatest value of ecosystem services per hectare are environments that are defined by water (Table 3.1). From this we know that the global value of water's role in producing ecosystem services is very large, but in any specific situation, we need to know the current and future local *marginal ecological-economic costs and benefits* associated with

Ecosystem Type	Annual Value of Ecosystem Services ($ per hectare)	Global Area (hectares $\times 10^6$)	Total Annual Value of Ecosystem Services ($ billion)
Estuaries	22,832	180	4110
Swamps/floodplains	19,580	165	3231
Sea grass/algal beds	19,004	200	3801
Tidal marsh/mangroves	9990	165	1648
Lakes/rivers	8498	200	1700
Coral reefs	6075	62	375
Tropical forest	2007	1900	3813

Table 3.1

Estimated Value of Ecosystem Services Provided by Leading Ecosystem Types

Source: Costanza et al., 1997.

various management or policy options. Accurately measuring these costs and benefits, rather than calculating the $33 trillion figure more accurately, is the methodological challenge for using ecological economics as a guide to decision making (Toman, 1998).

The *contingent valuation method* (CVM), for all its flaws, has proven useful in evaluating how individuals make trade-offs between marketable goods and services and non-market ecosystem services (Braden, 1997; Mitchell and Carson, 1989). CVM and other environmental economics methods, such as the property value (hedonic modeling) method, therefore have important roles to play in doing ecological economics, but ecological economics may provide a different interpretation of *willingness to pay* (WTP) and *willingness to accept* (WTA) compensation bids. For example, the high WTA bids often received by CVM researchers for diminishment in ecosystem services has sometimes been explained as risk or loss aversion, but may also be strong evidence of the high value people place on ecosystem services as a source of utility beyond that derived from marketable commodities. Secondly, while CVM treats ecosystem services as a source of individual utility similar to purchased goods and services, they of course are rarely individually owned but instead generally accrue to geographically defined communities over long periods of time. All of the time-honored debates about discount rates and the distinction between "consumers" and "citizens" therefore apply, with ecological economics generally favoring citizens with low discount rates. High WTA bids may also indicate ethical problems with receiving individual payment for diminishment of a community benefit. These vigorous debates over CVM illustrate the theoretical distinctions between neoclassical and ecological economics and provide a guide to better utilization of this valuable methodology in ecological-economic analysis.

Perhaps more powerful than CVM and other valuation techniques in the long run for evaluating management and policy options in a complex ecological-economic system are advancements in evolutionary algorithms, such as *genetic algorithms* (GAs). GAs have proven successful for decision support in a variety of water resources applications, including water supply system design (Murphy et al., 1993), groundwater management (Hilton and Culver, 2000), reservoir management (Esat and Hall, 1994; Nicklow and Bringer, 2001), and management of watershed land-use patterns for ecosystem services

(Bennett et al., 2004). GAs thus show promise for tackling problems of finding the highest and best multiple objective use of water in mathematically complex ecological-economic systems models where critical feedbacks within and among the various spheres of capital are taken into account. The NSF Biocomplexity in the Environment program may provide us with needed methodological advances to bring evolutionary algorithms and other methodologies into practice in water resources management for sustainability.

TOWARD SUSTAINABLE MANAGEMENT OF WATER RESOURCES

Ecological economics, in combination with perspectives borrowed from institutional economics, provides a guide to evaluating what *institutional reforms* are needed to make water resources management more sustainable. Much of the recent literature on reforming water resources management is consistent with sustainability. As examples of the application of ecological economics to policy, I will briefly discuss three current ideas for reforming water resources in terms of their relationship to sustainability: a human right to water, *integrated water resources management* (IWRM), and virtual water.

A Human Right to Water

Between 1970 and 2000, the proportion of people in developing countries who have access to potable water has increased from 30 percent to 80 percent. For sanitation the increase is from 23 to 53 percent (Lomborg, 2001). This represents an underreported success story of sustainable development. Nevertheless, the billion or more people still lacking safe drinking water and the 2 billion or more lacking wastewater services are precisely those who lack sufficient money income needed to generate the effective market demand for water that would make investment in delivery infrastructures profitable for private sector firms. These same societies also lack the financial capital to build infrastructures that can deliver safe drinking water as a public service. Yet the delivery of safe drinking water and wastewater management to populations that have lacked these basic human needs has the potential to cause a domino effect of development improvements by preventing over 2,000,000 deaths per year, mostly young children, at the hands of diarrheal diseases (Gleick et al., 2004). These improvements more than justify the investments despite their unprofitability in a narrow economic sense. One of the Millennium Development Goals adopted by the United Nations General Assembly is to halve by 2015 the proportion of people without sustainable access to safe drinking water (UNDP, 2003).

Japan leads all nations in aid directed to accomplish these goals (Glieck et al., 2004). As of this writing, the Bush administration had recently announced new commitments to improve the delivery of safe drinking water in order to bolster the U.S. program to reduce the impact of AIDS in Africa. More generally, the accessibility of safe drinking water greatly decreases the (unpaid) labor requirements of women to gather water and the incidence of gastrointestinal diseases in children. These effects improve the ability of young children to gain the full nutritional benefit of their limited food supplies, decreasing infant mortality. In turn, the practice of having large families to ensure that some children survive to provide support during old age becomes less important as a survival strategy among the poor. Therefore, fertility rates fall, freeing women for other important roles in earning income or in contributing to the social reproduction of the community. In other words, a lack of safe drinking water and basic wastewater services represents a human capital crisis and is the limiting factor in sustainable development.

The first 50 L per capita per day of potable domestic water used has very positive effects on human capital in the form of public health. Moreover, the volume of water that this represents is equivalent to only about 7 percent of mean rates of domestic use in the United States; domestic use is only 13 percent of all U.S. withdrawals, making 50 L per capita per day equivalent to only 1 percent of U.S. water withdrawals. It represents 3 percent of withdrawals in Bangladesh, but 35 percent in water-short Kenya. On this basis, Peter Gleick, at the Pacific Institute, and others have argued that this small but essential amount of potable water is a *right* rather than a *good*. While the manufactured capital needed to capture, treat, and deliver this water is substantial, the high public health values and high potential indirect effects on development justify national or international subsidization of delivery infrastructures to bring this water to most now lacking it, whether one conceives of the water delivered as a right or a basic good.

Integrated water resources management (IWRM) is based on the four 1992 Dublin principles:

Integrated Water Resources Management

1. Fresh water is a finite and vulnerable resource, essential to sustain life, development, and the environment.
2. Water development and management should be based on a participatory approach, involving users, planners, and policy-makers at all levels.
3. Women play a central part in the provision, management, and safeguarding of water.
4. Water has an economic value in all its competing uses and should be recognized as an economic good.

Principle 1 is discussed in some depth and principle 3 has been touched upon, but principles 2 and 4 need to be further explored because they are at the heart of the program of institutional reform that sustainability demands.

Watersheds, while long recognized as essential spatial units of hydrologic analysis, have increasingly been viewed in a number of countries as essential units of water governance. This trend has been driven by a change in the nature of water resources management problems away from engineering projects designed for water resources development, flood control, and waste treatment and toward integration of surface and groundwater management, management of polluted runoff, floodplain zoning and flood warning systems, management of water-based recreation, and protection and restoration of wetland and aquatic ecosystems and the services they deliver to local populations (Lant, 1998). This latter set of water resources management problems interfaces closely with land use and therefore with institutions of local social capital. In the United States, most of the thousand-plus recently-created watershed groups and initiatives lack the institutional capacity to manage the preceding list of issues. Nevertheless, institutions organized around relatively small-scale watersheds are very likely to grow as fora for stakeholder participation and to acquire legal authority in the process of meeting challenges such as development of total maximum daily load (TMDL) plans (Ruhl et al., 2003). Watersheds are geographically-defined units of natural capital, yet the social capital to manage these enormous assets is just beginning to develop. IWRM provides a way forward to accomplish this essential step toward sustainability.

Figure 3.2, adapted from Global Water Partnership (2000), shows how principle 4 is central to an ecological-economic approach to sustainability. The market value of water is only a portion of the economic value of water, to which must also be added the non-market values to human capital and ecosystem service values if the total ecological-economic value of water is to be identified. On the cost side, the fixed and variable cost of manufactured capital used to deliver water is the supply cost, but in order to find the total economic cost, the opportunity cost of allocating the water itself to its next best use, and any economic externalities (positive or negative) associated with this allocation must be added. The total ecological-economic cost, however, also includes any diminishment in ecosystem services associated with the allocation of water away from the site of its natural origin, a reduction in the quality of water returned, or the physical manipulation of aquatic environments associated with capturing the water. The highest and best ecological-economic use of water is the use with the greatest net value as shown in Fig. 3.2. As is apparent, this use often differs from current uses of water. Consequently, IWRM has been touted as an evolving framework for applying the concept of sustainability to the practice of water resources management.

One application of this approach is in managing in-stream flow in water-short regions by analyzing the relative marginal benefits of off-stream and in-stream uses of water (Fig. 3.3). Many studies have established that, in general, marginal benefits of water are highest for domestic uses, followed by commercial and industrial uses, irrigation of fruit and vegetable crops, and finally irrigation of grains, pastures, and other livestock feeds. Reallocation from lower to higher value uses, through markets or quasimarkets if these can be established, is advocated in order to increase the overall benefits of water use. However, when left in the stream, water also generates ecosystem services in the form of aquatic life and riparian zones that are often the most productive ecosystems in arid landscapes. While the marginal value of in-stream flow is likely declining in the

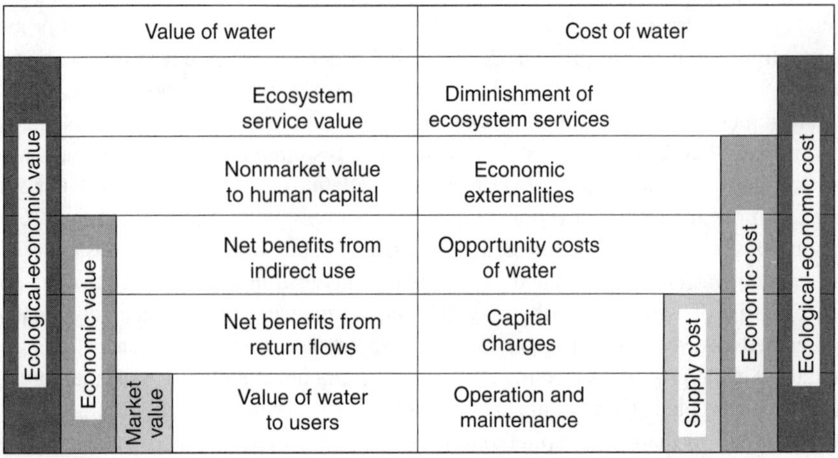

Figure 3.2

Comparison of market, economic, and ecological-economic value and cost of water.

Value of water		Cost of water	
Ecosystem service value		Diminishment of ecosystem services	
Nonmarket value to human capital		Economic externalities	
Net benefits from indirect use		Opportunity costs of water	
Net benefits from return flows		Capital charges	
Value of water to users		Operation and maintenance	

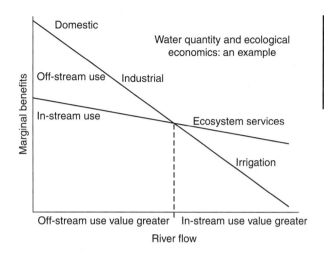

Figure 3.3
An ecological-
economic concep-
tualization of
optimal in-stream
flow.

majority of cases (each cubic meter of flow is less valuable than the last), even this simplified analysis, which leaves out thresholds that may be critical ecologically, points to
a level of flow where the marginal value of the ecosystem services produced by in-
stream flow begins to exceed the marginal benefits of off-stream uses, especially irrigation. This approach has applicability to the Klamath and Rio Grande examples cited
earlier where the marginal value of water left in these rivers for support of salmon or
endangered species is greater than the marginal value of crops irrigated with water that
was withdrawn from these rivers.

A similar logic can be applied to water quality using the case of fertilizer applications
(Fig. 3.4). While *yield* is maximized when the marginal value of fertilizer application is
zero, *profit* is maximized at a somewhat lower rate of fertilization, where the value of
marginal increases in yield equals the marginal cost (price) of fertilizer. However, fertilization has increasing marginal ecological opportunity costs in the form of groundwater
contamination with nitrates and eutrophication of waterways extending from the first
waterbody downstream of the field all the way to the hypoxic zone in the Gulf of
Mexico. These costs have considerable geographic variation depending on the hydrologic context of the agricultural land in question, such as the use of water in downstream
waterbodies and availability of filter strips, wetlands, and other sites for nutrient cycling.
When these costs are included in the analysis, the *ecological-economic optimum* level of
fertilization is lower than the profit-maximizing level; how much lower is site-specific.

Virtual Water

Virtual water is the amount of water used in the manufacturing or cultivation of specific
products, and therefore represents the water "embedded" within those products. For example, as a world average, paddy rice requires 2291 metric tons (m³) of water per ton of grain
produced. The figures for wheat (1394), soybeans (1789), and for maize/corn (909) are
somewhat less, but for cotton lint (8242), coffee (20,682), chicken meat (3918), and beef
(15,497), the figures are even higher. The cup of coffee that I drank while writing this
section required 140 L of water, the cotton T-shirt and leather shoes I am wearing embed
4100 and 8000 L, respectively, the glass of orange juice I had for breakfast, 170 L, and the

Figure 3.4

An ecological-economic conceptualization of optimal fertilization rates.

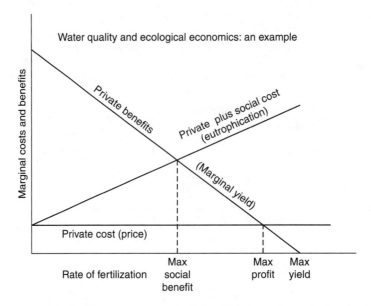

Water quality and ecological economics: an example

microchip that runs my computer, 32 L. If I have a tomato for lunch, that will require only 13 L, but if I have a hamburger, 2400 L of water will be used.

A 2004 report by the UNESCO International Hydrologic Program illustrates the relevance of virtual water in analyzing the "water footprint" of specific nations and the importance of trade in water-intensive products in redistributing water from humid to arid regions. World average per capita water use is 1243 m³ in the 1997 to 2001 period; 80 percent in the form of agricultural products. The United States consumes more water than any other large country at 2483 m³ per capita, 58 percent of which is agricultural. India consumes 980 m³ per capita, 86 percent of which is agricultural. China is similar at 702 m³ per capita, 87 percent of which is agricultural. Food trade is particularly important because it has the effect of moving "virtual" water from humid regions enjoying rain-fed agriculture to arid regions, thereby releasing water from irrigation. In 2001, 1625 Gm³ (cubic kilometers) of virtual water were traded internationally, 58 percent embedded in crops, 18 percent in livestock products, and 24 percent in industrial goods. This represents 15 percent of all agricultural water use and 34 percent of all industrial water use. Australia, Canada, United States, Argentina, and Brazil are the leading net virtual water exporters; Japan, Italy, United Kingdom, Germany, and South Korea are leading importers (Chapagain and Hoekstra, 2004). A number of countries rely on virtual water imports for over two-thirds of their effective water supply. These include arid countries such as Kuwait (87 percent), Oman (76 percent), Israel (74 percent), and Jordan (73 percent), but also affluent countries that import a large proportion of their raw materials such as the Netherlands (82 percent), Belgium-Luxembourg (80 percent), Switzerland (79 percent), and the United Kingdom (70 percent). Saudi Arabia is 53 percent dependent on virtual water, but could perhaps reduce unsustainable groundwater withdrawals and expensive desalination by using even greater food imports as a substitute.

Allan (2001) argues that regional differences in the total opportunity costs of using such vast quantities of water give humid regions a substantial comparative advantage over arid regions in food production. Arid regions, therefore, greatly benefit by importing virtual water in the form of food trade, and in fact are increasingly doing so, often behind the scenes, as the only economically sound means available to overcome regional water shortages. For example, Israel imports 87 percent, Jordan 91 percent, and Saudi Arabia imports 50 percent of their grain supply (Lomborg, 2001). Allan (2002), in applying the virtual water concept to the arid *Middle East/North Africa* (MENA) region, shows that MENA countries are now importing 50 million tons of grain annually and that these imports are the primary reason why political conflict over water (as opposed to conflict over other issues) has been minimal. In fact, the joint water committee governing the water resources of the Jordan River basin continued to hold meetings even during the height of the second *intifada* in 2001 and 2002. Wolf (2003) similarly points to the role water can play in maintaining lines of communication, even during times of political conflict, and the dominance of cooperation over conflict in the international management of water. By bringing to bear the international trading system in grain and other nonperishable agricultural products, the virtual water strategy makes it possible for arid regions such as MENA to meet their water needs by transporting, in a thousandfold more condensed form, water falling as rain and infiltrating into the fertile soils of the U.S. Midwest, the Canadian prairies, the Pampas of Argentina, or other regions where favorable conditions of climate, soil, and population density allow food production to exceed regional demand.

It is consistent with an ecological-economic approach to sustainability for populations living in arid regions to import much of their food from more humid regions and thereby preserve their scarce local water supplies for high value municipal and industrial use and for ecosystem services. The economic and ecological opportunity costs of allocating water to agriculture are much higher in arid than in humid regions, often even after transaction, transportation, and storage costs are accounted for. However, a few caveats must be offered. First, pursuing the virtual water strategy can undermine agricultural communities in arid food-importing regions (while augmenting those in humid food-exporting regions). Second, continued above world average rates of population growth in arid regions will exhaust even the power of the virtual water strategy to meet food and water needs. Arid regions must soon begin to follow the trend in the rest of the world toward lower total fertility rates, or be forced into massive and expensive desalination projects to meet even domestic and commercial water needs for which virtual water cannot substitute.

WHERE ECOLOGICAL ECONOMICS TAKES US

Water is never an end in itself. It is always a means to more fundamental ends. If we are to manage our water resources sustainably, what is it that we want to sustain? Fortunately, the answer is not that difficult. We want to sustain human welfare, widespread prosperity, peace, and ecosystem health, recognizing that sustaining each of these depends upon sustaining the others. We want to avoid, as competition for freshwater resources intensifies, sacrificing any of these for the sake of the others or for special interests.

Ecological economics, I have argued, provides us with the best normative and analytical guide to identifying sustainable paths and rectifying unsustainable paths by, essentially, expanding the meaning of "efficiency" to the larger concept of "sustainability" by including system interactions and nonmarket components such as ecosystem services and human capital (Fig. 3.1). What would ecological-economic analyses of water resources decisions tell us to do and what policy approaches would it identify as the best means to do it? In other words, if full ecological-economic cost and value of water (Fig. 3.2) were incorporated, how would water resources management change?

Ecological economics tells us that we need a global effort to continue to increase the proportion of people with potable water and basic sanitation from its current 80 percent and 53 percent, respectively. As is commonly pointed out in other terms, delivery of potable water is, like education, an investment in human capital with a very high rate of return, whether or not either water or education is a "right."

Ecological economics tells us that we need to arrest and reverse the steep decline in the health of many aquatic and coastal ecosystems through a program of ecological restoration that has only recently taken its first uncertain steps. The ecological improvements in Lake Erie and the Hudson River, for example, while partial, demonstrate the great benefits that could be derived if we ultimately prove to be successful in the Everglades, the Chesapeake Bay, and the Columbia and Penobscot Basin salmon runs. But the list of aquatic and coastal ecosystems whose services have been diminishing is much longer. Widespread bleaching of coral reefs is taking place as a result of a rise in ocean temperatures (World Resources Institute, 2000). The 20,000 watersheds on TMDL 303(d) lists do not begin to exhaust the set of watersheds where polluted runoff or past engineering greatly inhibits aquatic ecosystem health. Building momentum on this great task requires a synergy among traditional policy approaches, such as TMDL regulation and congressional funding for Army Corps of Engineers restoration projects, and new ones, such as an empowerment of watershed-based institutions and, especially, a change in water resource economics.

Incorporating full ecological-economic value and cost would, for example, overhaul the way water is used in agriculture in the United States and perhaps abroad. In rain-fed agriculture, the wettest lands currently used for crop production would become less profitable in that use than reallocating those lands to riparian zone protection and wetlands, the best multipurpose ecosystem service factories that we have. Carbon credits would induce farmers to "recarbonize" remaining croplands, with benefits not only to climate but to flood water retention in soils, future soil productivity, and the constructive utilization of livestock manure. Agrichemical use would become more expensive through taxation, tradable permits organized by watershed, or some other means. Reduced use would ameliorate the runoff and leaching of nitrogen, phosphorus, and pesticides into surface water and nitrates and pesticides into groundwater. The quantities of water needed to produce meat, especially beef, are enormous because of the inefficiency in converting water-consuming pastures and feed grains into meat. If the total ecological-economic costs of this water were incorporated into the price of the final product, the cost of meat production would increase significantly. The internalization of ecological costs of large-scale feedlot operations and the opportunity costs of using cropland for livestock feed rather than for crops for human consumption would raise the price even further. The resulting reduced demand for meat would decrease rates of heart disease (Willett and Stampfer, 2003), a further benefit to human capital.

In irrigated agriculture, farmers irrigating crops for livestock feed would find that leasing their water rights to utilities or to in-stream flows is far more profitable than irrigation in dry years and perhaps in all years. Farmers irrigating crops for human consumption would find that, like Israeli farmers, investments in efficient irrigation technology quickly pay for themselves in more expensive water saved. They may also find excellent export markets for their products in regions where very scarce water supplies need to be reserved for high-value domestic and industrial uses. Thus, in arid regions, confining irrigation to the highest value perishable crops and, in humid regions, reducing the proportion of meat in human diets to free water resources for virtual water export or for maintenance of local ecosystem services, are the cornerstones of sustainable water resources management. These would likely be the market outcomes of applying total ecological-economic cost to the price of water.

The water resources picture painted in these paragraphs is one that could be pushed into reality by an ecological-economics approach and is one that, more than the current picture, sustains human welfare, widespread prosperity, peace, and ecosystem health. Painting this picture would also, of course, be fraught with political challenges that the reader can readily identify. Do we want the picture? Are we up to the challenge?

ACKNOWLEDGMENTS

Portions of this chapter are reproduced with permission from Lant, C. L., 2004. Water Resources Sustainability: An Ecological Economics Perspective *Water Resources update* 127:20–30

REFERENCES

Allan, J. A., *The Middle East Water Question: Hydro-Politics and the Global Economy,* IB Tauris, London, 2001.

Allan, J. A., "Hydro-Peace in the Middle East: Why No Water Wars? A Case Study of the Jordan River Basin, *SAIS Review,* Vol. XXII, No. 2, (Summer-Fall 2002), pp. 5–72, 2002.

Bennett, D. A., N. Xiao, and M. P. Armstrong, "Exploring the Geographic Consequences of Public Policies Using Evolutionary Algorithms," *Annals of the Association of American Geographers,* Vol. 94 No. 4, pp. 827–847, 2004.

Braden, J. B. (ed.), "How Valuable is Valuation?" *Water Resources Update,* No. 109, pp. 1–4, 1997.

Bartelmus, P., "Accounting For Sustainable Development," *Environment, Growth, and Development,* Routledge, New York, 1994.

Capra, F., *The Web of Life: A New Scientific Understanding of Living Systems*, Anchor Books, New York, 1996.

Capra, F., *The Hidden Connections: Integrating the Biological, Cognitive, and Social Dimensions of Life into a Science of Sustainability,* Doubleday, New York, 2002.

Chapagain, A. K., and A.Y. Hoekstra, "Water Footprints of Nations," UNESCO- International Hydrologic Program, Institute for Water Education, available at: www.ihe.nl/downloads/projects/Report 16_Volume 1.pdf, 2004.

Costanza, R., "Visions, Values, Valuation, and the Need for an Ecological Economics," *Ecological Economics,* Vol. 51 No. 6, pp. 459–468, 2001.

Costanza, R. et al., "Modeling Complex Ecological Economic Systems," *Bioscience,* Vol. 43, No. 8, pp. 545–554, 1993.

Costanza, R. et al., "The Value of the World's Ecosystem Services and Natural Capital," *Bioscience,* Vol. 37, No. 6, pp. 407–412, 1997.

De Villiers, M., *Water: The Fate of Our Most Precious Resource,* Houghton Mifflin, Boston, 2000.

Esat, V., and M. J. Hall,"Water Resources System Optimization Using Genetic Algorithms," *Proceedings of the First International Conference on Hydroinformatics,* Rotterdam, The Netherlands, pp. 225–231, 1994.

Gleick, P. H. et al., *The World's Water 2004-2005: The Biennial Report on Freshwater Resources,* Island Press, Covelo, CA, 2004.

Global Water Partnership, Technical Advisory Committee, Integrated Water Resources Management, TAC Background Paper No. 4, 2000.

Gottfried, R. R., "The Value of a Watershed as a Series of Linked Multiproduct Assets," *Ecological Economics,* Vol. 5(2) pp.145–161, 1992.

Hawken, P., A. Lovins, and L. H. Lovins, *Natural Capitalism: Creating the Next Industrial Revolution,* Little, Brown, Boston, MA, 1999.

Hilton, A.C., and T. B. Culver, "Constraint Handling for Genetic Algorithms in Optimal Remediation Design," *Journal Water Resources Planning and Management,* ASCE, Vol. 126, No. 3, pp. 128–137, 2000.

Lant, C. L., "The Changing Nature of Water Management and its Reflection in the Academic Literature", *Water Resources Update*, Vol. 110, pp. 18–22, 1998.

Lant, C. L., S. E. Kraft, J. Beaulieu, D. Bennett, T. Loftus, and J. Nicklow, "Using Ecological-Economic Modeling to Evaluate Policies Affecting Agricultural Watersheds," *Ecological Economics,* 55(4):467–484, 2005.

Lomborg, B., *The Skeptical Environmentalist: Measuring the Real State of the World,* Cambridge University Press, Boston, 2001.

Mitchell, R. C., and R. T. Carson, *Using Surveys to Value Public Goods: The Contingent Valuation Method,* Resources for the Future, Washington, DC, 1989.

Mitsch, W. J., and J.G. Gosselink, *Wetlands,* 2d ed., Van Nostrand Reinhold, New York, 1993.

Murphy, L. J., A. R. Simpson, and G. C. Dandy,"Design of a Network Using Genetic Algorithms," *Water,* Vol. 20, pp. 40–42, 1993.

NSF Advisory Committee for Environmental Research and Education, *Complex Environmental Systems: Synthesis for Earth, Life and Society in the 21st Century: A 10-year outlook for the National Science Foundation,* National Science Foundation, Washington, DC, 2003.

Nicklow, J. W., and J. Bringer,"Optimal Control of Sedimentation in Multi-Reservoir River Systems Using Genetic Algorithms," *Proceedings of the 2001 Conference of the Environmental and Water Resources Institute, ASCE,* Orlando, FL, 2001.

Pearce, D., A. Markandya, and E. B. Barbier, *Blueprint for a Green Economy.* Earthscan, London, 1992.

Randall, A., "The Problem of Market Failure," *Natural Resources Journal,* Vol. 23(1) pp. 131–148, 1983.

Repetto, R., "Earth in the Balance Sheet: Incorporating Natural Resources in National Income Accounts," *Environment,* Vol. 34, No. 7, pp. 187–198, 1992.

Ruhl, J. B., C. L. Lant, S. E. Kraft, J. Adams, L. Duram, and T. Loftus, "Proposal for a Model State Watershed Management Act," *Environ. Law* (Lewis and Clark Law School), Vol. 33, No. 4, pp. 929–948, 2003.

Tietenberg, T., *Environmental and Natural Resource Economics,* 6th ed. Addison Wesley, Boston, 2003.

Toman, M., "Why Not to Calculate the Value of the World's Ecosystem Services and Natural Capital, *Ecological Economics,* Vol. 25 (1) pp. 57–60, 1998.

United Nations Development Program, *Human Development Report 2003,* United Nations, New York, 2003.

Willett, W. C., and M. J. Stampfer, "Rebuilding the Food Pyramid," *Scientific American*, pp. 64–71, January, 2003.

Wolf, A., K. Stahl, and M. F. Macomber, "Conflict and Cooperation Within International Rivers Basins: The Importance of Institutional Capacity," *Water Resources Update,* Vol. 125, pp. 31–40, 2003.

World Resources Institute, *World Resources 2000-2001: People and Ecosystems, the Fraying Web of Life,* World Resources Institute, Washington, DC, 2000.

Zilberman, D. (ed.), "Incentives and Trading in Water Resources Management, *Water Resources Update,* Vol. 121, 2002.

4

MANAGING FOR SUSTAINABILITY IN ARIZONA, USA: LINKING CLIMATE, WATER MANAGEMENT AND GROWTH

Jim Holway[1]
Associate Director
Global Institute of Sustainability
Professor of Practice, Civil, and Environmental Engineering
Arizona State University.

Katharine Jacobs[2]
Executive Director
Arizona Water Institute

---

[1] Previously served for 10 years as the Assistant Director of the Arizona Department of Water Resources.
[2] Previously served for 15 years as the Area Director of the Tucson Active Management Area for the Arizona Department of Water Resources.

INTRODUCTION

Arizona actively manages its water resources to ensure the long-term sustainability of the state's current water uses and future growth. This chapter describes the state's water management programs and draws lessons from Arizona's 25 years of experience implementing the 1980 *Groundwater Management Act* (GMA), in particular the efforts to ensure adequate supplies for new municipal growth. Linking water resources with planning for future growth is of critical importance throughout the rapidly growing arid southwestern United States, especially in the context of climate variability and change. Figure 4.1 shows Arizona's location within the southwestern United States and the Colorado River Basin, one of the major water supplies for the region.

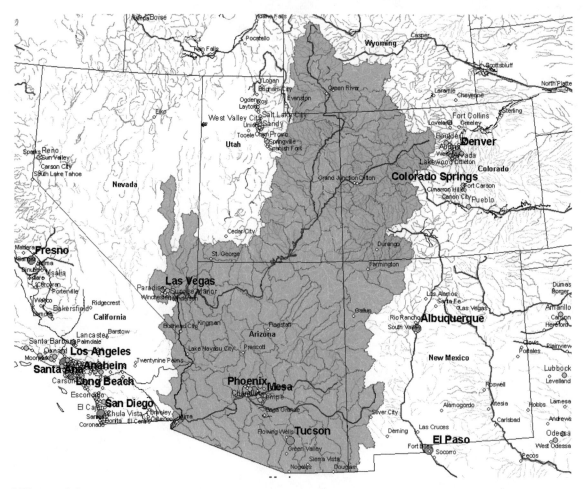

Figure 4.1

Location of Arizona and the Colorado River Basin in the southwest United States.

Arizonans have long noted the need for managing the state's groundwater resources. Groundwater levels have been declining in several basins in the central and southern parts of the state since the 1940s. A 1948 Critical Area Groundwater Code designated "critical" overdraft areas, but was ineffective in controlling the ongoing groundwater depletions. By the late 1970s there was growing recognition of the impacts of water level declines and resulting land subsidence in some areas. The U.S. Secretary of the Interior also declared that the long-desired *Central Arizona Project* (CAP), to import Colorado River water into central Arizona, would not be authorized unless Arizona took steps to reduce groundwater overdraft. A final catalyst to implementing effective groundwater controls was a lawsuit won by an agricultural irrigator to prevent the cities and mines from withdrawing groundwater from the vicinity of their irrigation wells and transporting it for use in other locations. These factors led to the adoption of the 1980 GMA, following a period of intense negotiation among a small group of stakeholders who were convened by then-Governor, Bruce Babbitt.

We will examine Arizona's programs in greater detail, after first exploring the state's water resources and reviewing new knowledge concerning the impact of climatic variability on western water supplies.

ARIZONA'S WATER RESOURCES AND CLIMATE

Arizona's Water Resources

Within the state's boundaries are three main physiographic provinces, the Colorado Plateau to the north, the Central Highlands above the uplifted divide known as the Mogollon Rim, and the Basin and Range province to the south (see Fig. 4.2). The Colorado Plateau province is characterized by numerous canyons and plateaus. It contains several large but not especially productive groundwater basins. The Central Highland area is characterized by a relatively narrow band of rugged mountains and generally high elevations. Groundwater availability is limited; the major watersheds, all tributary to the Gila River, supply water to the Phoenix area. This area is especially vulnerable to climatic variability, since there are no large groundwater storage basins. The Basin and Range province in the southern part of the state is characterized by parallel mountain ranges uplifted by tectonic activity, separated by broad alluvial valleys of materials eroded from the mountain fronts. The basins contain substantial groundwater supplies, frequently of very high quality, in aquifers thousands of feet deep with millions of acre-feet in storage (Arizona Department of Water Resources, 1994).

Groundwater supplies nearly 40 percent of the approximately 7.5 maf of water used in the state each year, with surface water, including diversions from the Colorado River, representing 55 percent of use. Approximately 74 percent of the water use in the state is for agricultural irrigation, though this percentage is expected to continue to decline over time with population growth and the urbanization of agricultural lands in central Arizona. Groundwater overdraft in central Arizona has created significant problems, such as increased well-drilling and pumping costs, declining water quality and subsidence of the land surface due to compaction of the aquifer materials after water is withdrawn.

The principal sources of surface water are the Colorado River and the Salt River. The Colorado River, the largest river in the Southwest at 1450 mi in length, drains

Figure 4.2
Physiographic regions and major rivers in Arizona.

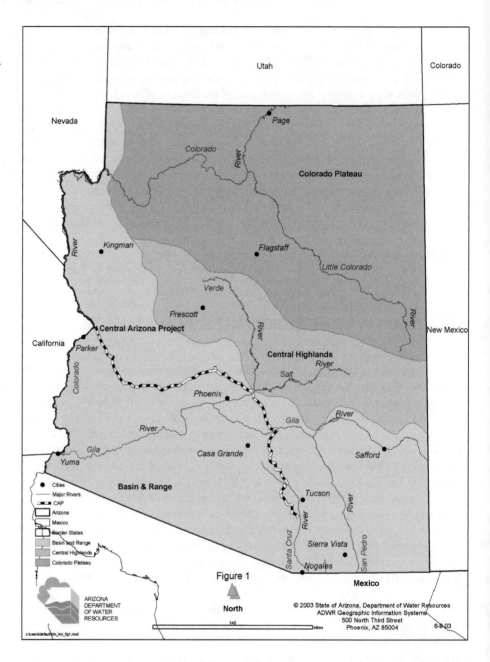

Figure 1

approximately 250,000 mi² with a drainage area extending over seven states as well as northwestern Mexico. Allocations of Colorado River water, which exceed the long-term average flow in the river, are managed by the U.S. Secretary of Interior and are divided among the Upper Basin states (above Lee's Ferry—Colorado, Utah, Wyoming, and New Mexico), the Lower Basin states (Arizona, California, and Nevada), and Mexico. The total volume allocated is 16.5 maf, of which 1.5 maf is allocated to Mexico. Of the Lower Basin allocation of 7.5 maf, Arizona has an allocation of 2.8 maf of Colorado River water.

Users along the river, primarily agricultural and Native American, are entitled to 1.3 maf with the remaining 1.5 maf imported into central Arizona through the CAP.[3]

The *Salt River Project* (SRP) has been delivering water to central Phoenix since 1903. It was the first multipurpose federal reclamation project, and currently delivers more than 1 maf of water to its water service area of 240,000 acres. Initially designed for agricultural irrigation, SRP now primarily delivers wholesale untreated water to municipal water providers. SRP is also one of the largest customers of excess CAP water during droughts on the Salt River system, since it buys CAP water to augment its supplies in shortage years.

Effluent use comprises approximately 7 percent of the state's water supply, with power plant cooling water at Palo Verde nuclear generating station west of Phoenix the largest single user. However, effluent is becoming an increasingly significant component of municipal water supplies and substantial investments have been made in advanced treatment and delivery systems to use reclaimed water for turf irrigation and aquifer recharge in many parts of Arizona.

The climate of Arizona varies dramatically with elevation. Conditions are generally dry (ranging from 3 to 11 in. of rainfall annually) and warm (average daytime temperatures of almost 90°F, with daytime summer temperatures commonly above 100°F) in the southern and central basins. At higher elevations, temperatures are lower and precipitation much higher, with average precipitation reaching 38 in. at Hawley Lake in the White Mountains. Annual precipitation averages 7 in. per year in the Phoenix and 10 in. per year in the Tucson metropolitan areas (Western Regional Climate Center, 2002). Precipitation is bimodal, with significant rainfall occurring in the winter months due to storm activity generated in the southern Pacific, and frontal storms originating in the Gulf of Mexico in the summer. Summer rainfall or "monsoon" activity is far more spatially heterogeneous than winter rainfall, with higher losses due to evapotranspiration. Thus, water supplies are more dependent on winter precipitation. The Phoenix metropolitan area is strongly affected by winter snowpack in the Salt-Verde river system (see Fig. 4.2 for the location of the Salt and Verde Rivers).

Climatic Conditions and Trends in Arizona

Although groundwater was historically the primary source of water supplies in Tucson, and a significant contributor to Phoenix area supplies as well, the completion of the CAP canal has permanently changed the water supply portfolio for the major metropolitan areas of the state. This means that central Arizona's water managers are now far more affected by short- to- medium-term climate conditions affecting either the Colorado River or Salt-Verde systems than they have been in the past. More than one-third of Arizona's total supply comes from the Colorado River. Of the water in the Colorado, 70 percent is generated in the Upper Basin. Therefore, Arizona's water supplies are significantly influenced by snowpack conditions in Colorado, Utah, and Wyoming.

---

[3] Arizona's 1.5 million acre-feet allocation for the Central Arizona Project also has the lowest priority on the river, thus making Arizona the state most vulnerable to shortages. Arizona accepted this lower priority in exchange for California's U.S. Congressional delegation dropping their opposition to funding of the CAP system. This lower priority has a significant impact on Arizona's water management policies and positions on Colorado River management controversies. (See Fig. 4.2 for the location of the CAP canal.)

The climate regime in the Upper Colorado River Basin is not always coincident with conditions in the Lower Basin because the influence of ocean temperatures in the south Pacific controls conditions more strongly in the southwestern United States than in the midcontinent. To the extent that the water supplies are not affected by the same climate regime, there are advantages to drawing supplies from a broad geographical area. On the other hand, a recent study for the SRP by Hirschboeck and Meko (2005) indicates that synchronous long-term drought is more common than the alternative.

Sea surface temperatures have been shown to be highly correlated with climate conditions, and are significant drivers of climate variability on a seasonal, annual, and in some cases, a decadal basis. The oceans store energy and release heat to the atmosphere. These ocean-atmosphere interactions result in persistent circulation patterns and storm tracks that affect precipitation across the United States. The *El Nino—Southern Oscillation* (ENSO) is the strongest and most important influence on interannual climate and weather variations in Arizona, particularly in winter (Andrade and Sellers, 1988). The two phases of ENSO are known as El Niño and *La Niña*. The episodes recur every 3 to 7 years on average, although instrumental and paleoclimate records indicate long-term variations in the recurrence interval (Stahle et al., 2000; Allan et al., 1996). In El Nino conditions, the western Pacific exhibits warmer than average temperatures, and this condition is associated with wetter than average winters in Arizona. La Niña, on the other hand, brings reliably dry winters to Arizona (McPhee et al., 2004). Investigators have linked La Niña with sustained drought in the southwestern United States. Some connections have been identified between ENSO conditions and the summer monsoons as well, though these connections are not as well documented as the impacts on winter precipitation.

Long-term variations in the Pacific and Atlantic Ocean conditions have recently been identified that appear to have a strong correlation with climate conditions in the United States on a decadal and multidecadal scale. The *Pacific Decadal Oscillation* (PDO) is characterized by persistent ENSO-like ocean temperature patterns that are pronounced in the north Pacific Ocean (Mantua and Hare, 2002). The multidecadal temperature conditions are associated with winter precipitation variations in the western United States, including below average winter precipitation and episodes of sustained drought in the Southwest. For example, the 1950s drought conditions (that continued through the mid-1960s) and a 20-years period of above average precipitation from the mid-1970s to the mid-1990s are correlated with ocean temperatures in the northern Pacific. The physical mechanisms behind the PDO are not currently known.

Multidecadal fluctuations in Atlantic Ocean temperatures, called the *Atlantic Multidecadal Oscillation* (AMO), are also associated with persistent dry conditions in the southwestern United States (Enfield et al., 2001), such as the 1950s drought. Recent research shows that the AMO in conjunction with the PDO can produce atmospheric circulation patterns conducive to persistent La Niña-like conditions and a higher frequency of drought years in the southwestern United States (McCabe et al., 2004). Multidecadal climate phenomena have significant implications for water management, particularly for surface water managers (Garfin et al., 2006). The fact that the climate system does not simply vary annually around a stable mean, but exhibits longer-term trends, has multiple implications for long-term water supply planning, conjunctive use of surface and groundwater, and reservoir management. It is hoped that as the mechanisms of decadal climate fluctuations are better understood, there will be enhanced predictive capacity to anticipate water supply conditions on an annual to multiannual basis.

Studies of selected tree species in mountainous areas in the western United States have shown that tree ring width is highly correlated with water supply availability if the trees are located in moisture-stressed situations. These records allow reconstructions of historic climatic conditions to roughly AD 1300, and indicate that sustained droughts of higher intensity than anything experienced in the past century were relatively common in the West. Such spatially-extensive droughts probably stretched simultaneously across major western river basins. Recent tree ring studies across the West have dramatically altered water managers' views of drought duration and intensity in the past, with corresponding concerns about the likelihood of future shortages.

The key lesson for water resources managers from paleohydrologic streamflow reconstructions is the danger in having a short-term record of streamflows when making long-term allocation decisions. The Colorado River water resources were overallocated by a significant margin in the 1920s in the context of the Colorado Compact and subsequent decisions that comprise the "law of the river." The allocation was based on one of the highest streamflow regimes in the last 500 years (Meko et al., 1995). Although the fact that the river was overallocated has been understood for decades, the true implications have only become apparent in the recent (post-1998) drought cycle, which occurred after substantial increases in demand for Colorado River water.

In addition to the challenges of climate variability, global warming is likely to further constrain the water supplies of the southwestern United States in the future[4]. Instrumental records in each of Arizona's seven climate divisions indicate that temperatures have been steadily increasing in Arizona (Garfin et al., 2006). An international assessment of climate change trends and impacts is conducted every 5 years by experts across the world who officially represent their governments. The most recent *International Panel on Climate Change* (IPCC) assessment suggests a combination of higher minimum and maximum temperatures, increased precipitation intensity, enhanced rates of evaporation, and increased precipitation variability in the southwestern United States (IPCC, 2001). The assessment also suggests that global warming is likely to increase the risk of extreme droughts and floods that occur with ENSO episodes in many regions. Accompanying the prediction of increased temperatures is a prediction of reduced snowpack in the mountains of the western United States, resulting from a combination of earlier snowmelt and an increased proportion of precipitation occurring as rain instead of snow. This effect, in combination with higher temperatures, means a shift in the hydrograph so that peak flows occur earlier in the year. This may increase the risk of flooding in the spring, but decrease surface flows during the summer. In the context of the major reservoirs on the Colorado River, the timing of runoff may not have a significant effect. However, effects on unregulated streams and on riparian habitat and aquatic species may be significantly negative.

[4] For a general discussion about the impacts of climate change on many different aspects of urban growth, see Ruth, 2006.

This is of particular concern in the context of trends toward reduced summer precipitation that have been observed in southern Arizona over the last century (Pool and Coes, 1999). Enhanced evaporation rates, combined with earlier snowmelt in the Upper Colorado River Basin would likely increase water supply vulnerability.

The IPCC consensus is that there are large uncertainties in the direction and magnitude of projected precipitation change in western North America. In contrast to inconsistent *precipitation* predictions, western United States and Colorado River Basin *temperature* projections are relatively consistent. Models show increased average temperatures in the Colorado River Basin in both summer and winter, with seasonal increases on the order of 2°C by 2050 and annual increases on the order of 4 to 5°C by 2099. The aforementioned increases are based on a "business as usual" greenhouse gas emission scenario.

Given the high sensitivity of the Colorado River Basin due to overallocation of water resources and high demands on the system, increased temperatures are likely to degrade Colorado River Basin system performance. Ramifications of degraded system performance include decreases in reservoir storage, hydroelectric power output, and deliveries to the CAP. Improved climate information has led to more and more concern about the reliability of water supplies in Arizona, in part due to Arizona's junior priority water rights under the 1968 Colorado River Basin Project Act. These factors underscore the importance of Arizona's long-term water management perspective and the assured water supply rules for municipal development. However, it should be noted that to date, Arizona has not explicitly considered the impacts of climate change on any of its programs. The only way to do so in the absence of clear projections regarding precipitation impacts is to make more conservative assumptions about supply (e.g., requiring a higher volume of backup supplies or assuming that more storage will be required to offset future shortages, because they are likely to be more frequent and/or of a higher volume) and also to include a global warming impact that may increase municipal use on the demand side of the *assured water supply* (AWS) calculations. Other southwestern states are dealing with similar issues of drought, variability in water availability, and rapid population growth. In the remaining part of this chapter we will focus on Arizona's water management experience and challenges for assuring adequate supplies and augmenting supplies through groundwater recharge and recovery programs.

ARIZONA'S WATER MANAGEMENT FRAMEWORK

Arizona's GMA focused almost exclusively on groundwater and did not affect the preexisting surface water management code, which remains a separate body of law. The hydrologic connections between surface water and groundwater are well documented, but Arizona's legal system does not recognize this fact (Glennon and Maddock, 1987). Like most other western states, surface water in Arizona is allocated based on prior appropriation, "first in time first in right." Under this system, the first user to put water to beneficial use has the highest priority right, and can divert his full allocation without regard for more junior users in the event of shortages in surface water flows. Prior to June 12, 1980, groundwater withdrawals in Arizona were generally subject only to the "reasonable use" doctrine, with little or no protection of prior users from new withdrawals.

The GMA:

1. Established grandfathered rights to groundwater for users that existed prior to June 12, 1980 and programs to reduce groundwater overdraft and the resulting water level declines

2. Supported completion of the CAP, a 330 mi long canal to bring Colorado River water to central and southern Arizona

3. Created the *Arizona Department of Water Resources* (ADWR) as the state water quantity management agency

The GMA established a regulatory framework for managing Arizona's groundwater, but focused most of the regulatory effort on parts of the state called *Active Management Areas* (AMAs), which were experiencing particularly acute groundwater overdraft problems. The three primary goals of the GMA were:

Groundwater Management Act Structure

1. To control the severe overdraft which was occurring in many parts of the state

2. To provide a means to allocate the state's limited groundwater resources to most effectively meet the changing needs of the state

3. To augment Arizona's groundwater through water supply development

The ADWR administers the provisions of the GMA in addition to other responsibilities related to flood control, Colorado River management, dam safety, and surface water rights administration.

The GMA statewide water management provisions are relatively limited, focusing on licensing of well drillers, well registration, notice of supply adequacy for new residential developments, and prohibitions on transportation of groundwater between most subbasins in the state.[5] Legislation passed in 2005 also requires water providers throughout the state to develop conservation, water supply and drought response plans, and for the first time to monitor and report their water use annually. In addition, comprehensive planning legislation, adopted as part of Arizona's growing smarter provisions in 2000, requires certain counties and municipalities to include a water resources element in their general plan.

Somewhat more restrictive management provisions apply to *Irrigation Nonexpansion Areas* (INAs), where no new land can be brought into agricultural production but there are no limits on nonirrigation uses of water. INAs were established in rural farming areas where groundwater overdraft problems and competition for groundwater supplies were significant but less severe than in AMAs. Two INAs, Douglas and Joseph City, were created by the GMA in 1980; the state agency (ADWR) established the Harquahala INA in 1982. Figure 4.3 shows the INAs and AMAs in Arizona.

---

[5] The limitations on groundwater transfers resulted from efforts by cities within the AMAs to buy "water ranches" in rural Arizona during the 1980s. The rural areas were concerned that water transfers would limit their economic future, and the legislature passed the Groundwater Transportation Act in 1990. This Act, and subsequent legislation in 1993, prohibits any transfer of groundwater across groundwater basin boundaries that is not expressly grandfathered within the legislation.

Figure 4.3

Arizona's active management areas and irrigation nonexpansion areas.

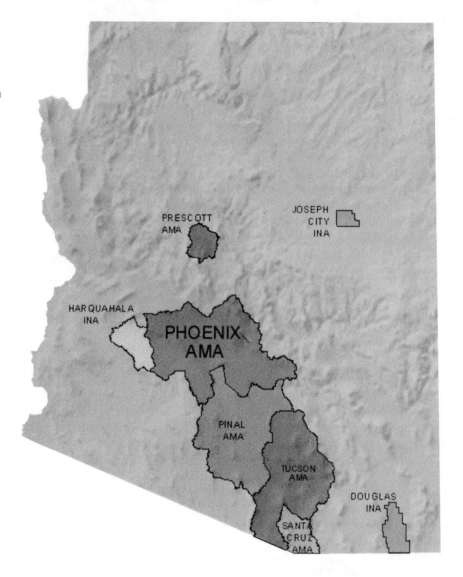

The majority of the groundwater management provisions of the GMA are focused on the AMAs. The code created four initial AMAs—Phoenix, Pinal, Prescott, and Tucson. A fifth AMA, the Santa Cruz AMA, was formed from the southern portion of the Tucson AMA in 1994. New AMAs and INAs can be designated by the state agency (ADWR), if necessary, to protect the water supply or on the basis of an election held by local residents of an area. Like the INAs, no new agricultural irrigation is allowed within AMAs. This limitation ties all farming activities to acreage that was irrigated prior to June 12, 1980. The boundaries of AMAs and INAs are generally defined by groundwater basins and subbasins rather than by the political lines of cities, towns, or counties.

The AMAs include over 80 percent of Arizona's population, over 50 percent of total water use in the state, and 70 percent of the state's groundwater overdraft, but only 23 percent of the land area. Total water demand is nearly 4 maf per year in the AMAs, with slightly over half of this being used for agricultural irrigation. Currently, overdraft across the five AMAs is approximately 500,000 ac-ft, or approximately 15 percent of total water use.

Arizona's GMA established a statutory management goal for each AMA. The goal for four of the five AMAs is based on the concept of safe-yield. "Safe-yield" means establishing and maintaining a long-term balance between the annual amount of groundwater withdrawn in an AMA and the annual amount of natural and artificial recharge in the AMA (A.R.S. § 45-561.12). The goal of achieving safe-yield by 2025, focused on eliminating groundwater overdraft, applies to the Phoenix, Tucson, and Prescott AMAs. In the Santa Cruz AMA, where significant international, riparian, and groundwater/ surface water interaction issues exist, the goal is to maintain safe-yield and prevent local water tables from experiencing long-term declines. In the Pinal AMA, where a predominantly agricultural economy exists, the goal is to allow the development of non-irrigation water uses, extend the life of the agricultural economy for as long as feasible, and preserve water supplies for future nonagricultural uses.

The safe-yield goal, as defined in the GMA, is not the same as the concept of sustainability,[6] which may consider impacts on surface flows and habitat in addition to basin-wide water balances. Safe-yield does not require that water levels remain stable throughout the basin; rather, it is a basin-wide accounting of total inflows and outflows to the aquifer. Awareness of the impacts of subsidence on infrastructure, particularly within urban areas of Arizona, and of the impacts of groundwater pumping on riparian areas has caused concern about the need to manage groundwater levels rather than only focusing on a basin-wide water-budget balancing of groundwater pumping and recharge. The Santa Cruz AMA, created in part to facilitate the conjunctive management of groundwater, surface water, and effluent does have the unique additional goal of maintaining water tables (and hopefully habitat and riparian flows in the Santa Cruz River) throughout the AMA.

Arizona's AMA groundwater management programs have four major regulatory components: the structure of the water rights system itself, conservation programs for major water users, the assured water supply program for new municipal development, and recharge and recovery programs facilitating conjunctive management of available supplies. The following program descriptions focus on the assured water supply and recharge and recovery efforts, but provide brief overviews of the other regulatory programs.[7] For areas outside of AMAs, comprehensive water resources planning mandates have been adopted in the last several years through county and municipality planning statutes and through recent modifications of the groundwater code. These provisions are also briefly discussed.

A series of five management plans must be adopted at specified dates between 1980 and 2025, to move the AMAs incrementally toward their management goals through demand management and supply enhancement (A.R.S. 45-563). Other programs are implemented

---

[6] We do not attempt to define sustainability in this chapter. For one of the most commonly referenced definitions of sustainability see World Commission on Environment and Development, 1987.

[7] The descriptions of Arizona's water management programs and the lessons learned from Arizona's experience, including demand management and planning efforts, are covered in greater breadth in an earlier article by the authors published in the Hydrogeology Journal (see Jacobs and Holway 2004).

through rule-making procedures that implement assured and adequate water supply, well-drilling construction and licensing, annual reports, water-measuring devices, capping of open wells, fees, and well-spacing and well-impact requirements.[8] The state agency (ADWR) also operates a number of planning, technical, and financial assistance programs to advance water management throughout the state.

Water Rights and Responsibilities in AMAs

The authority to withdraw groundwater from large wells (over 35 gal per minute) is controlled through a system of grandfathered rights that protects most groundwater users that were in place prior to 1980. New groundwater withdrawal permits are available under limited conditions.[9] In order to drill a large well, a separate permit is required and necessitates a demonstration that the new well will not create an unreasonable impact on any other well of record. In addition, water pumped from all large wells (35 gal per minute or larger) must be measured and reported on an annual basis to ADWR and users pay a withdrawal fee of $2 to $3 per ac-ft. The annual reports may be audited to ensure water-user compliance with the provisions of the GMA and management plans. Penalties of up to $10,000 a day may be assessed for deliberate violations, though most compliance activities result in stipulated settlements that may or may not include fines (A.R.S. 45-451 through 528).

Conservation Programs in AMAs

Mandatory conservation requirements are set for all large water users in the context of administrative law implemented through the series of 10-year management plans for each AMA (A.R.S. 45-564 through 575). The ADWR is required to include enforceable conservation requirements and a grants-based conservation assistance program in each management period for each AMA.

Agricultural groundwater rights holders with greater than 10 acres of land are given an annual allotment based on historic crops grown and an assumption of 80 percent irrigation efficiency. In addition, as indicated earlier, no new land can be brought into irrigation. Municipal water use is controlled through reductions in the average annual gallons per capita per day usage based on the conservation potential of all water companies serving more than 250 ac-ft. Industrial[10] users over 10 ac-ft are given allotments based on the use of the latest commercially available conservation technology, or are required to use specific conservation technologies. Alternative conservation programs based on use of approved best management practices are available for both agricultural and municipal water rights holders.

---

[8] The Arizona Department of Water Resources also has the authority to develop and publish substantive policies, in accordance with the State of Arizona Administrative Procedures, as necessary for additional guidance on regulatory program details not covered by statutes, rules, or management plans.

[9] Categories of available permits for new groundwater withdrawal authorities include such uses as general industrial uses, dewatering, mineral extraction, and electric power production. A very limited market has developed in Type II Nonirrigation Grandfathered Rights, which can be severed from the land and utilized by new water uses.

[10] Industrial users, for Groundwater Management Act purposes, are nonagricultural entities that have their own groundwater rights and do not receive service from municipal providers.

Assured Water Supply (AWS) Program in AMAs

The AWS program is designed to sustain the state's economic health by preserving groundwater resources and promoting long-term water supply planning within the state's five AMAs. The architects of Arizona's 1980 GMA recognized that the AMAs could not achieve safe-yield through conservation alone and that new growth, as well as some portion of existing municipal demand, would have to utilize more expensive renewable or imported supplies. To achieve this transition to renewable supplies the GMA required AWS rules to be adopted by 1995 (A.R.S. 45-576). These rules, finally adopted in February of 1995 on the third attempt, require all new subdivisions to demonstrate a secure 100-year supply of water, primarily from renewable water supplies, before a plan can be approved by the local government.

An AWS can be obtained in one of two different ways: first, designating water providers as holding an AWS for their entire service area, or second, the issuing of individual AWS certificates to a developer for their subdivision. Most of the larger cities, and a few private water companies, in the Phoenix and Tucson metropolitan areas are "designated" providers. These providers have successfully demonstrated to the state agency (ADWR) a portfolio of primarily renewable supplies adequate to serve their current demand, committed demands (for approved but not yet built lots), and at least 2 years of estimated new water commitments for the next 100 years. In cases where a water provider has not secured a designation, individual developers must obtain a certificate of AWS directly from the state agency for their subdivisions. Typically, new smaller communities on the growing edge of the urban areas, and development within unincorporated portions of the county are served by undesignated providers and developers must obtain their own certificates. Undesignated providers can continue to serve groundwater to their nonresidential water users and to residential users in lots which were approved prior to the 1995 adoption of the assured water supply rules.

Five separate criteria must be demonstrated for either a certificate or a designation;

1. Physical, legal, and continuous availability of water for 100 years
2. Water quality meeting state and federal drinking water standards
3. Sufficient financial capability to build and operate the necessary infrastructure
4. Consistency with the AMA management plan (principally conservation requirements)
5. Consistency with the AMA management goal (typically safe-yield)

The key criteria are consistency with the safe-yield goal and physical availability. Essentially, the subdivision needs to be supplied with a renewable or imported water supply. If the subdivision relies on groundwater, then any mined groundwater must be replenished, typically by joining the *Central Arizona Groundwater Replenishment District* (CAGRD). Subdivisions relying on groundwater and CAGRD membership must demonstrate that after 100 years of pumping, the depth to groundwater will not exceed 1000 ft below land surface.[11]

---

[11] The Pinal AMA, due to its different goal, has several significant differences in AWS rules. New subdivisions in the Pinal AMA can rely on groundwater for up to 125 GPCD for existing providers and 62.5 GPCD for new providers. In addition, the allowable depth to water is 1100 feet in the Pinal AMA. Pinal AMA interests are currently considering AWS rule changes which would significantly reduce the allowable groundwater use.

This program has forced major investments in the transition from overdrafted ground-water as the source of water supplies for urban areas toward the use of renewable or imported water supplies.[12] The AWS rules clearly demonstrate Arizona's commitment to ensuring a long-term secure water supply for its citizens living in the AMAs, and to making the investments required for infrastructure, treatment, and storage facilities.

Groundwater Recharge and Recovery Programs

Conjunctive management of all sources of water (groundwater, surface water, and effluent) facilitates a larger and more reliable supply. Storing excess surface water in aquifers during wet periods allows for recovery during droughts and increases the reliability of supplies. In addition, the supply of effluent, for which demand is greatest in summer for irrigation and production is greatest in winter due to seasonal residents, can be matched with demand through seasonal storage and recovery. Storage and recovery can also be used to further treat effluent through soil aquifer treatment. Conjunctively managing water supplies in this way is critical to the ability of many water providers to demonstrate an assured water supply.

Beginning in 1986, Arizona developed an innovative program to facilitate storing water underground in aquifers, and an associated recovery program that regulates withdrawal of the stored water (A.R.S. 45-801.01 through 898.01). The recharge and recovery program includes three types of permits:

1. Permitting of storage facilities to ensure hydrologic feasibility and quantify allowable capacity

2. Permitting of water storage for each individual storer at each facility and the type of water to be stored to ensure appropriate legal rights and water quality

3. Permitting of recovery wells to protect adjacent well owners and water users and to determine whether the wells are recovering the physical supplies actually recharged (though recovery is also permissible outside the area of hydrologic impact from the recharge)

Three principal means of conducting recharge are (1) constructed facilities such as recharge basins (ponds) (2) managed facilities that allow the water to run down a dry streambed and passively recharge (3) groundwater savings facilities where a farmer reduces groundwater pumping and takes delivery of an alternative supply, generating "credits" for a municipal provider or the CAP to pump the saved groundwater in the future. Recharge permits require consideration of hydrologic feasibility and prevention of unreasonable harm to other landowners and water users.

This program facilitates storage of surface water and effluent for future use, protection of rights to the stored water, and protection of water quality in aquifers. In addition to the importance of recharge credits for demonstrating assured water supplies and conjunctive

---

[12] The Arizona Municipal Water Users Association (AMWUA) representing most of the larger cities in the Phoenix metropolitan area (Avondale, Chandler, Gilbert, Glendale, Goodyear, Mesa, Peoria, Phoenix, Scottsdale, Tempe) has documented investments of $1.2 billion by their member cities in converting to the utilization of renewable supplies since the adoption of the 1980 GMA.

management of available supplies, recharge and recovery is also utilized by the *Arizona Water Banking Authority* (AWBA),[13] which stores excess Colorado River water for recovery during drought. Recharge and recovery also serves as a major storage and reliability feature for the CAP and its subsidiary, the CAGRD.

Central Arizona Groundwater Replenishment District   One of the most innovative and controversial institutions that has been developed in response to the AWS rules is the CAGRD. The CAGRD was created to provide access to renewable supplies for new developments that had no direct access to a CAP allocation or to surface water treatment plants and delivery infrastructure. The CAGRD is required to replenish in perpetuity all groundwater that is pumped by its members that is in excess of the groundwater that is allowed to be pumped under the AWS rules. This replenishment is required to take place within 3 years of the groundwater pumping. It has been very successful in attracting customers, perhaps more successful than anticipated. In part, the CAGRD is considered innovative because it is designed solely to support the AWS program by replenishing the groundwater use of its customers. Significant controversies relate to the ability of the CAGRD to store water in locations that are distant from the place where the groundwater was pumped (though it must be in a location where the water will be available for future recovery), to the fact that the CAGRD itself has more customers than were originally expected, and that the CAGRD, unlike individual providers or subdivisions seeking an assured water supply, is not required to demonstrate a 100-year supply. The CAGRD is currently dependent on the availability of surplus water for recharge.[14]

Arizona's Water Resources Planning Mandates
Within Arizona's AMA the 100-year water supply requirements for obtaining an assured water supply have generally been considered an adequate long-range planning requirement, at least for the larger cities which are designated providers. However, a number of efforts have been made to strengthen water planning requirements for areas outside of the AMAs. In 2000, Arizona's Growing Smarter requirements were modified to require a water resources element as part of the county and municipality comprehensive planning requirements. The water resources element is required for counties with a population of greater than 125,000; and for municipalities with more than 2500 population, unless they have fewer than 10,000 residents and are growing at a rate of less than 2 percent per year. This comprises 4 of 15 counties and 23 municipalities outside of the AMAs. The element must (1) identify known legally and physically available supplies, (2) identify demand resulting from growth projected in the general plan (generally 20 years of growth), and (3) identify how demand will be served by currently available supplies, or prepare a plan

---

[13] The Arizona Water Banking Authority was created in 1996 to store the currently excess AZ Colorado River supplies in order to firm municipal and industrial supplies during future anticipated droughts. The AWBA storage is also used to assist with Native American water rights settlements and for interstate water banking. The AWBA is funded through a combination of (1) property taxes from Maricopa, Pinal, and Pima counties which must be used to assist water management in the county from which the funds come (2) general fund appropriations which can be used for statewide benefit (3) groundwater withdrawal fees from the Phoenix, Pinal, and Tucson AMAs (4) funds from interstate banking. Additional information on the Arizona Water Banking Authority is available on the AWBA Web site at www.awba.state.az.us.

[14] Further information on the Central AZ Ground Water Replenishment District, and the Central AZ Water Conservation District, of which it is a part, is available through their Web site at www.cap-az.com

to obtain additional necessary water supplies (A.R.S. 9-461.05 and A.R.S. 11-821). Communities are not specifically required to go beyond the existing available information in preparing these comprehensive plan elements. Additional water resource planning requirements were adopted by the state legislature in 2005 (A.R.S. 45-342). All community water systems outside of AMAs must prepare (1) a water supply plan, (2) a drought preparedness plan, and (3) a water conservation plan. This legislation also required water providers outside of AMAs to begin reporting their annual water use. These plans will be due for large providers beginning in 2007 and for small providers 1 year later.

MANAGING FOR SUSTAINABILITY—ARIZONA'S EXPERIENCE

Arizona's water management programs have been shaped by the nature of available supplies in a semiarid state. Future water management efforts will likely be increasingly affected by an expanding understanding of the climate drivers that affect water availability. The state's programs have also been affected by the economics and technology of acquiring, treating, and distributing those supplies; the legal framework of land ownership and water rights; the growth and pattern of water demands; and the political nature of Arizona and its major historic water users. Management approaches in the larger central Arizona AMAs have been shaped by access to deep, although overdrafted, aquifers, and imported surface water supplies.[15] Many rural areas have limited groundwater and limited surface water rights. The high rate of population growth and the fast-paced changes in land and water uses throughout the state have resulted in unique management challenges. Arizona's approach has also been shaped by the state's politically conservative nature and resistance to government regulations and funding assistance. Fundamental choices made by Arizona in setting up water management programs included establishing regulatory programs in state-controlled AMAs, maintaining a dichotomy between groundwater and surface water management, and establishing a water rights structure within AMAs, which included grandfathering most existing groundwater uses. Given the politically conservative nature of Arizona, water managers from elsewhere are often surprised that Arizona has perhaps the most stringent and longest standing regulatory approach within the United States. The 1980 GMA was, in fact, an outgrowth of concerns about groundwater depletions, the threat of the federal government to not fund the CAP and legal challenges to groundwater transfers off the land from which it was pumped. The state regulatory structure provides parity among AMAs, but also allows for local input and implementation to tailor the management system to local conditions. This model has been successful, as has defining management areas based on hydrologic boundaries. The individual AMA's management plans provide the opportunity to accommodate the unique character of each AMA, though to date this has been used in only a limited way. The purpose of this section is to briefly review water management trends in Arizona since the adoption of the GMA and then to reflect on some of the policy choices and approaches taken by Arizona to deal with the connections between water management and growth.[16]

---

[15] In the Phoenix AMA, the ability of the Salt River Project to conjunctively manage and deliver approximately 1 million acre-feet of surface water, groundwater, and more recently CAP water, has shaped water management in that AMA.

[16] For a broader discussion of Arizona's lessons in the context of a comparative public policy framework, see Jacobs and Holway, 2004.

The last 25 years of Arizona's history has been a period of remarkable change and innovation. Due in part to Arizona's rapid rate of population growth and urbanization, and the dramatic diversification of the economy, Arizona has moved from a primarily resource-based economy (copper, cattle, cotton, and citrus) to an urbanized state more dependent on technology production, construction, and tourism. Nothing showcases this innovation and complexity better than the huge shifts in water management and water supply policy that have taken place.

Since the 1940s, the majority of water use in the AMAs was supported by groundwater, with the exception of the large surface water delivery system, the SRP, in the Phoenix area. Figure 4.4 illustrates the changes in water supplies utilized over the last two decades. The GMA charted a course for the municipal sector in the AMAs to move away from groundwater and toward renewable water supplies. This focus on the use of renewable supplies for the municipal sector was based on the expectation that municipal and industrial demand would continue to grow, while the demand of agriculture and mining would diminish over time. The transition to renewable water supplies was expected to be gradual, although substantial policy changes have been needed to facilitate the transition.

The original expectation was that in the early years of the CAP, agricultural entities would utilize all of the state's CAP allocation not yet needed by municipal, industrial, and Native American users. In fact, the costs associated with paying for the CAP water and the associated delivery systems made CAP water cost-prohibitive for agriculture initially, and major changes in pricing policy and water supply allocation have been made to respond to this problem.

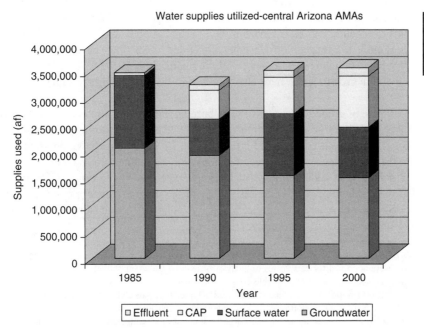

Figure 4.4

Water supplies utilized within central Arizona AMAs.[17]

---

[17] Data for this table comes from annual water uses reports submitted to the Arizona Department of Water Resources. The data shown for 2000 is actually made up of a composite of available information for 1998, 2000, and 2003 depending on available data for the three AMAs.

Municipal use of renewable supplies, although significant, also started more slowly than anticipated. However, primarily due to municipalities investments in utilizing renewable supplies, in the 15 years from 1985 to 2000, water supplies utilized switched from a 60 percent reliance on groundwater to a 60 percent reliance on renewable supplies. This conversion to renewable supplies will continue as grandfathered groundwater rights held by agricultural irrigators are urbanized and the new municipal use relies on renewable supplies. Recharge of CAP water and recovery from the aquifer has also been used extensively along with direct delivery for municipal use of CAP. The AWS rules have been one of the principal factors moving new municipal development to invest in surface water treatment plants and to utilize the more expensive SRP and CAP supplies.

The older cities in central Arizona (Phoenix, Tucson, Mesa, Glendale, Scottsdale, Tempe) began the conversion to use of renewable supplies well before the implementation of the 1995 AWS rules. Other rapidly growing cities converted to renewable supplies just in time to meet the rule requirements. Many of the newer cities and private water companies have chosen, at least for now, to pump groundwater and rely on the CAGRD for replenishment. Most of the larger cities in the Phoenix and Tucson metropolitan areas have also been designated as having an AWS; therefore they are required to reduce significantly their dependence on groundwater, typically to no more than 5 to 15 percent of their total supply. A number of these larger cities are using significantly less than this amount of groundwater during normal years and saving their groundwater for a backup supply during drought. Although the AWS rules apply only in the AMAs, these areas have captured the vast majority of the growth in Arizona over the last decade. Since the adoption of the AWS rules in 1995, over 1000 subdivisions representing between 200,000 and 300,000 lots received certificates of AWS through June 2005. Of all the new demand subject to the AWS, approximately 90 percent of the certificated lots between July 2003 and June 2005 have enrolled in the CAGRD and the CAGRD has additional commitments for certain designated service areas that have joined.[18] Designated providers, which comprise nearly all the major cities in the AMAs, but not necessarily the future growth areas, have also grown significantly during the same time period.

Recharge of groundwater aquifers has been an important component of demonstrating an AWS as well as for storing currently excess CAP to recover during drought shortages, and for effectively utilizing treated effluent, see Fig. 4.5. The recharge program has resulted in the development of 79 storage facilities with a combined capacity to store up to nearly 2 maf per year. Although large-scale *groundwater savings facilities* (GSF) received most of the recharge activity in the early years of the program, *underground storage facilities* (USF) now have a slightly greater capacity and are receiving storage levels similar to the

---

[18] Data from the Arizona Department of Water Resources indicates 1554 subdivisions representing 381,151 lots were approved for certificates or analysis of assured water supply between 1995 and 2005. However, these numbers include some double counting for reissued certificates and for lots issued an analysis of assured water supply (a preliminary determination typically requested by a few master planned communities). CAGRD data for the same time period indicates 975 subdivisions representing 191,412 lots with certificates of assured water supply joined the CAGRD. Looking at just fiscal years '04 and '05, for which better data exists from ADWR, approximately 90 percent of the lots issued certificates by ADWR joined the CAGRD. Therefore the actual number of new lots approved for certificates of assured water supply during this 10-year period is likely somewhere between 200,000 and 300,000.

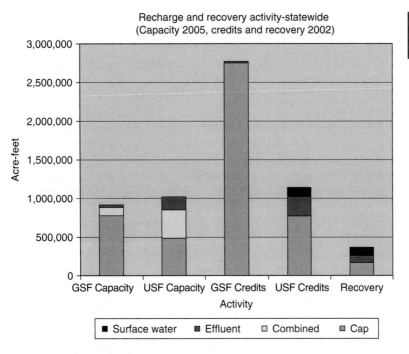

Figure 4.5

Recharge and recovery activity.

groundwater savings facilities. Of the nearly 500,000 ac-ft of recharge capacity capable of serving more than one water source (listed as combined in Fig. 4.5), the majority of this space is eligible to receive both CAP and other surface waters. Approximately 100,000 ac-ft of this combined capacity is also eligible to store effluent.

Water stored (shown as GSF credits and USF credits in Fig. 4.5), as of the end of 2002, was approximately 4 maf.[19] Of the 4 maf of water recharged through 2002, nearly half of the storage was done for the AWBA. Approximately 70 percent of the water stored has occurred through subsidized CAP use by agriculture in lieu of pumping groundwater (in GSF) and the remainder is from direct recharge (USF). However, as more underground storage facilities are brought on line, the proportion of storage through direct recharge is increasing.[20] Although the majority of recharge is being done with CAP water, over 260,000 ac-ft of effluent have also been recharged. Surface water, other than CAP water, typically must be recovered in the same year it was stored and is ineligible for long-term storage credits. Please note that Fig. 4.5 illustrates storage capacity data through June 2005, but recharge credits earned and recovered only through 2002, the latest comprehensive information available.

---

[19] The data in Fig. 4.5 is for statewide activities, however recharge capacities and credits earned are very small outside of the central Arizona AMAs.

[20] For example, the Arizona Water Banking Authority, which is the principal entity conducting recharge, has stored approximately 250,000 acre-feet per year in recent years and stored 59 percent of its 2004 recharge credits though direct recharge (in USFs). By contrast, in its second and third years of operation the AWBA stored only 32 percent of its credits in USFs.

Linking Water Supplies and Growth

Historically, throughout the United States, there has been a disconnect between water supply considerations and urban land use or growth management planning. Reasons for this disconnect range from the relative roles of federal, state, and local government, to the character of growth, to the differing nature of land use, and water resources planning (Lucero and Tarlock, 2003; Arnold, 2005; Babbit, 2005; Coulson, 2005; Hanak, 2005). Water is typically regulated at the state and federal level. Federal water management efforts, both regulations (mostly restricted to water quality and flood control) and investments in infrastructure, were primarily driven by reactions to a crisis or by an effort to facilitate regional economic development. Early twentieth-century federal dam building and irrigation projects to open up the West are a prime example of this federal role.[21] The federal government, as the major landowner in several western states, also plays a significant role through management of national forests, national parks, and other public lands.

Water quantity is typically regulated at the state level, if it is regulated at all. Water rights and quantity management are generally the responsibility of states with surface water and groundwater typically considered public resources subject to state law. A number of the western states, including Arizona within AMAs, issue various rights and permits to water users. By contrast, the owners of water delivery and treatment infrastructure are typically not the states but are local governments or private water companies and irrigation districts.[22]

Although land use management decisions are often integrally related to water issues, in the United States, regulation of land use and growth is generally the exclusive domain of local government (cities and towns, or the county if an unincorporated area) and actual land development investments are made by individual and corporate private property owners. Though there have been some recent regulatory efforts by state government on growth management, in general, these are strongly resisted by both state legislatures and local governments. Local land use decisions are often driven by efforts to manage nuisance land uses, satisfy local political elites, minimize public opposition, and increase the local tax base. Local comprehensive land use plans typically do not consider water supply adequacy, and for the most part are also weak and easily ignored or amended.[23]

In many communities, the local planners and the water managers do not interact. The character of urban growth, typically on the fringe of metropolitan areas in communities without significant resources or experience in dealing with growth, also hinders effective management. Additionally, because the boundaries of local jurisdictions rarely coincide

---

[21] Although the federal government does not directly participate in local land use decisions, numerous federal programs do have a major impact on land uses and the market which drives land use and land use changes throughout the country. Examples of major federal activities with a significant impact include agricultural price supports, programs to facilitate the home mortgage business, interstate highway planning and highway funding (Babbit, 2005).

[22] Certain major infrastructure projects in the west, such as the Central Arizona Project, are federally owned and operated by either regional districts or the federal government. In fact, recognition by the federal government of the need for major investments in dam building in the early 1900s for flood control and water supply initiated one of the most significant expansions of the role of the U.S. federal government (Holway and Burby, 1993).

[23] A number of states and local governments have however begun to strengthen general planning efforts by requiring consistency between planning documents and land use decisions such as zoning and by limiting the frequency with which general plans can be amended.

with aquifer and watershed boundaries, the areas benefiting from land use decisions and those bearing the cost of water resources impacts frequently are different, so the water resources costs of some land use decisions may not be recognized by the local government approving development. Finally, water managers traditionally have avoided involvement with land use decisions, focusing instead on acquiring the water and building the infrastructure to meet the expected water demands. In the arid southwest, it is generally accepted that water will continue to move uphill toward development and money. So far, water management decisions for the most part have continued to be left to the water management experts without significant involvement by others. Recent efforts in Arizona to integrate long-range water and land use planning and to consider water adequacy prior to approving new developments will be discussed later.

The majority of land in Arizona is state, federal, or Native American lands with only 16 percent of the state in private ownership.[24] However, most water uses occur on these private lands and the rights of private property owners are vigorously defended in Arizona. It is the decisions and investments of multiple water users and providers (cities, farmers, irrigation districts, private water companies, industries, and individuals) that most strongly affect how water is used in Arizona. An effective approach for state programs is to influence the individual behaviors and investment decisions that collectively determine how water is actually managed. Different types of programs, both regulatory and nonregulatory, are needed depending on the decisions that need to be affected. By providing regulatory certainty, a clear water rights system and the grandfathering of existing users, Arizona's GMA encouraged investments in conservation and use of renewable supplies. Establishment of the water rights structure protected existing users and assisted private markets to function. Creation of such a water rights structure, though perhaps not essential for a regulatory program to operate, is fundamental to the operation of Arizona's regulatory demand and supply management programs.

Arizona's innovations to ensure sustainable water supplies sufficient for continued urban growth include adoption of conservation programs and investments in utilization of renewable supplies by cities; the conservation and AWS provisions of the GMA; the recharge and recovery program, groundwater replenishment district and AWBA; water rights settlements with Native American communities; new requirements for inclusion of a water resources element in comprehensive plans by larger or fast-growing local governments throughout the state; and recent changes which will require water supply, drought, and conservation planning by water providers throughout the state. Arizona is in a unique position relative to the linkages between water and growth. Within the AMAs, new development is subject to the rigorous state AWS requirements. In the non-AMA, primarily rural areas, however, Arizona has one of the weakest programs in the country. In these areas, weak state rules are generally considered to prevent local governments from making stronger regulatory linkages between growth and adequate supplies. Though it is too soon to evaluate the impact of the new water supply planning requirements, these will, at a minimum, raise awareness of water resources issues and likely increase the call to improve available information. One result of the recent requirement to prepare a water

Arizona's Assured Water Supply Program and Water Planning Mandates

---

[24] Such a high percentage of federal public lands is common in only a few western states. In most of the states the vast majority of land is privately owned.

resources element for local general plans was a heightened awareness of the lack of water resources information in much of rural Arizona and of the disconnection between local governments and private water companies.

Adoption of Arizona-style AWS requirements will, in general, be politically acceptable only where alternative supplies are available to meet the requirements. Use of these types of rules to require investments in renewable supplies for new growth is much more likely to be accepted than using such requirements to halt growth. In addition, given the intensely political nature of local land use decisions, placing the authority for requiring adequate water supplies at a level of government higher than that responsible for approving subdivisions may be critical to a successful program. On the other hand, advocates for improving the integration of water and growth decisions want water adequacy and allocation decisions to be linked to land use decisions at the local level. An appropriate state role may be to further encourage local authority while serving as a final stopgap, with monitoring and compliance authority, to make sure water adequacy was demonstrated prior to final subdivision approval. Programs to encourage conversion from groundwater to renewable supplies and regulations requiring new growth to use renewable water supplies are the cornerstone of Arizona's efforts to reduce overdraft in the AMAs. A couple of key lessons to highlight include the decisions being targeted by the AWS program, the institutional and ownership issues involved in recharge, the role of the CAGRD, and the incorporation of climatic uncertainty.

The objective of the AWS program is to ensure new municipal development has a secure and renewable supply of water that will not exacerbate groundwater mining. The relevant decision makers are developers who want to build, landowners who hold vacant land, and local jurisdictions that approve new subdivisions.[25] The AWS program features a strong regulatory approach, with control at the state level, to prohibit local governments from permitting the subdivision of land unless the requirement for a secure 100-year water supply is met. Arizona's AWS program also includes significant compliance authority. This authority includes regular reviews of annual reports and the ability to revoke AWS designations or to halt the issuance of new certificates which would prevent the plating or sale of new subdivision lots. The state agency (ADWR) has used this authority on several occasions to force compliance with the AWS criteria.

Implementation of the AWS rules would not have been politically feasible in Arizona without providing a convenient mechanism for most residential developers to continue building, particularly those at or beyond the edge of the urban area without ready access to renewable supplies. The CAGRD, by committing to replenish groundwater used by its members, provided this mechanism and allowed adoption of the AWS rules. The AWS rules were also dependent on the passage of recharge and recovery statutes. These statutes provided the critical protection that an entity storing water in the aquifer could retain access to that water and could recover the water anywhere in the same active management area and legally consider the water to be from the source recharged, surface water (CAP) or, effluent, rather than groundwater.[26] These provisions both protect ownership of recharge credits and

---

[25] For a discussion of the potential impact of assured water supply type requirements on housing prices and availability (see Hanak, 2005).

[26] This is important because conservation requirements generally are not applied to effluent and AWS rules require use of nongroundwater supplies.

facilitate some limited markets for transferring recharge credits. The recharge statutes also put in place a regulatory structure for permitting recharge facilities. The recharge and recovery programs combine to allow aquifer space to be used for storage of excess waters and later recovery.

Arizona's AWS program may not adequately consider the state's climatic variability, particularly in the context of the likelihood of longer and more intense drought cycles in the future. CAP municipal and industrial allocations are treated as though they are 100 percent reliable. Even with the firming activities of the AWBA, these supplies are subject to shortage. The state may also have assumed a greater volume of SRP supplies than will actually be available over the long term when accounting for drought periods. The implication of the vulnerability to drought of these supplies is that more backup supplies should be secured. The current AWS rules do allow additional groundwater pumping during drought periods, but in areas where physical availability of groundwater is limited sufficient supplies may not be available or set aside as part of the AWS approval. Recent improvements in the understanding of climatic variability may help tighten the evaluation methods and standards for future revisions of the AWS rules.

It is also likely that additional adjustments will be made to the rules to ensure the reliability of the required 100-year supplies. This may involve long-term dry-year option agreements with agriculture, through which municipal supplies can be "backed up" by contracts with agricultural users.

Under the "physically and continuously available" criteria of the AWS program the state's evaluation of groundwater availability is probably conservative, though hydrologic uncertainties regarding the volumes of available groundwater and administrative issues may affect the long-term program's effectiveness. Both the hydrology and the overall groundwater allocations relative to available supplies will need to be revisited periodically to evaluate whether the program is functioning properly.

A final lesson from the Arizona experience comes from the recognition that comprehensive water management programs grow and evolve over many years. The GMA, with the creation of a long-term goal and a series of 10-year management plans, put in place an incremental approach to reaching safe-yield and ensured an ability to respond to changing conditions.

Continued growth, both within and outside of AMAs will certainly place stress on the resiliency of the water management system and the long-term sustainability of Arizona's water supplies. Population projections indicate the Central Arizona AMAs could increase from the current population of nearly 5 million to 15 million by 2100. However, the central Arizona AMAs are likely to be able to sustainably serve this level of growth by utilizing a combination of (1) their extensive (though already overdrafted) groundwater reserves as backup supplies during drought, (2) current and new infrastructure for importing additional supplies, (3) requirements that new growth secure renewable or imported supplies, and (4) potential access to renewable supplies currently serving agricultural and Native American water right holders[27] (Holway, Newell and Rossi, 2006). However, if necessary investments are not

---

[27] Since 1978, six major Native American water rights settlements have been reached in central Arizona, including the largest ever Indian water rights settlement, the 2004 Arizona Water Rights Settlement Act. Provisions of these settlements allow some water to be leased to cities.

made, if regional water and land use managers do not properly prepare for future droughts, if political or economic forces restrict the ability to purchase and transfer water supplies, or if other unforeseen events occur, then a future of ongoing water crises could occur. Areas of the southwestern United States without both significant water storage potential and water importation infrastructure will face significant growth limits and possibly insurmountable challenges to long-term sustainability.

CONCLUDING THOUGHTS

Arizona's water management efforts within AMAs, although heavily regulatory, have largely been successful. The state has reduced its reliance on groundwater and increased use of more expensive and sustainable surface water supplies. The legal framework and management approaches in place have provided the assurances of stable supplies and the certainty necessary to encourage investments in Arizona's future. As the west continues to grow and face new water management challenges, Arizona's water managers will need to address a number of new challenges, including the need to (1) work cooperatively with water users and interested parties throughout the state and the Colorado River Basin to secure the future supplies for urban growth;[28] (2) forge regional partnerships within urban areas to develop coordinated long-range aquifer management strategies which incorporate conjunctive use and both water quantity and water quality needs; (3) modify the CAGRD as necessary to ensure it remains a viable component of the region's water supply mechanisms and to ensure it is not forced to grow beyond its ability to secure supplies for replenishment and its appropriate role; (4) improve the understanding of climatic variability and future global climate change and incorporate this knowledge into long-range water management planning, almost definitely requiring more investment in storage and backup supplies to offset future CAP shortages; (5) modify the state's regulatory framework to evolve with the changing nature of water uses, supplies, and understanding of climatic variability and to facilitate the necessary water management programs and investments; (6) learn from the efforts of other tribal, state and local governments; (7) look beyond the 2025 AMA safe-yield goals and commit to long-term objectives; and (8) address environmental quality, ecosystem health, and quality-of-life concerns as they relate to water management.[29] The continued rapid growth of Arizona's urban areas will require building on past water management successes and further investing in the infrastructure and water management capacity necessary to ensure long-term sustainability.

[28] On February 3, 2006, all seven Colorado River Basin States signed a letter to the U.S. Secretary of Interior proposing certain shortage guidelines and coordinated management strategies to be included within the scope of the Environmental Impact Statement for the Department of Interior's proposed Colorado River Reservoir Operations. The letter, The Seven Basin States' Proposal and the Basin States Final Draft Agreement are available through the following websites:
U.S. Bureau of Reclamation—Lower Colorado River Region website, accessed June 8, 2006. Information on seven basin states shortage sharing negotiations at http://www.usbr.gov/lc/region/programs/strategies.html and Arizona Department of Water Resources website, accessed June 8, 2006 at http://www.azwater.gov/dwr/Content/Hot_Topics/Colorado_River/Status.htm
[29] See Kenney 2005 for a discussion of lessons the American west can draw from international water management experiences.

REFERENCES

Allan, R., J. Lindesay, and D. Parker, *El Niño Southern Oscillation and Climatic Variability,* CSIRO, Collingwood, VIC, Australia, 1996.

Andrade, E. R., and W. D. Sellers, "El Niño and Its Effect on Precipitation in Arizona and Western New Mexico," *Journal of Climatology,* Vol. 8, 4, pp. 403–410, 1988.

Arizona Department of Water Resources, *Arizona Water Resources Assessment*, 1994. Arizona Department of Water Resources, Phoenix, Arizona.

Arnold, C. A. (ed.), *Wet Growth: Should Water Law Control Growth?,* Environmental Law Institute, Washington, DC, 2005.

Babbit, B. E., *Cities In The Wilderness: A New Vision Of Land Use In America,* Island Press, Washington, DC, 2005.

Connall, Jr., and D. Desmond. "A History of the Arizona Groundwater Management Act," *Arizona State Law Journal,* No. 2, pp. 313–343, 1982.

Coulson, S. E., "Locally Integrated Management of Land-Use and Water Supply: Can Water Continue To Follow The Plow?" Masters Thesis, Urban and Regional Planning, University of Colorado, Denver, 2005.

Enfield, D. B., A. M. Mestas-Nuñez, and P. J. Trimble, "The Atlantic Multidecadal Oscillation and its Relation to Rainfall and River Flows in the Continental U.S.," *Geophysical Research Letters,* Vol. 28, No. 10, pp. 2077–2080, 2001.

Garfin, G. M., M. Crimmins, and K. L. Jacobs, "Drought, Climate Variability, and Implications for Water Supply and Management," in *Arizona Water Policy: Water Management Innovations in an Urbanizing Arid Region.* B. Colby and K. Jacobs (eds.), Resources for the Future, Washington, DC, 2006.

Glennon, R., and T. Maddock, *The Concept of Capture: The Hydrogeology and Law of Stream/Aquifer Interactions,* Rocky Mountain Mineral Law Foundation, Westminster, Colorado, P. 89, 1987.

Hanak, E., *Water For Growth: California's New Frontier,* Public Policy Institute of California, San Francisco, CA, 2005.

Hirschboeck, K. K., and D. M. Meko, *A Tree-Ring Based Assessment of Synchronous Extreme Streamflow Episodes in the Upper Colorado & Salt-Verde-Tonto River Basins.* University of Arizona Laboratory of Tree-Ring Research, Tucson, AZ, available at http://fpcluster.ccit. arizona.edu/khirschboeck/srp.htm, 2005.

IPCC, *Climate Change 2001: The Scientific Basis. Contribution of Working Group I to the Third Assessment Report of the Intergovernmental Panel on Climate Change, Cambridge University Press, NY,* 2001.

Holway, J. M., and R. J. Burby, "Reducing Flood Losses: Local Planning and Land Use Controls," *Journal of the American Planning Association*, Vol. 59, No. 2, pp. 205–216, 1993.

Holway, J. M., P. Newell and T. S. Rossi, "Water and Growth: Future Water Supplies for Central Arizona," Discussion Paper, Global Institute of Sustainability, Arizona State University, Tempe, Arizona 2006.

Jacobs, K. L., and J. M. Holway, "Managing for Sustainability in an Arid Climate: Lessons Learned from 20 Years of Groundwater Management in Arizona, USA," *Hydrogeology Journal,* Vol. 12, pp. 52–65, 2004.

Kenney, D. S. (ed.) *In Search of Sustainable Water Management: International Lessons for the American West and Beyond.* Edward Elgar. Cheltenham, UK 2005.

Lucero, L., and A. D. Tarlock, "Water Supply and Urban Growth in New Mexico: Same Old, Same Old or a New Era?" *Natural Resources Journal*, Vol. 43, 3 pp. 803–805, 2003.

Mantua, N. J., and S. R. Hare, "The Pacific Decadal Oscillation," *Journal of Oceanography,* Vol. 58, 1, pp. 35–44, 2002.

Mayo, A., "A 300-Year Water Supply Requirement: One County's Approach," *Journal of the American Planning Association,* Vol. 56, No. 2, pp. 197–208, 1990.

McCabe, G. J., M. A. Palecki, and J. L. Betancourt, "Pacific and Atlantic Ocean Influences on Multidecadal Drought Frequency in the United States," *Proceedings of the National Academy of Sciences,* Vol. 101, No. 12, pp. 4136–4141, 2004.

McPhee, J., A. Comrie, and G. Garfin,. *Drought and Climate in Arizona: Top Ten Questions & Answers,* Final Report, CLIMAS, Tucson, available at. http://www.ispe.arizona.edu/climas/learn/drought/DroughtQ&A.pdf, 2004.

Meko, D., C. W. Stockton, and W. R. Boggess, "The Tree-Ring Record of Severe Sustained Drought," *Water Resources Bulletin,* Vol. 31, No. 5, pp. 789–801, 1995.

Pool, D. R., and A. L. Coes, "Hydrogeologic Investigations of the Sierra Vista Sub-watershed of the Upper San Pedro Basin, Cochise County, Southeast Arizona," U.S. Geological Survey Water Resources Investigations Report 99-4197, Tucson, Arizona 1999.

Ruth, M. (ed.) *Smart Growth and Climate Change: Regional Development, Infrastructure and Adaptation.* Edward Elgar, Cheltenham, UK 2006.

Stahle, D. W., E. R. Cook, M. K. Cleaveland, M. D. Therrell, D. M. Meko, H. D. Grissino-Mayer, E. Watson, and B. H., Luckman, "Tree-Ring Data Document 16th Century Megadrought Over North America," *Eos,* Vol. 81, No. 12, p. 212, 2000.

Western Regional Climate Center, Comparative Data, 2002.

World Commission on Environment and Development, *Our Common Future* (Brundtland Report) Oxford Univ. Press, New York, 1987.

5 OPTIMIZATION MODELING FOR SUSTAINABLE GROUNDWATER AND CONJUNCTIVE USE POLICY DEVELOPMENT

Richard C. Peralta
Department of Biological and Irrigation Engineering
Utah State University
Logan, Utah

A. Mark Bennett III
Arkansas Natural Resources Commission
Little Rock, Arkansas

Ann W. Peralta
Peralta and Associates, Inc.
Hyde Park, Utah

Robert N. Shulstad
College of Agricultural and Environmental Sciences
University of Georgia
Athens, Georgia

Paul J. Killian
STS Consultants Ltd.
Green Bay, Wisconsin

Keyvan Asghari
Department of Civil Engineering
Isfahan University of Technology
Isfahan, Iran

INTRODUCTION

Worldwide, many aquifers have been mined, causing significant declines in groundwater levels. Without reductions in groundwater pumping, unacceptable changes in head or water quality, economic hardship, or possible irreversible aquifer damage can occur. Decision makers are frequently called upon to declare management policies for such areas. The case study described illustrates explicitly formulating, evaluating, and comparing alternative policies, so that the best one can be selected. It involves using simulation and simulation/optimization modeling.

Simulation models are needed to predict physical system response to stimuli. Stimuli can include groundwater pumping, aquifer recharge, diversions from streams, return flows to streams, flows across boundaries, and the like. Some stimuli are manageable (directly controllable by man) and some are not.

Simulation/optimization (S/O) models can help identify the best set of manageable stimuli values (the best management strategy). Such models predict the best possible outcomes from tested policies. This is valuable, even if those ideal outcomes cannot be totally achieved in practice, for social, institutional, or other reasons.

SIMULATION/OPTIMIZATION MODELS

Introduction

An S/O model calculates a strategy for maximizing achievement of user-defined management objectives, subject to user-specified restrictions. An S/O model couples a simulation module that predicts the results of management, with an optimization module that computes the mathematically best management strategy for the user-posed optimization problem. An example groundwater pumping strategy consists of a set of spatially, and perhaps temporally, distributed groundwater extraction rates.

Using an S/O model is different than using a normal simulation model (here termed an S model), such as MODFLOW (McDonald and Harbaugh, 1988) and MT3DMS (Zheng and Wang, 1998). A groundwater S model user must prepare a pumping strategy and input it into an S model. Then the S model predicts how the modeled physical system will respond to the strategy. S models do not calculate optimal management strategies. Using them to develop strategies requires trial and error and will only produce the mathematically best strategy for simple problems.

Again, S/O models directly compute optimal strategies for user-specified optimization problems. The S/O model's optimization module calculates the optimal strategy. S/O models include S modules to predict system responses, and ensure that an evolving strategy satisfies user-specified constraints. An S/O model's predictive ability is as accurate as the incorporated S model.

Methods of Simulating within S/O Models

Common methods for predicting system response to stimuli within S/O models include embedding physical process equations directly as constraints; employing convolution (superposition) equations containing influence coefficients (dirac delta functions or discrete kernels) as substitute simulators; employing response surfaces (examples include regression equations, interpolators, and artificial neural networks) as substitute simulators; and calling simulation models directly. Different methods are better for different situations.

Within S/O models, the most common ways for computing optimal solutions are either classical *operations research* (OR) or *heuristic* (HO) methods. Common OR methods include simplex, gradient search, branch and bound, and outer approximation techniques. Common HO (or evolutionary) methods include *genetic algorithm* (GA), *simulated annealing* (SA), *tabu search* (TS), and hybrid methods combining features of multiple techniques. Techniques vary in their utility for different types of simulated and optimized systems.

Methods of Optimizing within S/O Models

There are many possible combinations of simulator(s) and optimizer(s). Again, some combinations are better than others for particular situations. Selected examples are mentioned here. Simple field optimization problems can be addressed via relatively simple and inexpensive S/O models. These utilize analytical flow or transport equations as simulators and usually employ classical OR optimizers. Analytical equations are appropriate for relatively homogeneous field scale groundwater or conjunctive water management problems—for situations in which one would commonly use an analytical equation, such as those by Theis, Theim, Glover-Balmer, and others (Peralta, 1999) for prediction. For such simple situations, available capabilities often do not justify using numerical (finite difference or finite element) S models. Peralta and Wu (2004) present sample applications. System nonlinearities, including unconfined aquifer transmissivity, are often addressed automatically using cycling (also termed *successive linear optimization*).

Coupled Simulators and Optimizers

Optimizing water quantity management for significantly heterogeneous physical systems usually requires using numerical flow S models (finite difference, finite element, or analytical element) as simulators. Example S models are MODFLOW (McDonald and Harbaugh, 1988), and SWIFT (Reeves and Cranwell, 1981; Reeves et al., 1986a, b). S/O models for such problems usually use classical OR optimizers, and address system nonlinearities using cycling. Many journal papers describe using such techniques. Peralta and Shulstad (2004) applied S/O modeling to compare sustained yield water policy alternatives for different legal-institutional-hydrogeologic settings.

Optimizing water quantity and quality management for heterogeneous groundwater systems often requires using numerical flow and transport models as simulators. For some problems classical OR optimizers are adequate, but for most extremely nonlinear problems, heuristic optimizers or hybrids are better. For example, HO is most common for designing optimal *pump and treat* (PAT) systems for remediating groundwater contamination. Hegazy and Peralta (1997), Peralta (2001), and Peralta et al. (2003) provide many linear and nonlinear groundwater contamination remediation examples, using OR and HO optimizers such as are in the SOMOS software (SSOL and HGS, 2001; Peralta, 2003).

SUSTAINABLE GROUNDWATER AND CONJUNCTIVE USE UNDER RIPARIAN RIGHTS—REASONABLE USE DOCTRINE

Introduction to Grand Prairie S/O Application

The Grand Prairie is an important rice, soybean, and aquaculture producing area in southeastern Arkansas (Fig. 5.1). Historically, most of the region's irrigation water has come from a quaternary aquifer that is part of the Mississippi plain alluvial aquifer

Figure 5.1

The Grand Prairie study area.

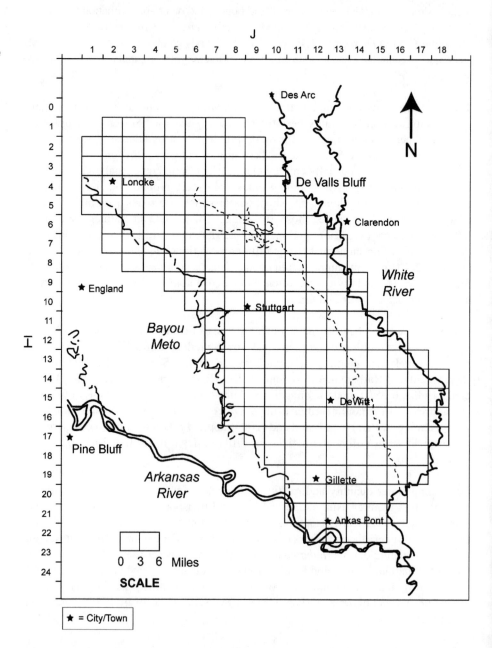

Groundwater levels have dropped in the Grand Prairie for most of the last 100 years, causing extensive and deep potentiometric surface depression (Fig. 5.2). As water levels in the alluvial aquifer have dropped, irrigators have begun to use water from the Sparta aquifer in greater quantities. The tertiary Sparta aquifer is the main source of drinking water in the area. For some time, the public has been concerned about economic problems that would result from water shortages or excessively costly irrigation water (ANRC, 1990). Water users, water managers, and those involved in policy formation asked several questions about Grand Prairie water use. First, what across-the-board proportionate reduction in current groundwater use would be necessary to achieve a safe sustained yield of groundwater? An Arkansas landowner has "a common and correlative right to the use of this underground water upon his land to the full extent

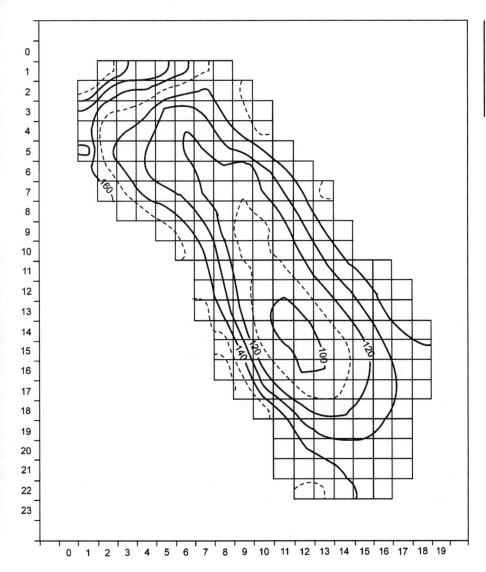

Figure 5.2

Recent Grand Prairie potentiometric surface (ft above mean sea level).

of his needs if the common supply is sufficient, and to the extent of a reasonable share there of, if the supply is so scant that the use by one will affect the supply of other over-lying users" (Jones v. OZ-ARK-VAL Poultry Co., 228 Ark. at 81, 306 S.W. 2d at 115). Groundwater pumping corresponding to a sustained yield groundwater extraction strategy might well be defined as "reasonable use" (Peralta and Peralta, 1984a, b). Arkansas water law is in transition. The Arkansas Supreme Court has been inclined to defer to state agency decisions concerning water issues [351 Ark. 289, 92 S.W.3d 47 (2002)]. However, in the absence of clear authorizations to the ANRC from the General Assembly, groundwater scarcity could result in court mandated reductions in ground-water withdrawals (Peralta and Peralta, 1984 a, b).

The second Grand Prairie question was whether implementing on-farm conservation measures could reduce water need (demand) sufficiently to permit developing an acceptable sustained yield strategy that would not require diverting river water to non-riparian farm lands. Agriculture, including aquaculture, uses almost all of the quaternary groundwater withdrawn in that region. Rice and aquaculture producers used an average of 2 and 7 ac-ft (2.57 and 8.64 10^3m^3) of water per acre per year, respectively, and are sometimes criticized because of their high consumptive water use.

The third Grand Prairie question arose after the White River-Grand Prairie irrigation district was formed under Act 114 of 1957 (Ark. Stat. Ann. 12-1401, et seq.). The district's right to exist and to divert river water to the Grand Prairie was affirmed by the Arkansas Supreme Court in 1984 (Frank Lyon, Jr. et al. v. White River-Grand Prairie Irrigation District, 281 Ark. 286. 664 S.W.2d 441), but surface water storage and distribution systems did not exist to all parts of the region. Assuming that surface water can be used to replace current quaternary groundwater use within areas where it can be delivered, could all current regional water demand be satisfied in the long term? In other words, could a sustained yield strategy be developed substituting a combination of groundwa-ter, diverted river water, and on-farm water conservation measures to replace current quaternary groundwater use?

These questions were answered by the presented S/O modeling process. The S/O model computed optimal water allocation strategies. Each strategy included the spatial distribu-tion of river water and quaternary groundwater that should be used annually, and the dis-tribution of unsatisfied water demand—demand that *cannot* be met by either source under that strategy. Here water demand is defined as the average recent rate of water withdrawn for agricultural and aquacultural uses from the quaternary aquifer. Water demand can be reduced by implementing water conservation measures. This study addressed the water demand of acreages that were dependent on quaternary groundwater, and assumed that acreages being supplied with water from other sources would continue to do so.

Also estimated were the future net economic return from continuing groundwater-supported agriculture in the Grand Prairie (based on projected potentiometric heads, and the like), and the economic consequence of a particular allocation strategy (the annual sum of all changes in net economic return from the unmanaged, unconstrained maintenance of current pumping). This includes changes in returns from production and changes in costs, including cost of production, water supply, and lost opportunity resulting from not fully satisfying water demand. A sustained yield scenario requires the switch of some aquacultural or irrigated crop acreages to nonirrigated soybeans.

To answer the posed questions, four alternative policy scenarios (and optimization problem formulations) were addressed. Optimal sustained groundwater yield strategies and the predicted annual economic consequences for the 10 years following strategy implementation are presented for each scenario. Only 10 years of predictive simulation was performed because that was the duration of the model calibration era.

Two objective functions were used: minimization of unsatisfied water demand and minimization of the common percentage of reduction in groundwater use needed to achieve a sustained yield (an objective not previously examined in the literature). Also considered are the reduction in water demands achieved through on-farm water conservation measures and the use of diverted river water. An economic optimization objective function was not used because Arkansas' water users cannot be forced to adhere to a regionally economically optimal strategy.

In subsequent sections, technical and nontechnical topics are separated to avoid unnecessary repetition and to aid water management strategy comparisons. Thus, the section on "Optimization Problem Formulations" describes computer simulation and optimization methods, without detailing input values and assumptions. The section on "Application and Results" describes application of the methods (including inputs), and results of those techniques.

In overview, the section on "Model A—Minimizing the Common Proportion of Reduction in Pumping Needed to Achieve a Sustained Yield (Used for Scenario I)" summarizes the equations representing the first optimization problem that is solved. The section on "Model B—Minimizing Unsatisfied Water Demand (Used in Scenarios II-IV)" summarizes equations of the other three solved optimization problems (values used as bounds in those equations are presented later in the section on "Application and Results"). The section on "Economic Postprocessor" discusses the economic postprocessor used to predict economic impacts of each computed optimal strategy. The section on "General Hydrologic Assumptions and Constraints" describes sufficient hydrogeology to explain boundary conditions, and values generally used within bounds and constraints of the section on "Optimization Problem Formulations" optimization problems. The section on "Alternative policy scenarios and results" describes results. For each of the four posed optimization problems, this includes S/O model-developed strategies, and economic postprocessor output.

Model A—Minimizing the Common Proportion of Reduction in Pumping Needed to Achieve a Sustained Yield (Used for Scenario I)

Optimization Problem Formulations

By what proportion must *all* groundwater users in a region reduce their current withdrawals in order to achieve a sustained yield? In aquifers with rapidly declining water levels, the Arkansas courts could limit each overlying landowner to a proportionate or prorated share of the available supply [Hudson v. Dailey, 156 Cal. 617, 105 p. 748 (1909); Jones v. OZ-ARK-VAL Poultry Co., 228 Ark. 76, 306 S.W. 2d 111 (1957); Lingo v. City of Jacksonville, 258 Ark. 63, 522 S.W.2d 403 (1975)]. Model A addresses this possibility indirectly by maximizing the common proportion, χ, of current groundwater withdrawals that can be pumped from each cell in a sustained yield setting. (By this is meant the largest X for which a steady state flow solution exists that does not violate any bounds or constraints.) The percentage by which current withdrawals need to be reduced equals $(1 - X)$ times 100. Assuming that current groundwater withdrawal represents the upper limit on pumping in any cell, the optimization problem formulation is:

$$\text{Max } X \qquad (5.1)$$

Subject to

$$p^U(\hat{e}) \, X = p(\hat{e}), \text{ for all pumping cells} \qquad (5.2)$$

$$0.0 \leqslant X \leqslant 1.0 \qquad (5.3)$$

$$0.0 \leqslant p\,(\hat{e}) \leqslant p^U(\hat{e}) \qquad (5.4)$$

$$q^{zL}(\hat{o}) \leqslant q^z(\hat{o}) \leqslant q^{zU}(\hat{o}), \text{ for all specified-head cells} \qquad (5.5)$$

$$h^L(\hat{o}) \leqslant h(\hat{o}) \text{ , for all variable-head cells} \qquad (5.6)$$

where L and U = lower or upper bounds, respectively, on variables

$\hat{e}, (\hat{o})$ = indices identifying finite difference cells within the groundwater model $(L^3 T^{-1})$

$p(\hat{e})$ = groundwater pumping extraction at cell $(\hat{e})$. Flow leaving the aquifer is positive in sign $(L^3 T^{-1})$

$q^z(\hat{o})$ = groundwater flow entering or leaving the study area through a specified-head cell $(L^3 T^{-1})$

Other constraint equations link pumping decision variables to head and boundary flow state variables.

Model B—Minimizing Unsatisfied Water Demand (Used in Scenarios II-IV)

Minimizing annual unsatisfied water demands is sometimes the same as maximizing total annual groundwater pumping, z. Model B, similar to formulations by Aguado et al. (1974), Alley et al. (1976), and Elango and Rouve (1980) for small systems, and recently by many more authors for real systems is:

$$\text{Max } Z = \sum p(\hat{a}) \qquad (5.7)$$

Where $(\hat{a})$ identifies a cell at which diverted river water is unavailable, subject to constraint Equations 5.4 to 5.6. This formulation works because there was sufficient river water available to satisfy demand in cells to which it could be reasonably conveyed via new canals.

Economic Postprocessor

If no optimal strategy is implemented, the net economic return of agricultural production relying on the alluvial aquifer would average \$8.33 10^6 per year during the following decade. This was estimated using unit costs and revenues reported by Harper et al. (1989). These same values are used in the economic postprocessor to estimate the economic impact of implementing each policy. Although no longer current, these costs effectively demonstrate the methodology. Considered factors include those from crop budgets (Table 5.1) and the variable cost per unit volume of groundwater. The variable cost of 1 ac-ft (1.23 10^3m^3) of groundwater in a cell was calculated by multiplying an

Table 5.1

Economic
Parameters

Crop Parameters		Crops			
		Aquaculture	Irrigated Rice	Nonirrigated Soybean	Soybean
Yield	lb/ac	1100	4410	2400	1620
	(kg/ha)	(1233)	(4942.9)	(2690)	(1815.7)
Value	$/lb ($/kg)	1.12	0.1056	0.1046	0.1046
		(2.47)	(6.232)	(0.231)	(0.231)
Water Requirement	ac-ft (10³m³)	7(8.64)	2(2.47)	0.4(0.693)	0.0(0)
Fixed Costs	$/ac	227.23	117.75	119.28	90.43
	($/ha)	(560.9)	(289.1)	(294.7)	(223.4)
Variable Costs, Except Water Cost	$/ac	604.37	245.57	171.56	165.96
	($/ha)	(1493.4)	(606.8)	(423.9)	(410.1)

Water Cost Parameters		
Pumping plant energy, repair, and lubrication	$/ac-ft-ft	0.18
	($/10³m³-m)	(0.48)
Pumping plant maintenance	$/ac-ft	1.65
	($/10³m³)	(1.34)
Arkansas River water	$/ac-ft	17.00
	($/10³m³)	(13.82)
White River water	$/ac-ft	31.00
	($/10³m³)	(25.2)

estimated seasonal average total dynamic head for wells in that cell by pumping plant energy, repair and lubrication costs, and adding the pump maintenance cost. The procedure described by Peralta et al. (1985) was used to estimate the average seasonal total dynamic head for a well from projected head at the center of the cell. They computed dynamic daily drawdown curves of 500 gpm wells pumping to supply the water demand of 50 acres of rice for average and drought conditions in the Grand Prairie. They illustrated how much saturated thickness was needed in the springtime to assure that drawdown does not exceed two-thirds of the saturated thickness during the following summer (for power efficiency reasons one does not normally want the drawdown to exceed two-thirds of the design saturated thickness). They demonstrated that 25 ft is the minimum saturated thickness needed for efficient operation under drought conditions.

After optimal strategies were computed, standard methods were used to estimate the average change in net return that would result soon after strategy implementation. The change is the sum of the changes in gross receipts, fixed costs, variable costs exclusive of the costs of obtaining water, and the variable costs of the water.

The change in net return from current assumed return is presented, rather than an estimated net return, because several costs are not included in available crop budgets.

This cost-benefit analysis approach allows omitting the cost of land, the value of the labor of the farmer and his family, and general farm overhead.

Assumptions in computing the change in return resulting from strategy implementation are as follows. If, for a particular sustained yield strategy, there is inadequate water in a cell to satisfy demand, nonirrigated soybeans will replace aquaculture or an irrigated crop. Irrigated soybean acreages are the first to be switched to dryland soybeans, followed by aquacultural and rice acreages. The fixed expenses for the original crop will continue for a while, even after a crop change is implemented. A producer needing to switch crops has adequate surplus capacity in his machinery set to produce nonirrigated soybeans. After five years, the considered fixed cost is the average of that of the original crop and nonirrigated soybeans.

The change in gross returns and variable costs resulting from strategy implementation is based on the crop acreages that the strategy can support with water plus any nonirrigated soybean acreages made necessary by inadequate water supply. The change in the variable cost of water in a cell is the difference between the cost of groundwater in an unmanaged future and the sum of the costs of optimal groundwater and diverted river water. The unit cost of groundwater in a cell was estimated using the total dynamic head of representative wells derived from the transient water table elevations that will result from either continued current pumping or optimal pumping.

The cost of diverted river water differed with source (Fig. 5.3, Table 5.1). Values include delivery to identified cells and $3/ac-ft ($2.44/10^3m^3) to move the water to fields in the cells (U.S. Army Corps of Engineers, 1984a, b); and personal communication, Dwight Smith). Economies of scale were neglected.

Application and Results

General Hydrologic Assumptions and Constraints

Because of the impermeable layer overlying the aquifer in internal cells, vertical recharge of the aquifer in the Grand Prairie is small (Sniegocki, 1964; Griffis, 1972; Peralta et al., 1985). Recharge occurs at peripheral specified-head cells and comes from either rivers or extensions of the aquifer outside the area. To avoid the risk that an optimal pumping strategy would cause boundary heads to drop in the field, the amount of recharge induced through any of those cells is limited during optimization. It is assumed that if that limit is not exceeded, actually implementing the optimal strategy will maintain current boundary heads. In the model, the greatest annual recharge permitted to occur in any peripheral constant head cell (most negative rate of flow entering that cell from outside the study area), was the greatest value calculated to occur based on the springtime hydraulic gradients of the recent decade. The use of constant head-constrained flux boundary conditions was justified previously for similar areas (Yazdanian and Peralta, 1986a, b).

Groundwater pumping can occur at all interior cells. The lower limit on groundwater withdrawal in any internal cell was zero. This assured that a feasible strategy could be computed for each scenario. Two sets of upper bounds on pumping were used, depending on the scenario. In the first set, the upper limit for a particular cell was the average use of quaternary groundwater. This totaled 286,000 ac-ft (353 10^6m^3) (Table 5.2).

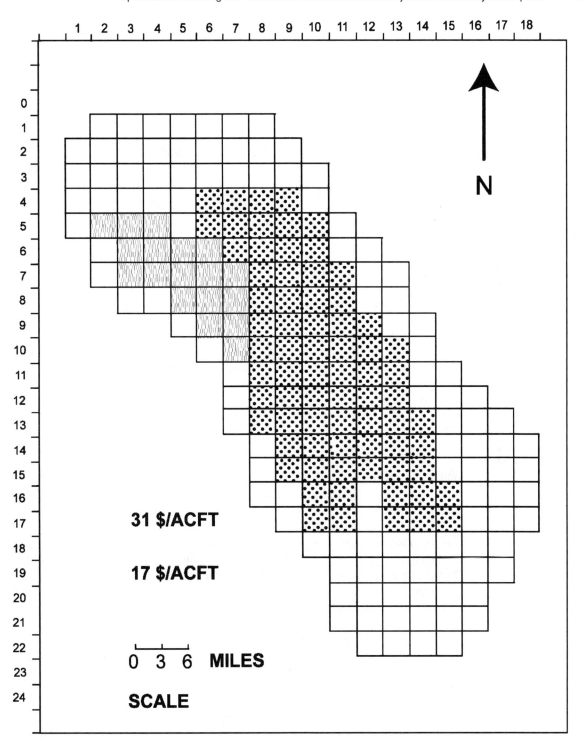

Figure 5.3
Cells to which river water can be diverted and the unit cost of that water ($/ac-ft).

Table 5.2

Optimal Strategies
and Short-Term
Annual
Consequences of
Strategy
Implementation

		Scenarios				
		Base	I	II	III	IV
Water demands	1000 ac-ft	286	286	286	253	253
Groundwater use	1000 ac-ft	286	38	118	115	62
River water use	1000 ac-ft	0	0	0	0	160
Unmet water demand	1000 ac-ft	0	248	168	138	31
Change in net economic return from that of unmanaged groundwater use	1000$(%)	NA	−6985	−4066	−2634	−1948

To convert from acre-feet to cubic meter multiply by 1.23E03.

A different set of upper bounds on pumping was used when assuming water demand can be reduced by conservation measures. By implementing measures suggested by Harper et al. (1989), total water demand can be reduced to 253,000 ac-ft ($312\ 10^6 m^3$). Through a survey of literature and water users, Harper concluded that the following measures would not reduce yields or increase production expense. For example, assume that 20 percent of the rice acreage in the region was generally maintained at a flood depth of 6 to 8 in. (15 to 20 cm). On those acreages, 6 in. of water could be saved each year, without adversely affecting yields, by changing to a 2 to 4 in. (5 to10 cm) flood depth (Ferguson, 1970). Also, 29 percent of the soybean acreage was furrow irrigated. A 35 percent reduction in water use could be obtained for those acres by irrigating only alternate furrows. In addition, it was assumed that aquacultural consumptive use could be reduced by 20 percent from 7 to 5.6 ft (2.133 to 1.71 m) per year. Thus, the three assumed conservation measures were the reduction in rice flood depth for 17,000 acres (6880 ha), the conversion to alternate furrow irrigation of soybeans for 13,000 acres (5261 ha), and the 20 percent reduction in consumptive water use for 13,800 aquacultural acres (5585 ha).

The lower bounds on potentiometric surface head in Equation 5.6 would assure at least 20 ft (6.1 m) of saturated thickness in all internal cells. Peralta et al. (1985) determined 20 ft to be the minimum saturated thickness needed for a representative 500 gpm (1893 lpm) well irrigating 50 acres (20.2 ha) of rice to remain operable throughout a climatically average pumping season.

The ability to predict head response to pumping came from calibrated finite difference models (Griffis, 1972; Peralta et al., 1985). Average aquifer hydraulic conductivity is 270 ft/day (82.3 m/day) and effective porosity is 0.3 (Engler et al., 1945). Peralta et al., calibrated for a 10-year period.

Alternative Policy Scenarios and Results

Table 5.2 displays historic (base) and optimal sustained yield strategies and economic consequences for four scenarios. Table 5.3 displays the base crop acreages supported by quaternary alluvial groundwater. It also contains the acreages that would exist for each

	Scenarios				
	Base	I	II	III	IV
Aquaculture	13,800	2,600	5,300	5,900	10,700
Rice	86,300	56,800	66,200	67,000	79,900
Irrigated soybean	45,700	23,900	32,000	32,000	41,200
Nonirrigated soybean	NA	62,500	42,300	40,900	14,000

Table 5.3

Crop Acreage
Consequences of
Strategy
Implementation

To convert from acres to hectares multiply by 0.4047.

crop under each scenario if the crop-switch rules described previously were invoked. Cropping in accordance with these acreages results in the changes in net return shown in Table 5.2.

To develop Table 5.2 optimal strategies, the S/O models attempt to satisfy water demands either from groundwater alone (scenarios I–III) or from conjunctive use of groundwater and diverted surface water (scenario IV). Dixon and Peralta (1986) demonstrated that there are adequate divertible water resources available in the Arkansas River and White River for this purpose. U.S. Army Corps of Engineers (1984a, 1984b) investigations indicate the cells to which surface water can be realistically diverted (Fig. 5.2). In all scenarios, water demand that cannot be met by either groundwater or surface water is considered to be unsatisfied.

Scenario I assumed no improvement in water conservation practice and the absence of diverted river water. The correlative rights doctrine is assumed and pumping reductions are proportionally egalitarian. The smallest common proportion of reduction in current pumping in all cells, X, needed to achieve a sustained yield (and satisfy all imposed bounds on heads and flows) was 0.86. Only 14 percent of current pumping, 40,000 ac-ft ($49.3 \ 10^6 m^3$), could be withdrawn per year from the aquifer. The 86 percent unsatisfied demand would result in an average annual reduction in net economic return of $6,985,000—83 percent of the base return. A policy incorporating this objective, representing one possible application of the correlative rights doctrine, would have a serious impact on the region. A difficulty with this scenario would be the requirement to reduce everyone by the same proportion. This limitation on freedom would prevent a more favorable solution from being computed.

Scenario II assumed no increase in water conservation measures and the unavailability of river water. With the objective of minimizing unsatisfied demand, 119,000 ac-ft ($147 \ 10^6 m^3$) of groundwater could be withdrawn, resulting in 169,000 ac-ft ($208 \ 10^6 m^3$) of unsatisfied demand. The 49 percent reduction in economic return is severe. However, the freedom to let some pump at a greater proportion X than others would reap hydrologic reward.

Scenario III differed from scenario II only in the implementation of conservation measures. Groundwater use under scenario III was 3000 ac-ft ($3.7 \ 10^6 m^3$) less than that of scenario II, but because water demands were greatly reduced, unsatisfied water demands were significantly less. The reduction in net return from the base strategy was 32 percent. When compared with scenario II, this corresponded to an improvement in net return of \$43 for each ac-ft (1.23 10^6m^3) of reduced water demand (\$1,432,000 / 33,000 ac-ft). Assuming a one-third reduction in net return was probably still unsatisfactory, the region would need to look beyond water conservation for a solution.

Simultaneous availability of river water for diversion and use of conservation measures distinguished scenario IV. The total water provided is greater than those for the other three optimization scenarios, and groundwater use would decline. Opportunity to fully satisfy water demands in some cells with river water would allow a spatial redistribution of groundwater pumping. There were still 31,000 ac-ft (38.2 10^6m^3) of unsatisfied demand and a 23 percent reduction in base net return.

In summary, sustained yield strategies were computed for all four policy alternatives. However, no tested policy could sustainably provide enough water to satisfy all average water demands. Scenario IV, a combination of water conservation and river water diversion, would be regionally most beneficial. Agency selection of any of scenarios II to IV would be preferable to court order of scenario I. The scenario I policy, a strict correlative rights doctrine implementation, with proportionate reductions in use, maximizes social equity but would be hydrologically and economically inefficient.

Some additional computer processing was performed. Sensitivity analysis showed that optimal strategy rankings were insensitive to assumed aquifer parameters and boundary conditions. Other predictive simulations showed that one could expect there to be more available groundwater after 10 years if the scenario IV strategy were immediately implemented, than if current pumping continued (a comparison between implementing scenario IV versus no management). No-management transient simulation predicted decreasing saturated thickness in many cells. The transient simulation assumed (1) there could be no quaternary groundwater withdrawal (100 percent unsatisfied demand) in a cell having less than 10 ft (3.05 m) of saturated thickness (Peralta et al., 1985) (2) only 50 percent of groundwater demands could be pumped (50 percent unsatisfied demand) in a cell having 10 to 20 ft (3.05 to 6.1m) of saturated thickness. In that case, after 10 years there could be at least 31,400 ac-ft (38.7 10^6m^3) of unsatisfied demand. One could expect that figure to increase with time if no management strategy were adopted. On the other hand, if a strategy for scenario IV were implemented promptly, unsatisfied demand would not be expected to increase above 31,000 ac-ft (38.2 10^6m^3).

Reported results were invaluable for assessing the desirability of tested policies for sustained yield management. The relative ranking of computed strategies is unlikely to change with data refinements or with more extensive and detailed simulation models applied over greater areas. Further detailed evaluation and planning would be desirable before implementing a particular strategy. For example, more exacting analysis of river water diversion dynamics would be necessary before implementing scenario IV.

Regardless of the desirability of further data, the presented procedure reasonably estimated the best effect of water policy decisions. Partially as a result of this public information effort, the federal government through the U.S. Army Corps of Engineers and the state of Arkansas shared the planning costs. The construction costs are being funded—65 percent by the federal government and 35 percent by the State of Arkansas and the White River Regional Irrigation Distribution District. Long-term water purchase contracts will be used to provide operation and maintenance cost of the facilities and to repay part of the 35 percent state and district share. The policy we tested has been partially adopted, and is being tailored to best allow sustainable production while protecting or enhancing environmental quality (USACE, 1999).

CONCLUSIONS

Steady-state groundwater optimization and economic evaluation aid in assessing how water policy can affect groundwater sustained yields and economic return. This is valuable for policy analysis before policy implementation. Here, we evaluated alternative policies for an irrigated region in which historic pumping rates were not physically sustainable. Thus, some reduction in water use and short-term economic loss would result from implementing any sustained yield strategy. The goal of this effort was to determine which policy would result in the most acceptable strategy with the least economic harm. We computed optimal strategies for four policies.

Court imposed application of a strict interpretation of the correlative rights doctrine was the worst-case scenario. Grave economic losses would result from even the smallest across-the-board proportional (86 percent) reduction in groundwater pumping needed to achieve a sustained yield.

The least economic hardship (23 percent reduction in net return) would result if conservation measures are practiced and surface water is diverted from adjacent rivers. If either of these actions is omitted, at least a one-third total reduction in net return would result. Improved conservation alone will not solve the problem.

Study results were a guide for the planning and policy-development process. They are not detailed strategies proposed for implementation. Improved knowledge of system parameters, a larger model area, finer model spatial discretization, and calibration of a current MODFLOW model (instead of our simulator of 20 years ago) is unlikely to change the relative ranking of the strategies.

The demonstrated comparisons of alternative sustained groundwater yield and conjunctive use policies included preparing optimization problem formulations for alternative policies; computing the best spatially distributed sustainable groundwater extraction strategy for each policy; and evaluating short-term economic consequences. After the procedure was applied to the Grand Prairie region of Arkansas (Peralta et al., 1985), results were presented at public meetings in Arkansas. The clearly illustrated alternative potential outcomes helped mobilize local, state, and federal efforts to achieve sustainable conjunctive use for irrigated agricultural production in the Grand Prairie and elsewhere in the state. Subsequently, the Arkansas General Assembly passed necessary legislation (Act 1051 of 1985, Ark. Code Ann. §15-22-301; Act 154 of 1991, Ark. Code Ann. §§ 15-22-901 to 914; Act 1050 of 1999, Ark. Code Ann. §§ 15-22-1201 to 1218), the Grand Prairie Area Demonstration Project was established (USACE, 1999), and the U.S.G.S. performed MODFLOW calibration and S/O model optimization (Reed, 2003; Czarnecki et al., 2003). The ANRC has contracted with the USGS for additional modeling to assist in determining acceptable spatially distributed sustainable groundwater extraction rates. Upon completion of the study, the ANRC will probably begin the process of formally adopting pumping rates.

Arkansas Act 1051 of 1985 (Ark. Code Ann. §15-22-301) moved the state toward the possible application of a sustained yield strategy in the Grand Prairie. The Act requires registration of groundwater withdrawals with the Arkansas Natural Resources Commission (formerly the Arkansas Soil and Water Conservation Commission, see Act 1243 of 2005), except from domestic wells or wells pumping less than 50,000 gallons per day. The Act specifically provides for nonriparian surface water transfers (Ark. Code Ann. §§ 15-22-301 to 302). The *Arkansas Groundwater Protection and Management Act* (AGPMA) followed in 1991 (Act 154 of 1991, Ark. Code Ann. § 15-22-901 to 914). Under the AGPMA the ANRC is empowered to designate Critical Groundwater Areas, defined by the Arkansas' Water Plan as areas where the "quantity of groundwater is rapidly becoming depleted or the quality is being degraded," and to "promulgate rules and regulations for groundwater classification and aquifer use, well spacing, issuance of groundwater rights within critical groundwater areas [Ark. Code Ann. §15-22-904(1)]. By 1998, the Grand Prairie was designated as a critical area for the quaternary alluvial aquifer and for the tertiary Sparta aquifer. In general, water levels have continued to decline more or less as projected by our 1985 work (Smith, 2006).

Regulation has not been proposed for the Grand Prairie or for any other area in the state. Even after critical area designation, the AGPMA imposes cumbersome restrictions on ANRC groundwater management activities, requiring "a new process involving lengthy legal proceedings, additional notice and public hearings" before regulations can be imposed (www.anrc/arkansas.gov/critical_groundwater_designation_fact_sheet.pdf). Further action by the General Assembly is needed to facilitate groundwater management in critical areas, especially during times of extreme shortage (Perkins, 2002).

The Grand Prairie Area Demonstration Project is designed to allow continued irrigation pumping from the alluvial aquifer at its long-term sustainable yield, while reserving the Sparta aquifers for municipal and industrial uses. Among project features are a pumping station to divert excess surface water from the White River, the construction of a surface water distribution system of canals, pipelines, and channel alterations, channel structures, increased on-farm irrigation storage, conservation, and water reuse (U.S. Army Corps of Engineers, 1999). In spite of features to benefit waterfowl and to reestablish areas of native prairie grasses, some environmental groups tried (unsuccessfully) to slow or stop the project (Arkansas Wildlife Federation v. U.S. Army Corps of Engineers, 431 F.3d 1096). The Grand Prairie Area Demonstration Project offers hope that the alluvial and Sparta aquifers will be able to sustain beneficial uses indefinitely. Without the project, the alluvial aquifer may no longer be able to sustain irrigated agriculture beyond 2015 (U.S. Army Corps of Engineers, 1999).

S/O modeling is a truly powerful tool. When data and simulators were relatively simple, it was applied carefully to part (the Grand Prairie) of an extensive alluvial aquifer by constraining boundary recharge rates. That S/O modeling provided long-term guidance that is still valid 20 years later. Applied now with newer simulators and optimizers to a much larger portion of the alluvial aquifer, S/O modeling can help with quantifying regulated limits on groundwater pumping to assure sustainability. Two S/O models available to perform these tasks are GWM (Ahlfeld et al, 2005) and SOMOS (SSOL and HGS, 2001).

ACKNOWLEDGMENTS

This work was funded primarily by the Winthrop Rockefeller Foundation and the Arkansas Agricultural Experiment Station. The Utah State University Agricultural Experiment Station, Arkansas Natural Resources Commission, and University of Georgia also assisted.

REFERENCES

Aguado, E., I. Remson, M. F. Pikul, and W. A. Thomas, "Optimal Pumping in Aquifer Dewatering," *Journal of Hydraulics, ASCE,* 100(HY7), pp. 860–877, 1974.

Ahlfeld, D. P., P. M. Barlow, and A.E. Mulligan, "GWM-A Ground-Water Management Process for the U.S. Geological Survey Modular Ground-Water Model (MODFLOW-2000)," U.S. Geological Survey Open-File Report 2005-1072, p. 124, 2005.

Alley, W. M., Aguado, E., and I. Remson, "Aquifer Management Under Transient and Steady-State Conditions," *Water Resources Bulletin,* Vol. 12, No. 5, pp. 963–972, 1976.

Arkansas Natural Resources Commission (ANRC), "State Water Plan," Executive Summary, Little Rock, Arkansas, 1990.

Czarnecki, J. B., B. R. Clark, and T. B. Reed, "Conjunctive-Use Optimization Model of the Mississippi River Valley Alluvial Aquifer of Northeastern Arkansas," U.S. Geological Survey Water-Resources Investigations Report 03-4230, p. 29, 2003.

Dixon, W. D., and R. C. Peralta, "Potential Arkansas and White Rivers Water Available for Diversion to the Grand Prairie," *Special Report in the Arkansas State Water Plan,* Arkansas Soil and Water Conservation Commission, Little Rock, Arkansas, p. 21, 1986.

Elango, K., and G. Rouve,"Aquifers: Finite-Element Linear Programing model," *Journal of the Hydraulics Division, ASCE,* 106(HY10), pp. 1641–1658, 1980.

Engler, K., D. Thompson, and R. Kazman, "Groundwater Supplies for the Rice Irrigation in the Grand Prairie Region, Arkansas," *Bulletin No. 457, Agricultural Experiment Station,* University of Arkansas, Fayetteville, Arkansas, 1945.

Ferguson, J. A., "The Effect of Rice Flood Depth on Rice Yield and Water Balance," *Arkansas Farm Research,* Vol. 19, No. 3, p. 4, 1970.

Griffis, C. L., "Groundwater-Surface Water Integration Study in the Grand Prairie of Arkansas," *Arkansas Water Resources Research Center Publication No. 11,* University of Arkansas, Fayetteville, Arkansas, 1972.

Harper, J. K., R. C. Peralta, and R. N. Shulstad, "On-Farm Reservoir Construction in the Grand Prairie Region of Arkansas: And Engineering Economic Analysis," *Report FS 89-40,* University of Georgia, Division of Agricultural Economics, p. 52, 1989.

Hegazy, M. A., and R. C. Peralta, "Feasiblity Considerations of an Optimal Pumping Strategy to Capture TCE/PCE Plume at March AFB, CA," *Prepared for Earth Technology*

Corporation. Report SS/OL 97-1 Systems Simulation/Optimization Laboratory, Department of Biological and Irrigation Engineering, Utah State University, Logan, UT, p. 41, 1997.

McDonald, M. D., and A. W. Harbaugh, "A Modular Three-Dimensional Finite-Difference Groundwater Flow Model," *U.S. Geological Survey Techniques of Water-Resources Investigations*, Book 6, Chap., A1, 1988.

Peralta, R. C. "Conjunctive Use of Ground Water and Surface Waters for Sustainable Agricultural Production," *FAO of the United Nations Consultancy Report,* p. 158, 1999.

Peralta, R. C., "Remediation Simulation/Optimization Demonstrations," in Seo, Poeter, Zheng, and Poeter (eds.), Proceedings of MODFLOW and Other Modeling Odysseys, IGWMC, Golden, Colorado, pp. 651–657, 2001.

Peralta, R. C., "SOMOS Simulation/Optimization Modeling System," Proceedings of MODFLOW and More 2003: Understanding through Modeling, IGWMC, Golden, CO, pp. 819–823, 2003.

Peralta, R. C., I. M. Kalwij, and S. Wu, "Practical Simulation /Optimization Modeling for Groundwater Quality and Quantity Management," *MODFLOW & More 2003: Understanding through Modeling*, IGWMC, Golden, CO, pp. 784–788, 2003.

Peralta, R. C., and R. Shulstad, "Optimization Modeling for Groundwater and Conjunctive Use Water Policy Development," Proceedings of FEM-MODFLOW International Conference, IAHS, Karlovy Vary, Czechoslovakia, September 2004.

Peralta, R. C., and A. W. Peralta,"Arkansas Groundwater Management Via Target Levels," *Transactions of the American Society of Agricultural Engineers,* Vol. 27, No. 6, pp. 1696–1703, 1984a.

Peralta, R. C., and A. W. Peralta, "Using Target Levels to Develop a Sustained Yield Pumping Strategy in Arkansas, a Riparian Rights State," *Special Report in the Arkansas State Water Plan,* Arkansas Soil and Water Conservation Commission, Little Rock, Arkansas, 1984b.

Peralta, R. C., and S. Wu., "Software for Optimizing International Water Resources Management," Proceedings of EWRI 2004 World Congress, ASCE, Salt Lake City, U.S.A., June 2004.

Peralta, R. C., P. Killian, and K. Asghari, "Effect of Rules and Laws on the Sustained Availability of Groundwater," Phase I Project Completion Report for the Winthrop Rockefeller Foundation, Little Rock, Arkansas, p. 32, 1985.

Peralta, R. C., A. Yazdanian, P. Killian, and R. N. Shulstad, "Future Quaternary Groundwater Accessibility in the Arkansas Grand Prairie 1993," *Bulletin No. 877,* Agricultural Experiment Station, University of Arkansas, Fayetteville, Arkansas, 1985.

Perkins, G. A., "Arkansas Water Rights: Review and Considerations for Reform," University of Arkansas Little Rock Law Review, 25 U, ARK Little Rock L. Review 123, 2002.

Reed, T. B., "Recalibration of a Ground-Water Flow Model of the Mississippi River Valley Alluvial Aquifer of Northeastern Arkansas, 1918–1998, with Simulation of Water Levels Caused by Projected Ground-Water Withdrawals through 2049," U.S. Geological Survey Water-Resources Investigations Report 03-4109, p. 58, 2003.

Reeves, M., and R. M. Cranwell, "User's Manual for the Sandia Waste-Isolation Flow and Transport Model (SWIFT) Release 4.81," NUREG/CR-2324 and SAND81-2516, Sandia National Laboratories, Albuquerque, New Mexico, 1981.

Reeves, M., D. S. Ward, N. D. Johns, and R. M. Cranwell, "Theory and Implementation for SWIFT II, The Sandia Waste-Isolation Flow and Transport Model for Fractured Media," NUREG/CR-3328 and SAND83-1159, Sandia National Laboratories, Albuquerque, New Mexico, 1986a.

Reeves, M., N. D. Johns, and R. M. Cranwell, "Data Input Guide for SWIFT II, The Sandia Waste-Isolation Flow and Transport Model for Fractured Media," NUREG/CR-3162 and SAND83-0242, Sandia National Laboratories, Albuquerque, New Mexico, 1986b.

Smith, E., Personal communication, 2006.

Sniegocki, R. T., "Hydrogeology of a Part of the Grand Prairie Region, Arkansas," *Geological Survey Water-Supply Paper 1615-B,* U. S. Dept. of the Interior, Washington, DC, 1964.

Systems Simulation/Optimization Lab. and HydroGeoSystems Group (SSOL and HGS), "Simulation/Optimization Modeling System (SOMOS) Users Manual," SS/OL, Bio. & Irrig. Eng. Dept., Utah State Univ., Logan, Utah, p. 457, 2001.

U.S. Army Corps of Engineers, "Eastern Arkansas Region Comprehensive Study; Grand Prairie Region and Bayou Meto Basin, ArkansasP; Grand Prairie Area Demonstration Project; General Reevaluation Report; Vol. 1, Main Report and Final Environmental Impact Statement," USACE Memphis District, p. 427, 1999.

U.S. Army Corps of Engineers, "Interbasin Transfer Arkansas River to Bayou Meto and Lower White Basin," *Arkansas State Water Plan Special Report,* Arkansas Soil and Water Conservation Commission, Little Rock, Arkansas, 1984a.

U.S. Army Corps of Engineers, "Interbasin Transfer, Lower White to Bayou Meto," *Arkansas State Water Plan Special Report*, Arkansas Soil and Water Conservation Commission, Little Rock, Arkansas, 1984b.

Yazdanian, A., and R. C. Peralta, "Maintaining Target Groundwater Levels Using Goal-Programming: Linear and Quadratic Methods," *Transactions of the ASAE,* Vol. 29, No. 4, pp. 995–1004, 1986a.

Yazdanian, A., and R. C. Peralta, "Sustained-Yield Groundwater Planning by Goal Programming," *Groundwater,* Vol. 24, No. 2, pp. 157–165, 1986b.

Zheng, C., and P. P. Wang, "MT3DMS: A Modular Three-Dimensional Multispecies Transport Model for Simulation of Advection, Dispersion and Chemical Reactions of Contaminants in Groundwater Systems," Departments of Geology and Mathematics, *Univ. of Alabama. Technical Report,* Prepared for Waterways Experiment Station, U.S. Army Corps of Engineers, 1998.

LEGAL CITATIONS

Act 114 of 1957, Ark. Code Ann. §§ 21-14-4, et seq. as amended.
Act 217 of 1969, Ark. Code Ann. § 15-22-503.
Act 1051 of 1985, Ark. Code Ann. §§ 15-22-301 to 304.
Act 154 of 1991, Ark. Code Ann. §§ 15-22-901 to 914.
Act 1050 of 1999, Ark. Code Ann. §§ 15-22-1201 to 1218.
Arkansas Wildlife Federation v. United States Army Corps of Engineers, 431 F.3d 1096 (2005).
Arkansas Soil and Water Conservation Commission v. City of Bentonville, 351 Ark. 289, 92 S.W.3d 47 (2002).
Frank Lyon, Jr. et al v. White River-Grand Prairie Irrigation District et al., 281 Ark. 286, 64 S.W.2d 441 (1984).
Harris v. Brooks, 225 Ark. 436, 283 S.W. 2d 129 (1955).
Hudson v. Dailey, 156 Cal. 617, 105 p.748 (1909).
Jones v. OZ-ARK-VAL Poultry Co., 228 Ark. 76, 306 S.W. 2d 111 (1957).
Lingo v. City of Jacksonville, 258 Ark. 63, 522 S.W.2d 403 (1975).

6 MULTIOBJECTIVE ANALYSIS FOR SUSTAINABILITY—CONJUCTIVE USE PLANNING OF GROUNDWATER AND SURFACE WATER

James McPhee
Department of Civil and Environmental Engineering
University of California
Los Angeles, California
Departamento de Ingeniería Civil, Facultad de Ciencias Físicas y Matemáticas
Universidad de Chile
Santiago, Chile.

William W-G. Yeh
Department of Civil and Environmental Engineering
University of California
Los Angeles, California

INTRODUCTION

As population in urban centers around the world increases and economic development raises standards of living, domestic and industrial water demands grow steadily. Mounting pressure on existing water supply sources is compounded by uncertainty regarding the future reliability of water supply due to the not-yet quantified effects of global climate change. Conjunctive use of groundwater and surface water can counter these effects and enhance the reliability of water supply systems by taking full advantage of the almost unlimited storage capacity of groundwater systems. By controlling the total water resources of a region, conjunctive use planning of groundwater and surface water can increase the efficiency, reliability, and cost-effectiveness of water use, particularly in river basins with spatial or temporal imbalances in water demands and natural supplies. Rarely do regions and times of high rainfall and runoff coincide with regions and times of extensive water development and demand. Rather, periods of lowest stream flow and groundwater recharge usually coincide with the largest demand, or vice versa. Integrated management of surface water and groundwater can reduce these deficiencies by using groundwater to supplement scarce surface water supplies during the drier seasons. During the periods of medium or high runoff, surface water can then be used to satisfy the water demands and to recharge the aquifers using spreading basins, abandoned stream channels, and wells.

Conjunctive use of surface water and groundwater requires knowledge about past and current availability of water sources from a hydrologic point of view; the potential for recharging aquifers with a suitable amount of water; and the behavior of water under the ground surface and associated risks of contamination, together with the potential for natural treatment of reclaimed water. Most importantly, relevant questions applicable to conjunctive use involve how to operate such systems, that is, when to recharge, when and how much to extract from the aquifer, and how to value the benefits associated with capacity expansion (Philbrick and Kitanidis, 1998).

It follows that conjunctive use requires accurate knowledge about the physical systems as well as decision-making tools for adequately managing recharge and pumping operations. Knowledge about physical systems can be achieved with adequately planned field data collection campaigns, as well as with numerical modeling. On the other hand, optimization or system analysis techniques can provide invaluable help in identifying good solutions to problems that generally are complex, multidimensional, and unstructured in nature.

GOVERNING EQUATIONS

With respect to conjunctive use, we are interested in how much water can be recharged into the aquifer. Surface water sources can be local or external, and can consist in excess runoff from precipitation or reclaimed water. The most common methods for aquifer recharge consists of some form of detention pond that slows surface water flow in order to increase the time and surface of contact between surface water and the soil's top layer, and hence to augment seepage into the ground. From the above, relevant equations pertain to the movement of water from the soil surface into the ground. There are several models of infiltration, with varying degrees of complexity. Some of the more common expressions are the Horton infiltration model and the Green-Ampt model. The Horton equation expresses the infiltration capacity f_p as:

$$f_p = f_c + (f_0 - f_c)e^{-kt} \tag{6.1}$$

where f_c is a final or equilibrium capacity, f_0 is an initial infiltration capacity, and k is a constant parameter representing the rate of decrease in infiltration capacity. Despite its simple form, the Horton's model has limited applicability due to the difficulty of estimating f_0 and k (Viessman and Lewis, 2003). The Horton model may be attempted, however, if one has observations of infiltration rates at different times.

The Green-Ampt model is based on Darcy's law. Since its inception in 1911, it has gone through modifications that have enabled the model to simulate conditions of excess water at the surface at all times, steady rainfall input, and unsteady precipitation (e.g., Mein and Larson, 1973; Chu and Marino, 2005). The original form of the Green-Amp model relates the infiltration rate, f_p, to the distance L from the soil surface to a wetting front formed by water seeping into the ground:

$$f_p = \frac{K_S(L + S)}{L} \tag{6.2}$$

where S is the capillary suction at the wetting front and K_S is the hydraulic conductivity of the wetted soil. A more convenient form of the Green-Amp model is obtained by recognizing that the cumulative infiltration, F, is equivalent to the product of L and the initial moisture deficit, $\theta_s - \theta_i = $ IMD. Considering that $f_p = dF/dt$, making the appropriate substitutions, and integrating dF/dt with initial conditions, $F = 0$ at $t = 0$, one obtains

$$F - S \times \text{IMD} \times \log_e\left(\frac{F + \text{IMD} \times S}{\text{IMD} \times S}\right) = K_s t \tag{6.3}$$

Equation (6.3) is useful for conjunctive use planning because it allows for a determination of infiltration at any time. The parameter that is the hardest to estimate is the capillary suction potential, S, because it presents a wide range of variability for different soil types under different moisture conditions.

Another mechanism for incorporating surface water to the subsurface is mechanical recharge through injection wells. The dynamics of this process, however, are

Surface Water

determined by the groundwater governing equation and are thus discussed in the next section.

Groundwater

The governing equation for groundwater flow is:

$$S_s(\mathbf{x})\frac{\partial h(\mathbf{x}, t)}{\partial t} = \nabla \cdot [K(\mathbf{x})\nabla h(\mathbf{x}, t)] \pm q_s(\mathbf{x}, t) \qquad \mathbf{x} \in \Omega \qquad (6.4)$$

subject to the initial condition

$$h(\mathbf{x}, 0) = h_0(\mathbf{x}) \qquad \mathbf{x} \in \Omega \qquad (6.5)$$

and the generalized boundary condition

$$K(\mathbf{x})\nabla h(\mathbf{x}, t)\mathbf{n} = \alpha[H - h(\mathbf{x}, t)] + Qx \in \Gamma \qquad (6.6)$$

where $\mathbf{x}$ = position vector
$h(\mathbf{x}, t)$ = hydraulic head as a function of position and time [L]
$K(\mathbf{x})$ = hydraulic conductivity [L/T]
Ω = flow domain of the aquifer
Γ = domain's boundary
$S_s(\mathbf{x})$ = specific storage [L]$^{-1}$
$q_s(\mathbf{x}, t)$ = internal sink/source term [L^3/T]
t = time
$\mathbf{n}$ = unit vector normal to Γ pointing out of the aquifer
H = prescribed boundary head [L]
Q = prescribed boundary flux [L/T]
α = parameter controlling the type of boundary condition

In Eqs. 6.4 to 6.6 and hereafter, "$\nabla\cdot$" and "∇" are the divergence and gradient operators, respectively.

Modeling Issues

In surface water and groundwater, the dynamics of water movement within a natural environment are strongly influenced by quantities—permeability, storage coefficients, streambed roughness, and the like—that are not directly observable at every point of the model's domain. Furthermore, it is most likely that these parameters have different values at each location in the three-dimensional space, thus yielding a numerical model with potentially infinite degrees of freedom. Therefore, *parameterization* constitutes an important step in solving the governing equations. Parameterization (Carrera and Neuman, 1986a, 1986b, and 1986c, Yeh, 1986) involves discretizing the model domain into a finite number of parcels in one, two, or three dimensions, and assigning constant parameter values to these parcels. The appropriate parameter values assigned to each of the parcels have to be calibrated using historical observations, a process better known as *inverse modeling*. Inverse modeling has been a subject of intense research for the last forty years, and many important issues and methods related to it are discussed in Yeh (1986), Madsen et al. (2002), and Doherty and Johnston (2003), among others. Because there are many sources of error (e.g., observation error, model structure error, and conceptual model error) the solution of the inverse problem is seldom unique, and therefore the predictions made with calibrated models always are associated to some degree of uncertainty. Uncertainty in model predictions, hence, should always be accounted for, or at least kept in mind, when evaluating conjunctive use plans.

MANAGEMENT OBJECTIVES IN CONJUNCTIVE USE PROJECTS

Typically, a conjunctive use project is implemented to augment the available water that can be supplied for one or more uses. Agricultural, domestic, or industrial water demands have different characteristics regarding spatial and time distributions. Aside from short-term variability considerations, a conjunctive-use project should be able to store water in the aquifer when hydrologic conditions generate a situation of surplus. The stored water then can be pumped from the ground during subsequent periods of scarcity. Although the potential benefits of groundwater as a supply stabilizer (i.e., through improving reliability and reducing the impact of drought) is widely recognized, conjunctive use projects have yet to acquire a preponderant role, due to the difficulty in making decisions related to how much and when to recharge to or pump from the aquifer (Philbrick and Kitanidis, 1998). When analyzing water supply, it is important to stress the concepts of *severity of drought* and *hedging*. Severity of a drought is a concept that attempts to capture the notion that water deficits can be damaging due to two major factors: the *magnitude* of the deficit (or the value of the deviation between observed and average flow conditions for a given time period) and the *duration* of the deficit. In a very simplified manner, the severity (S) of a drought k can be defined in terms of duration (D) and magnitude (M) as:

Supply Objectives

$$S(k) = D(k) \times M(k) \qquad (6.7)$$

The definition of what constitutes a drought, however, is a more complicated matter that depends on the particular field of study. For example, it is possible to define hydrologic droughts (stream flow deficit), agricultural droughts (soil moisture deficit), and hydrometeorological droughts (precipitation deficit), among others (Chow et al., 1988). In fact, a large deficit that lasts a short period can be much more damaging than a moderate deficit that extends for a longer time. The phenomenon discussed has been well documented and is related to the fact that losses exhibit a nonlinear behavior with respect to water scarcity. If available hydrologic forecasts show that an extended period of droughts and water shortage is unavoidable, from an operational point of view often it is preferable to *hedge* the water supply, reducing deliveries moderately (by establishing shifts, reducing pipe pressures, and so forth) in order to save water and thus "distribute" the deficits in time. Both the study of past drought events and the operational need of making hedging decisions in real time require the adoption of a drought index. A drought index provides a quantitative measure so that comparisons can be made among events, hedging mechanisms activated, and in general so that decisions can be made based upon objective criteria. Several drought severity indexes have been proposed by regulatory agencies around the world. Hsu (1995) presents some shortage severity indexes proposed by the U.S. Army Corps of Engineers (*shortage index*, SI) and the Japan Water Resources Development Public Corporation (*deficit percent day index*, DPD). The SI is a lumped ratio of the total deficit in a given period to the design supply in that period, and does not account for the temporal distribution of the deficits within the analysis period. The DPD, on the other hand, is equal to the sum of the product of the percent daily deficit rate and the number of contiguous days the deficit event lasts, for all deficit events within an analysis period. In an attempt to incorporate the strengths of both indexes, Hsu (1995) proposed a *generalized shortage index* (GSI), defined as:

$$\text{GSI} = \frac{100}{N} \sum_{i-1}^{N} \left(\frac{\text{DPD}a_i}{100 \times \text{DY}_i} \right)^k \qquad (6.8)$$

where N = number of sample years

k = coefficient (usually takes a value of 2)

$\text{DPD}a_i$ = sum of all DPDs in year i

DY_i = number of days in the year

Compared to SI, an index such as GSI is more sensitive to the intensity and duration of the deficit, implying a stricter objective for conjunctive use planning.

Quality Objectives

A conjunctive use project ultimately will blend water from different sources, and thus with potentially widely varying quality levels. If one is to minimize the required treatment prior to distribution, then it would be desirable to blend high quality with poor quality water in a proportion that yields a final product requiring only marginal treatment, such as disinfection. Sometimes this may not be achievable, in particular in arid and semiarid areas, where surface water often presents high levels of dissolved solids due to multiple reclamation cycles. In most cases, it is assumed that subsurface sources show higher quality than surface sources, and therefore the injection of water into the soil should be carried out carefully in order to avoid excessive contamination of groundwater. For K sources of water with various quality levels C_k, $k = 1, \ldots, K$, the mass balance equation yields the resulting quality after blending:

$$C_{\text{res}} = \frac{\sum\limits_{k=1}^{K} C_k \times Q_k}{\sum\limits_{k=1}^{K} Q_k} \tag{6.9}$$

Another set of quality objectives relates to preserving the natural characteristics in coastal aquifers subject to intense pumping. Reduced groundwater piezometric head in the vicinity of the ocean may reverse the natural flow direction toward the ocean with the potential for seawater to reach the domestic pumping wells. Salt water–contaminated wells are extremely difficult to rehabilitate, therefore caution should be exercised when designing a pumping plan in coastal areas. The problem of seawater intrusion can be modeled realistically as a variable-density groundwater flow and transport problem (Simmons et al., 2001), but such a formulation poses significant computational challenges. An alternative approach involves solving only a flow problem and replacing salt concentration constraints with velocity and piezometric gradient constraints in order to ensure that extraction wells are shielded from seawater (Emch and Yeh, 1998).

Economic Objectives

Economic objectives in conjunctive use typically involve minimizing some discounted form of installation and operation costs associated with injecting, pumping, treating, and distributing water to the different users that benefit from the project. Additionally, shortage costs should be considered. Shortage costs are often difficult to estimate and therefore benefit/cost analyses generally take into consideration only the capital and operating costs of the alternatives. A feasible mechanism for estimating shortage cost is presented by Philbrick and Kitanidis (1998), who propose equating these costs with consumers' willingness to pay for domestic water use. Willingness to pay is estimated by assuming that the elasticity of demand for water is constant, and by adopting nominal values for the price and average annual supply. Pumping costs usually depend on the capacity of the installed wells, but also on the state of the system given that the required energy for pumping is a function of the total lift. Recharge costs, on the other hand, are highly variable depending on the methods used in implementing the recharge program. Recharge methods may include surface spreading, injection, and enhanced natural recharge, that is, replacing groundwater pumping with surface resources in wet years (Philbrick and Kitanidis, 1998). All these mechanisms are influenced strongly by highly variable and uncertain parameters such as soil type, infiltration rate, and subsurface geology.

Environmental Objectives

During the past decades the human race has learned, sometimes at dear cost, that engineering projects, which alter nature, carry with them environmental impacts that cannot be overlooked. The natural environment in which a conjunctive use project is embedded needs to be characterized and understood in a way such that impacts can be predicted, quantified, and mitigated. In semiarid regions, fragile riparian habitats depend on groundwater discharge to surface streams in periods of low flow (U.S.G.S., 1999). Altering the natural cycles of stream flow, either by transferring surface resources from one stream to another, or by excessive pumping of the aquifer, can affect negatively the delicate balance existing in these ecosystems. In other cases, increased concentration of solutes otherwise absent from the natural environment can affect negatively flora and fauna inhabiting aquatic systems, like wetlands. The environmental objectives to be considered in a conjunctive use project will, hence, depend strongly on the unique features of the project. However, some typical objectives or constraints include minimizing the drawdown observed in an aquifer after some period of operation (McPhee and Yeh, 2004), constraining the movement of contaminant plumes in the subsurface (Shafike et al., 1992), and controlling salinity in surface streams (Louie et al., 1984).

Sustainability

In recent years, increased public awareness regarding the impacts of water resources projects have modified the paradigms under which these projects are carried out. For example, in many areas of the world, the construction of new multipurpose reservoirs is a highly controversial subject. Together with environmental concerns, the attention given to the concept of sustainability has been increasing when evaluating the benefits of almost any human project. Sustainability as a concept lacks a well-established definition, but a loose interpretation of the term involves the satisfaction of our current needs while at the same time the preservation of the capacity of future generations to satisfy their own. The *American Society of Civil Engineers* (ASCE), through the work of a task committee, has proposed some indicators to measure sustainability (ASCE and UNESCO 1998). Efficiency, survivability, and sustainability are three among several planning objectives that focus on future conditions. In order to define these concepts, suppose that for each possible decision to be made today (k), a welfare outcome $W(k,y)$ results for each period y from now into the future. Also, suppose that the minimum welfare for survival is W_{min}. A decision will be *efficient* if it maximizes the present value of current and future welfare outcomes. As the rate of discount, r, increases, the welfare outputs occurring in the distant future become less important compared to welfare outputs occurring near the present. Efficiency, then, favors projects with immediate benefits over projects that ensure a steady stream of benefits for a long time. On the other hand, low interest rates favor projects that are less likely to survive economically, and that are less likely to incorporate investment in environmental protection.

A project or decision, k, is considered *survivable* if its welfare outcomes are greater or equal than the minimum survival welfare, W_{min}, for every period y in the future. Finally, an alternative is considered *sustainable* if the average (over some finite time period) welfare for future generations is no less than the average welfare available to previous generations. A related interpretation states that an alternative is sustainable if it does not imply long-term decreases in the welfare available to future generations, that is, $W(k, y + 1) \geq W(k, y)$. The length of the time period y should be such that natural variations in availability of a resource, like water, are averaged over the period of analysis.

Additional definitions of sustainability include concepts such as *reliability, resilience,* and *vulnerability*. Application of these concepts requires identification of suitable criteria that can be quantified numerically or "linguistically" (e.g., "poor," "good," "excellent"), as well as expressed as a function of water resources time series. If upper and lower bounds of acceptability are defined for the evaluation criteria, it is possible to define the (a) *reliability statistic* as the ratio of the number of satisfactory values to the total number of simulated periods (b) *resilience statistic* as the ratio of the number of times a satisfactory value follows an unsatisfactory value to the total number of unsatisfactory values (c) *vulnerability statistic* as a measure of the duration of failure.

As the contents covered in this section suggest, sustainability indicators can be defined in various ways. Depending on the particularities of each problem, some indicators will favor alternatives that perform poorly regarding other measures of sustainability. This, together with the multiplicity of other management objectives that have to be taken into account, points strongly toward the multicriteria nature of the decision-making process for conjunctive use.

MULTIOBJECTIVE OPTIMIZATION

Definition

In general, a multiobjective optimization problem can be expressed as follows:

$$\min Z(\mathbf{q}) = [Z_1(\mathbf{q}), Z_2(\mathbf{q}), ..., Z_F(\mathbf{q})]$$

subject to: (6.10)

$$g_c(\mathbf{q}) \leq 0, \qquad c = 1, 2, ..., m$$

$$\mathbf{q} \in Ad$$

where F = number of objectives
$\mathbf{q}$ = n-dimensional vector of decision variables
$Z_f(\mathbf{q}), f = 1, ..., F$ = objective function
Ad = admissible decision variable set

Equation (6.10) represents a vector optimization, and we can say it is a *convex vector optimization problem* if the constraint functions g_c are convex and Z is convex in Ad. The primary source of confusion associated with vector optimization arises when we want to compare feasible solutions. Let x and y be two feasible solutions. We would like to say that $Z(x) \leq Z(y)$ means that x is better than or equal to y. The problem is that, depending on the particular form of the individual components of the vector objective Z, $Z(x)$

and $Z(y)$ need not be comparable. That is, neither is better than the other. For example, how does one compare a water storage/distribution system that increases supply reliability to a municipal water district in 5 percent with respect to another, but costs $100 million more?

Returning to the general case, if there is a feasible x such that $Z(x) \leq Z(y)$ for all feasible y, then we say that $Z(x)$ is the *optimal value* of the problem. In general, however, it is much more likely that such a point does not exist. In such cases, it is possible to find an *efficient* or *nondominated* solution (Cohon, 1978), which is defined as the solution in which no improvement can be made with respect to one objective without harming at least one of the other objectives. The set of efficient solutions is also called the *Pareto* set. Formally, a decision vector $\mathbf{q}^* \in Ad$ is Pareto optimal if there does not exist another $\mathbf{q} \in Ad$ such that $Z_f(\mathbf{q}) \leq Z_f(\mathbf{q}^*)$ for all $f = 1, \ldots, F$ and $Z_j(\mathbf{q}) < Z_j(\mathbf{q}^*)$ for at least one index j. An objective vector is Pareto optimal if the corresponding decision vector is Pareto optimal.

A standard technique for obtaining Pareto optimal points in Eq. (6.10) is the weighting method. Let $W = (w_1, w_2, \ldots, w_F)$ be a positive vector in $\mathbf{R}^F$. Then it is possible to build the *scalar* optimization problem

Solution Methods—Mathematical Programming

$$\min Z(\mathbf{q}) = \sum_{f=1}^{F} w_f^* Z_f(\mathbf{q})$$

subject to: (6.11)

$$g_c(\mathbf{q}) \leq 0, \qquad c = 1, 2, \ldots, m$$

$$\mathbf{q} \in Ad$$

and let x be an optimal solution. Then x is Pareto optimal for the vector optimization problem (Boyd and Vandenberghe, 2004). It follows that by varying the weighting vector W it is possible to find different Pareto optimal solutions for the vector optimization problem. Some authors, however, have pointed out that the weighting coefficient method as described earlier may fail to generate all efficient solutions if the efficient frontier is nonconvex (Watkins and McKinney, 1997).

Another alternative for finding Pareto optimal solutions is the so-called constrained method (Cohon, 1978; Louie et al., 1984; Shafike et al., 1992). This is performed by optimizing one of the objectives $Z_f(\mathbf{q})$ while incorporating the others in the constraint set and parametrically varying the objectives in the constraint set from their lower bounds to upper bounds. The entire methodology for obtaining the Pareto via the constrained method set is summarized in the following steps:

1. Solve each objective subject to common constraints, without regard to the other objectives.

2. Generate a payoff table which contains the upper and lower bounds for each objective. It is assumed that the lowest possible satisfaction with respect to one objective is obtained when optimizing some of the other conflicting objectives. In order for this to happen, the objectives should be competing.

3. Generate the Pareto set by solving a series of optimization problems.

Solution Methods— Multiobjective Evolutionary Algorithms

Evolutionary algorithms, in particular *genetic algorithms* (GA), have become popular tools for finding optimal solutions to complex optimization problems. GA uses principles derived from the evolution of biological organisms in order to obtain better-performing solution sets to an optimization problem, starting from an initial population. The bibliography on the basic principles and applications of GA to water resources systems is extensive, and a detailed explanation of the mutation, crossover, and reproduction processes that constitute the basis of GA is beyond the scope of this work. A flow chart of a simple GA is shown in Fig. 6.1. In their most basic form, GA are intended to deal with unconstrained optimization because there is no way to incorporate a priori knowledge regarding the feasibility of a particular solution obtained from the random

Figure 6.1

Flow chart of simple genetic algorithm.

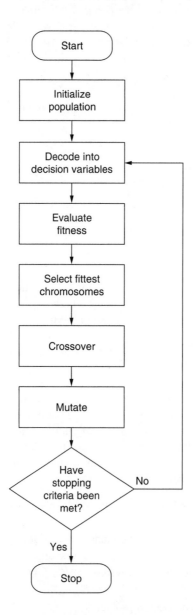

mechanisms that generate a "child" solution from a pair of "parent" individuals. However, some have attempted constrained optimization by adding a penalty term to the fit function of nonfeasible individuals. In this way, nonfeasibility is passed from generation to generation as an undesirable trait that should tend to be avoided by the algorithm. A number of variations of the classical GA have been proposed in order to solve multiobjective optimization problems. The initial variations essentially involved coding into the GA the weighting and constrained methods described earlier. This failed to take advantage of one of the main strengths of GA, which is to sample the solution space by several individuals at a time. More contemporary versions of the GA benefit from this multiple sampling by incorporating the degree to which a solution is dominated by others in the fitness value assignment. This process, also known as Pareto sorting, is at the core of one the most popular multiobjective algorithms, the *nondominated sorting genetic algorithm II* (NSGA-II) (Deb et al., 2002). NSGA-II converges to the Pareto front by assigning fitness values to *subsets* of solutions based on domination sorting. Solution s_1 is said to dominate solution s_2, if s_1 is not outperformed by s_2 along any criteria, while at the same time outperforms s_2 with respect to at least one criterion. Another desirable feature of multiobjective algorithms that NSGA-II incorporates involves maintaining a sufficient spread of the Pareto front as the algorithm converges by using a crowded-comparison operator. NSGA-II also can handle constrained optimization by means of binary tournament selection that incorporates the concept of domination. Under this approach, solution s_1 is said to be constrained-dominant over solution s_2 if any of the following conditions are met:

1. Solution s_1 is feasible and solution s_2 is not.
2. Solutions s_1 and s_2 are both infeasible, but solution s_1 has a smaller overall constraint violation.
3. Solutions s_1 and s_2 are feasible and solution s_1 dominates solution s_2.

Examples of application of NSGA-II to water resources management can be found in Reed et al. (2004) and Dorn and Ranjithan (2003).

LINKING SIMULATION AND OPTIMIZATION

Evaluation of the impact of a particular conjunctive use plan on the environment, as well as its potential for reaching the stated management goal, is often carried out with the aid of a simulation model. As described before, management goals may relate to flow of water in the surface and subsurface of the ground, transport of contaminant in water bodies as well as in the unsaturated and saturated zones, sediment transport, and ecosystem preservation. Numerical simulation tools can be linked to optimization routines in order to provide the search algorithm with information about how well the candidate alternatives perform. If gradient-based optimization algorithms are used, the quantities of interest will be the changes of the objective function with respect to unit variations of the decision variables. If combinatorial or evolutionary algorithms are used, in general only the values of the objective function are required, and no gradient values need to be computed. For gradient-based methods, a commonly used technique for linking groundwater simulation with optimization requires adopting a first-order Taylor's series expansion to express head as a linear function of pumping and recharge rates:

$$h_s(\mathbf{q}_0 + \Delta\mathbf{q}) = h_s(\mathbf{q}_0) + \sum_v \frac{\delta h_s}{\delta q_v}\Delta q_v. \tag{6.12}$$

From Eq. (6.12), it can be seen that the key elements in linking simulation and optimization are the influence coefficients (Becker and Yeh, 1972), $\delta h_s/\delta q_v$, where h_s is the hydraulic head of groundwater at location $s = 1, \ldots, H$, and q_v is the vth element of the vector $\mathbf{q}$ of manageable pumping or recharge decision variables. The influence coefficients are computed using the simulation model. If one uses the perturbation method or the sensitivity equation method, $N + 1$ simulation runs are required to obtain the coefficients for N decision variables (Yeh, 1986).

The embedding approach (Ahlfeld and Heidari, 1994) embeds the discretised set of finite-difference or finite-element equations of the simulation model directly in the constraint set of the optimization model. This approach, however, is practical only for cases in which the number of computational nodes in the simulation model is limited. Fredericks et al. (1998) present a decision support system for conjunctive use in which they link a networklike surface water accounting model with MODFLOW (McDonald and Harbaugh, 1998; Harbaugh et al., 2000) using discrete kernel/response functions. Both the response function and the influence coefficient method assume that the behavior of the aquifer is linear with respect to excitations produced by pumping wells, surface recharge, and so forth. This assumption, however, is valid for confined aquifers or unconfined aquifers when the water table depression does not exceed 10 percent of the saturated thickness of the unconfined aquifer. If expected drawdown after aquifer operations is greater than this threshold, an iterative method should be employed in which response functions are recomputed each time a new value of the excitation variables is assigned. For additional information on simulation-optimization approaches, refer to Gorelick (1983), Willis and Yeh (1987), Yeh (1992), and Ahlfeld and Heidari (1994).

MULTICRITERIA DECISION MAKING

Multicriteria decision making involves using analytical methods to help the decision maker select a preferred project among multiple alternatives. Because more than one objective is being considered, alternatives that are optimal in the sense that they minimize all objective functions may not exist. Therefore, solutions that provide an acceptable compromise between objectives have to be sought. Ranking of alternatives may be performed under several premises. Stansbury et al. (1991) used *composite programming* to evaluate and rank water transfer options in a context of agricultural development. Shafike et al. (1992) used *compromise programming*, *ELECTRE II* and *multicriterion Q-analysis* to rank a set of alternatives for groundwater management. They reduced the decision search space to those alternatives providing an efficient trade-off between three groundwater management objectives. Duckstein et al. (1994) used compromise programming, *ELECTRE III*, *multiattribute utility function*, and *UTA* (*utilité additive*) to search for the Pareto set of pumping policies that provide the optimal trade-off between total pumping rate, costs, and risk.

Compromise programming belongs to the category of distance-based methods for project ranking. Each project alternative constitutes an n-dimensional vector, where n is the number of objectives. A norm may be used to measure the distance between the alternative and the ideal optimal point, defined as the one that maximizes simultaneously all n objectives. On a normalized, 0–1 scale, this is equivalent to the distance to the point with all coordinates equal to one.

Distance-Based Methods: Compromise Programming

ELECTRE III (Roy 1968; 1991) operates by outranking the possible alternatives through pair-wise comparison of a finite set of alternatives evaluated on a consistent family of criteria. Given two alternatives, ELECTRE III defines indifference, weak-preference and strict-preference relations between them, using a set of indifference, preference and veto thresholds. Among its interesting features, ELECTRE distinguishes itself from other multiple-criteria solution methods because it is noncompensatory. This means that a very bad score on a particular criterion cannot be outweighed or balanced by several good scores in other criteria. A further characteristic of ELECTRE is that it allows for incomparability among criteria. Incomparability, which should not be confused with indifference, occurs when it is not possible to determine if alternative a is superior to alternative b or vice versa.

Outranking Techniques: ELECTRE III

ELECTRE introduces the concept of an indifference threshold q, such that the preference relations between a and b are defined as follows (assume objectives are being maximized):

$a\mathbf{P}b$ (a is preferred to b) $\Leftrightarrow$ $Z(a) > Z(b) + q$

$a\mathbf{I}b$ (a is indifferent to b) $\Leftrightarrow$ $|Z(a) - Z(b)| \leq q$

To model the zone in which the decision maker shifts from indifference to strict preference, a buffer zone or zone of weak preference is defined by introducing the preference threshold, p. The double threshold model is then:

$a\mathbf{P}b$ (a is strongly preferred to b) $\Leftrightarrow$ $Z(a) - Z(b) > p$

$a\mathbf{Q}b$ (a is weakly preferred to b) $\Leftrightarrow$ $q < Z(a) - Z(b) \leq p$

$a\mathbf{I}b$ (a is indifferent to b) $\Leftrightarrow$ $|Z(a) - Z(b)| \leq q$

The method's ultimate goal is to build an outranking relation $\mathbf{S}$, meaning that $a\mathbf{S}b$ if and only if "a is at least as good as b" OR "a is not worse than b." The test to accept the assertion $a\mathbf{S}b$ is implemented using two principles:

- A concordance principle, which requires that most objectives, considering their relative importance, support the assertion
- A nondiscordance principle, which requires that among those objectives that do not support the assertion, none of them is strongly against the assertion

Q-analysis was developed by Atkin (1974, 1977) as an approach for studying the structural characteristics of social systems in which two sets of indicators are related to each other. Q-analysis has been used as a multicriterion decision-making tool in different disciplines (Hiessl et al., 1985; Duckstein and Nobe, 1997). The applications found in the literature relate to the selection of the best project out of a relatively small number of alternatives. The method requires a normalized data matrix $\mathbf{A}$ and a weight

Multicriterion Q-Analysis (MCQA)

vector **w**. Element a_{ij} in **A** rates the alternative i for criterion (in this case, objective) j. MCQA I uses a *project satisfaction index* (PSI) and a *project comparison index* (PCI). MCQA II additionally includes a *project discordance index* (PDI). The PSI of an alternative is independent from other alternatives, whereas the PCI and PDI of an alternative are dependent on the objective values for the other alternatives. The PSI for alternative d_i is a measure of how well d_i satisfies the criteria, and is defined as:

$$\mathbf{PSI}(i) = \sum_{j,k} \alpha(k) \cdot w_j \cdot b_{ij}^k \tag{6.13}$$

where $\alpha(k)$ is a threshold level, w_j is a weighting factor, and b_{ij}^k is a binary variable that takes values of one if alternative i satisfies objective j, at least at the threshold level and zero otherwise. PCI ranks the alternatives by comparing q-connectivity. PDI ranks the alternatives by comparing discordance q-connectivity. For a detailed description of this and other concepts, see Duckstein and Nobe (1997). The last step in the methodology is to rank the alternatives using the *project-rating index* (PRI). MCQA I utilizes PRI1, while MCQA II uses PRI2. A lower value of PRI indicates a better alternative. PRI1 and PRI2 are defined as:

$$\mathbf{PRI}1(i)_p = [(1 - \mathbf{PSI}(i))^p + (1 - \mathbf{PCI}(i))^p]^{1/p} \tag{6.14}$$

$$\mathbf{PRI}2(i)_p = [(1 - \mathbf{PSI}(i))^p + (1 - \mathbf{PCI}(i))^p + (\mathbf{PDI}(i))^p]^{1/p} \tag{6.15}$$

The application of MCQA as well as ELECTRE III requires pair-wise comparison of all the alternatives being considered.

Ordered Weighted Averaging

Ordered weighted averaging (OWA)(Yager, 1988), on the other hand, presents a simpler, yet conceptually powerful method for ranking alternatives. OWA works by assigning normalized weights to the components of a sorted vector of criteria. Formally, a mapping $F: I^n \rightarrow I$ (where $I = [0, 1]$) is an OWA operator if there exists a weighting vector $W = (W_1, W_2, \ldots, W_n)$, associated with F, such that:

$$W_i \in (0,1)$$
$$\sum_i W_i = 1 \tag{6.16}$$

where

$$F(a_1, a_2, \ldots, a_n) = W_1 b_1 + W_2 b_2 + \cdots W_n b_n \tag{6.17}$$

In Eq. (6.17), b_i is the ith largest element in the collection $a_1, a_2, \ldots, a_n$. It is important to stress that the weights are associated with a particular ordered position rather than a particular element. This property allows OWA to exhibit properties such as symmetry. More importantly, OWA is capable of extending the quantification of alternatives beyond the traditional binary logic propositions "there exists" and "for all," to intermediate quantifiers associated with natural language, such as "almost all," "few," "many," "most," and the like (Yager, 1988).

CASE STUDY: SAN PEDRO RIVER BASIN

This section presents a case study of a multiobjective simulation-optimization model developed for studying surface water-groundwater management options for the *Upper San Pedro River Basin* (USPRB) (McPhee and Yeh, 2004). The USPRB is located in the semiarid borderland of southeastern Arizona and the state of Sonora, Mexico. From an ecological point of view the basin, and particularly its upper portion, is an ecosystem of unparalleled importance in the region. The riparian corridor is a key factor in the journey of migratory birds, harboring the richest, densest, and most diverse inland population—with 385 identified species—in the continental United States. Additionally, 82 species of mammals, 43 kinds of reptiles and amphibians, and populations of cottonwood-willow and mesquite trees also inhabit this ecosystem (Kingsolver, 2000). Of particular relevance is the *San Pedro Riparian National Conservation Area* (SPRNCA). The SPRNCA is the United States' first riparian reserve, established by the U.S. Congress in 1988 in an effort to preserve the rare riparian habitat from damage due to increasing water demands in the surrounding area. Water flow paths in the USPRB begin in the mountain ranges that run parallel to the course of the river. Snowmelt reaches the aquifer primarily through mountain-front recharge, and subsequently groundwater flows toward the center of the basin, where it seeps into the stream of the San Pedro River and is responsible for maintaining winter baseflow. Increased pumping in areas of the Basin has yielded cones of depression that impair the natural flow of groundwater to the surface, and therefore some reaches of the San Pedro River have changed their hydrologic regime from perennial to intermittent.

General Information

McPhee and Yeh (2004) employed two models to study this problem. They used a MODFLOW model developed at the University of Arizona (Goode and Maddock, 2000) to obtain the sensitivity coefficients that link pumping and recharge rates at more than 3000 locations with changes in groundwater head. These sensitivities are input into a management model, which in turn consists of a linear optimization problem. At a basin-wide scale, spatial variability of the demand motivates the definition of operational sectors (users), $j = 1, \dots, U$. User j may receive water from any of the water sources, $i = 1, \dots, S$. Treatment plants, $k = 1, \dots, T$, receive effluents and return flow from the operational sectors and afterwards the treated water is discharged or recycled at discharge facilities, $l = 1, \dots, D$. Demand at the operational sectors may be offset by the implementation of water conservation projects, $m = 1, \dots, W$. Three management objectives are considered in the formulation: minimizing the net present value of mitigation costs, maximizing aquifer yield and minimizing drawdown at selected locations. Minimizing the net present value of groundwater depletion mitigation cost requires searching for alternative sources of water outside of the basin, as well as modifying temporal and spatial patterns of pumping and recharge. McPhee and Yeh (2004) select a subset of conservation and mitigation strategies identified by the *Upper San Pedro Partnership* (USPP, a stakeholder organization). These include, among others, opening new well fields in a neighboring basin, importing surface water from the Central Arizona Project and implementing recharge wells along the most threatened areas in the SPRNCA. *Maximizing aquifer yield* requires taking into account the sum of pumping rates at every well location deemed as "manageable." In this case, *minimizing drawdown at selected locations* acts as a surrogate objective of the ultimate goal, which is to keep critical stretches of the San Pedro River from becoming intermittent. The objective function

Models

selected to represent this goal is the l^1-norm of the differences between a target head and the simulated head at a set of selected locations.

The three objectives presented in McPhee and Yeh (2004) and their associated tradeoffs serve as guidelines for decision makers in analyzing future options for development and ecological conservation in the basin. Although for the San Pedro Basin the major current concern is to protect the riparian habitat, the yield objective is useful for assessing the yield potential of the aquifer and the influence of development patterns over the ecosystem. For a given groundwater exploitation pattern, it is expected that drawdown can be mitigated by increasing the cost of recharging the aquifer or by offsetting excess demand with more expensive imported water. Likewise, at any given cost level the amount of groundwater to be extracted will be linked to the resulting drawdown.

In McPhee and Yeh's (2004) formulation, the decision variables include whether to implement a mitigation project or not and the amount of water that should be imported from that source. Therefore, the problem as stated is a *mixed-integer nonlinear programming problem* (MINLP). MINLP problems are extremely difficult to solve, and in general global optimality cannot be guaranteed. The problem is linearized by making some assumptions regarding the discrete variables. First, for a long-term policy analysis, the mitigation variables are made continuous and constant over the entire planning horizon. Additionally, all mitigation projects are assumed to start at the beginning of the planning horizon and continue to the end of the planning horizon. As a result the need for integer variables is eliminated.

Computational requirements make it impractical to include as independent decision variables the pumping rates at each time period for each of the 3470 pumping wells identified in the original San Pedro simulation model. McPhee and Yeh (2004) preprocess the available data in order to reduce the dimension of the *decision space*. The first step is to differentiate those wells that may be subject to management from those that, due to legal or practical constraints, would be very difficult to manage. Wells selected for management purposes are those classified as public, irrigation, and industrial, while domestic, stock, commercial, and institutional wells are assigned their historical pumping rates. Public wells are owned by municipal or private water supply utilities, and industrial wells are owned by a few users. In both cases it is assumed that management can be reached through negotiation between user organizations, cities, and the well owner. In the case of irrigation wells management options are more limited, generally consisting of land purchases that permit a change in the use of land and hence the shut-off of the wells. Irrigation wells are included in the management model in order to evaluate the ecological yield of certain land purchases. After this first selection process 507 irrigation wells, 32 industrial wells and 112 public wells are left for consideration as decision variables. Further clustering follows in order to associate the pumping rates from several wells to a single pumping parameter (Hill et al., 2000). Pumping variables are grouped according to use, owner, and location, yielding 128 well decision variables. Finally, a sensitivity analysis allows discriminating among those well parameters that influence more significantly groundwater heads in the basin, resulting in only 48 pumping decision variables.

Stakeholder Involvement

Sustainable water resources management is an interdisciplinary problem that requires the coordinated effort of a wide range of specialists, decision makers and stakeholders. For this reason generating alternatives become an iterative process in which simulation

Table 6.1
Payoff Matrix,
Case A

Objective Being Optimized	Objective Being Evaluated		
	Drawdown (m)	Cost ($)	Yield (m³/day)
Drawdown	6.92	2.17E+08	−21,813
Cost	10.81	0.0	−97,698
Yield	11.11	0.0	−135,857

Table 6.1
Payoff Matrix,
Case A

and optimization provide information to decision makers. At the same time decision makers drive the analysis through expressing their views in terms of objective functions and constraints. In the San Pedro River Basin case study, McPhee and Yeh (2004) benefit from stakeholders' participation through the USPP (http://www.usppartnership.com), consisting of 20 agencies and organizations with interest in the San Pedro River and its future. The USPP funds several research efforts that focus on the sustainability of development in the basin, and is a major factor in coordinating policy and science. The model of participation that the USPP has established should constitute a significant guideline for conjunctive use plans and watershed management.

McPhee and Yeh (2004) present two cases that exemplify different assumptions regarding the way decision variables can be dealt with. In case A, it is assumed that the entire planning horizon is represented by a single management period in such a way that the optimal pumping rates obtained for existing wells subject to management are constant. For the same reason, mitigation projects always are assumed to start at the beginning of the planning horizon. Therefore, linear variables can characterize the operational rates at which these projects are implemented. In case B, pumping rates at existing wells subject to management are allowed to vary with time. The planning horizon in case B (20 years) is divided into five four-year management periods during which pumping rates remain constant. No scheduling of mitigation projects is allowed. This implies that, if selected, a mitigation project will start at the beginning of the planning horizon.

Simulation-Optimization Results

Tables 6.1 and 6.2 show the payoff matrices for cases A and B, respectively. The range between the lower and upper bounds of objective functions 1 and 2 (cost and yield) is discretized into 36 points that are introduced as constraints to generate the nondominated set. No major changes in the drawdown objective in terms of the best and worst values are observed when allowing pumping rates to change with time. However, in terms of aquifer yield, greater flexibility in the pumping policy allows distributing pumping in a way that results in a slightly better drawdown objective while at the same time extracting more water from the aquifer.

Objective Being Optimized	Objective Being Evaluated		
	Drawdown (m)	Cost ($)	Yield (m³/day)
Drawdown	6.88	3.10E+08	−28,050
Cost	11.04	0.0	−135,289
Yield	11.15	0.0	−135,857

Table 6.2
Payoff Matrix,
Case B

Figure 6.2

Drawdown versus
yield tradeoff.

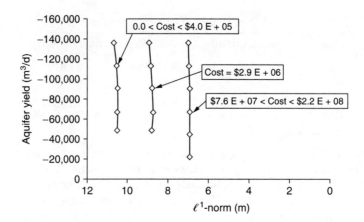

Figures 6.2 and 6.3 illustrate the two-way tradeoff between the drawdown objective and the yield and cost objectives, respectively. Less than 36 points are visible in each non-dominated set because some points overlay others. Figure 6.2 suggests that the drawdown objective as defined is somewhat independent from aquifer yield, if recharge is implemented along the locations selected as drawdown state variables. Therefore, three levels of satisfaction are observed:

1. Average drawdown is 6.9 m when mitigation cost is 7.6×10^7 or above.
2. Average drawdown is around 8.8 m when mitigation cost is equal to 2.9×10^6.
3. Average drawdown is around 10.8 m when mitigation cost is less than 4.0×10^5.

The values in the boxes in Fig. 6.2 indicate the parameterized values of mitigation costs to which each series of points correspond. When more than a cost value appears in the box, it is because alternatives corresponding to those parameterized cost values lay over the same projected position in the plane formed by the "yield" and "drawdown" objectives. Figure 6.3 shows the tradeoff between the drawdown objective and mitigation costs. This figure is a cross section of Fig. 6.2, and the independence from

Figure 6.3

Drawdown versus
cost tradeoff.

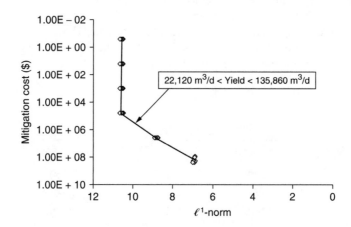

aquifer yield can be seen here as well. McPhee and Yeh (2004) point out that the apparent independence between drawdown and yield objectives results from the recharge wells being positioned too close to the locations at which drawdown is evaluated, therefore "shielding" these locations from effects caused by pumping in farther away locations.

Each point of the nondominated sets computed for cases A and B represents a specific policy alternative that is to be compared with the others in order to decide management actions. These alternatives have in common the fact that they are *efficient*, that is, no improvement can be achieved for a particular objective without diminishing the satisfaction associated with the other objectives. The problem of selecting which alternative remains open because all alternatives are equivalent from an efficiency point of view. Alternative selection is a complex problem that involves human interaction and conflict resolution. As such, it is very hard, if not impossible to model. Under these conditions, a modeler's primary goal should be to present the results of the analysis as clearly as possible, and to facilitate negotiation and incremental approaches to alternative selection using computational tools (e.g., Thiessen and Loucks, 1992; Cai et al., 2004; McPhee and Yeh, 2004)

CONCLUDING REMARKS

Conjunctive use planning of groundwater and surface water emerges today as one of the most viable means for accomplishing sustainable and reliable water supply in many parts of the world. Recent studies in climate change suggest that the intensity of the hydrologic cycle may be increasing, resulting in longer periods of drought and more intense storm events. By making use of the comparatively unlimited storage capacity of aquifers to store excess stream flow during wet periods or seasons, it is possible to reduce the risk of supply shortages and increase the efficiency of existing supply systems, usually at a fraction of the economic and environmental cost associated with surface water reservoirs and other conveyance and storage facilities. Because evaluation of conjunctive use planning involves elements of groundwater and surface water hydrology, water chemistry, ecology, systems analysis, and the like, it is clear that we face a multidisciplinary task that requires careful estimation of the different potential effects (environmental, public health, existing infrastructure) of conjunctive use projects. Numerical modeling tools are paramount components of the analysis of the impacts of conjunctive use projects, and optimization techniques provide invaluable aid in filtering out less effective alternatives from a space of potential solutions that may otherwise be very hard to assess. It has been observed that the success of conjunctive use planning is hindered by the difficulty in deciding when and how much to recharge, as well as when and in what quantity to pump groundwater from the aquifer. If all available tools of analysis can be integrated efficiently, and if planning is carried out in concert with stakeholders as well as operators and decision makers, conjunctive use planning of groundwater and surface water should provide a viable solution to the problem of increasing water demands and possibly decreasing water supply in the future.

ACKNOWLEDGMENTS

This material is based on work supported in part by NSF under EAR-9876800 (SAHRA) and EAR0336952.

REFERENCES

Ahlfeld, D. P., and M. Heidari, Applications of Optimal Hydraulic Control to Groundwater Systems," *Journal of Water Resources Planning and Management-ASCE*, Vol. 120, No. 3, pp. 350–365, 1994.

ASCE, and UNESCO, *Sustainability Criteria for Water Resource Systems*, American Society of Civil Engineers, Reston, Virginia. (1998).

Atkin, R. H., *Mathematical Structure in Human Affairs*, Heinemann, London, 1974.

Atkin, R. H., *Combinatorial Connectives in Social Systems*, Birkhauser, Basel, 1977.

Becker, L., and W. W. G. Yeh, "Identification of Parameters in Unsteady Open-Channel Flows," *Water Resources Research*, Vol. 8, No. 4, pp. 956–965, 1972.

Boyd, S. P., and L. Vandenberghe, *Convex Optimization*, Cambridge University Press, Cambridge, UK, New York, 2004.

Cai, X. M., L. Lasdon, and A. M. Michelsen, "Group Decision Making in Water Resources Planning Using Multiple Objective Analysis," *Journal of Water Resources Planning and Management-ASCE*, Vol. 130, No. 1, pp. 4–14, 2004.

Carrera, J., and S. P. Neuman, "Estimation of Aquifer Parameters under Transient and Steady-State Conditions. 1. Maximum-Likelihood Method Incorporating Prior Information," *Water Resources Research*, Vol. 22, No. 2, pp. 199–210, 1986a.

Carrera, J., and S. P. Neuman, "Estimation of Aquifer Parameters under Transient and Steady-State Conditions. 2. Uniqueness, Stability, and Solution Algorithms," *Water Resources Research*, Vol. 22, No. 2, pp. 211–227, 1986b.

Carrera, J., and S. P. Neuman, "Estimation of Aquifer Parameters under Transient and Steady-State Conditions. 3. Application to Synthetic and Field Data," *Water Resources Research*, Vol. 22, No. 2, pp. 228–242, 1986c.

Chow, V. T., D. R. Maidment, and L. W. Mays, *Applied Hydrology*, McGraw-Hill, New York, 1988.

Chu, X. F., and M. A. Marino, "Determination of Ponding Condition and Infiltration into Layered Soils under Unsteady Rainfall," *Journal of Hydrology*, Vol. 313, No. 3–4, pp. 195–207, 2005.

Cohon, J. L., *Multiobjective Programming and Planning*, Academic Press, New York, 1978.

Deb, K., A. Pratap, S. Agarwal, and T. Meyarivan, "A Fast and Elitist Multiobjective Genetic Algorithm: NSGA-II," *IEEE Transactions on Evolutionary Computation*, Vol. 6, No. 2, pp. 182–197, 2002.

Doherty, J., and J. M. Johnston, "Methodologies for Calibration and Predictive Analysis of a Watershed Model," *Journal of the American Water Resources Association*, Vol. 39, No. 2, pp. 251–265, 2003.

Dorn, J. L., and S. R. Ranjithan, "Evolutionary Multiobjective Optimization in Watershed Water Quality Management," *Evolutionary Multi-Criterion Optimization, Proceedings*, Vol. 2632, pp. 692–706, 2003.

Duckstein, L., and S. A. Nobe, "Q-Analysis for Modeling and Decision Making," *European Journal of Operational Research*, Vol. 103, No. 3, pp. 411–425, 1997.

Duckstein, L., W. Treichel, and S. Elmagnouni, "Ranking Groundwater-Management Alternatives by Multicriterion Analysis," *Journal of Water Resources Planning and Management-ASCE*, Vol. 120, No. 4, pp. 546–565, 1994.

Emch, P. G., and W. W. G. Yeh, "Management Model for Conjunctive Use of Coastal Surface Water and Ground Water," *Journal of Water Resources Planning and Management-ASCE*, Vol. 124, No. 3, pp. 129–139, 1998.

Fredericks, J. W., J. W. Labadie, and J. M. Altenhofen, "Decision Support System for Conjunctive Stream-Aquifer Management," *Journal of Water Resources Planning and Management-ASCE*, Vol. 124, No. 2, pp. 69–78, 1998.

Goode, T., and T. Maddock, "Simulation of Groundwater Conditions in the Upper San Pedro Basin for the Evaluation of Alternative Futures," Arizona Research Laboratory for Riparian Studies, The University of Arizona, 2000.

Gorelick, S. M., "A Review of Distributed Parameter Groundwater-Management Modeling Methods," *Water Resources Research*, Vol. 19, No. 2, pp. 305–319, 1983.

Harbaugh, A. W., E. R. Banta, M. C. Hill, and M. G. McDonald, "MODFLOW-2000, the U.S. Geological Survey Modular Ground-Water Model," *00-92*, U.S. Geological Survey, 2000.

Hiessl, H., L. Duckstein, and E. J. Plate, "Multiobjective Q-Analysis with Concordance and Discordance Concepts," *Applied Mathematics and Computation*, Vol. 17, No. 2, pp. 107–122, 1985.

Hill, M. C., E. R. Banta, A. W. Harbaugh, and E. R. Anderman, "MODFLOW-2000, the U.S. Geological Survey Modular Ground-Water Model—User Guide to the Observation, Sensitivity, and Parameter-Estimation Processes and Three Post-Processing Programs," *00–184*, U.S. Geological Survey, 2000.

Hsu, S. K., "Shortage Indexes for Water-Resources Planning in Taiwan," *Journal of Water Resources Planning and Management-ASCE*, Vol. 121, No. 2, pp. 119–131, 1995.

Kingsolver, B., "The Patience of A Saint," National Geographic Magazine, April 2000, pp. 80–97.

Louie, P. W. F., W. W. G. Yeh, and N. S. Hsu, "Multiobjective Water-Resources Management Planning," *Journal of Water Resources Planning and Management-ASCE*, Vol. 110, No. 1, pp. 39–56, 1984.

Madsen, H., G. Wilson, and H. C. Ammentrop, "Comparison of Different Automated Strategies for Calibration of Rainfall-Runoff Models," *Journal of Hydrology*, Vol. 261, No. 1–4, pp. 48–59, 2002.

McDonald, M. G., and A. W. Harbaugh, "A Modular Three-Dimensional Finite Difference Ground-Water Flow Model," Techniques of water-resources investigations of the U.S. Geological Survey, U.S. Government Printing Office, Washington, DC, 1998.

McPhee, J., and W. W. G. Yeh, "Multiobjective Optimization for Sustainable Groundwater Management in Semiarid Regions," *Journal of Water Resources Planning and Management-ASCE*, Vol. 130, No. 6, pp. 490–497, 2004.

Mein, R. G., and C. L. Larson, "Modeling Infiltration during a Steady Rain," *Water Resources Research*, Vol. 9, No. 2, pp. 384–394, 1973.

Philbrick, C. R., and P. K. Kitanidis, "Optimal Conjunctive-Use Operations and Plans," *Water Resources Research*, Vol. 34, No. 5, pp. 1307–1316, 1998.

Reed, P. M., and B. S. Minsker, "Striking the Balance: Long-Term Groundwater Monitoring Design for Conflicting Objectives," *Journal of Water Resources Planning and Management-ASCE*, Vol. 130, No. 2, pp. 140–149, 2004.

Roy, B., "Ranking and Choice in Pace of Multiple Points of View (Electre Method)," *Revue Francaise D Informatique De Recherche Operationnelle*, Vol. 2, No. 8, pp. 57–75, 1968.

Roy, B., "The Outranking Approach and the Foundation of ELECTRE Methods," *Theory and Decision*, Vol. 31, No. 1, pp. 49–73, 1991.

Shafike, N. G., L. Duckstein, and T. Maddock, "Multicriterion Analysis of Groundwater Contamination Management," *Water Resources Bulletin*, Vol. 28, No. 1, pp. 33–43, 1992.

Simmons, C. T., T. R. Fenstemaker, and J. J. M. Sharp, "Variable-Density Groundwater Flow and Solute Transport in Heterogeneous Porous Media: Approaches, Resolutions and Future Challenges," *Journal of Contaminant Hydrology*, Vol. 52, No. 1–4, pp. 245–275, 2001.

Stansbury, J., W. Woldt, I. Bogardi, and A. Bleed, "Decision Support System for Water Transfer Evaluation," *Water Resources Research*, Vol. 27, No. 4, pp. 443–451, 1991.

Thiessen, E. M., and D. P. Loucks, "Computer-Assisted Negotiation of Multiobjective Water-Resources Conflicts," *Water Resources Bulletin*, Vol. 28, No. 1, pp. 163–177, 1992.

U.S.G.S., "Hydrogeologic Investigations of the Sierra Vista Subwatershed of the Upper San Pedro Basin, Cochise County, Southeast Arizona," *99-4197*, U.S. Geological Survey, 1999.

Viessman, W., and G. L. Lewis, *Introduction to Hydrology*, Prentice Hall, Upper Saddle River, NJ, 2003.

Watkins, D. W., and D. C. McKinney, "Finding Robust Solutions to Water Resources Problems," *Journal of Water Resources Planning and Management-ASCE*, Vol. 123, No. 1, pp. 49–58, 1997.

Willis, R., and W. W. G. Yeh, *Groundwater Systems Planning and Management*, Prentice Hall, Englewood Cliffs, NJ, 1987.

Yager, R. R., "On Ordered Weighted Averaging Aggregation Operators in Multicriteria Decision-Making," *IEEE Transactions on Systems Man and Cybernetics*, Vol. 18, No. 1, pp. 183–190, 1988.

Yeh, W. W. G., "Review of Parameter-Identification Procedures in Groundwater Hydrology—the Inverse Problem," *Water Resources Research*, Vol. 22, No. 2, pp. 95–108, 1986.

Yeh, W. W. G., "Systems-Analysis in Groundwater Planning and Management," *Journal of Water Resources Planning and Management-ASCE*, Vol. 118, No. 3, pp. 224–237, 1992.

7

Uncertainties and Risks in Water Resources Projects

Yeou-Koung Tung
Department of Civil Engineering
Hong Kong University of Science & Technology
Hong Kong, China

INTRODUCTION

In the planning and development and design stages of a water resource project there exist a variety of uncertainties which render the ability of the system to achieve its intended goals uncertain. In other words, decisions made under uncertainties would always be subject to potential failure.

Failure of a hydrosystem engineering project would generally incur damages or adverse consequences. For example, failure of a flood defense system could result in direct and indirect economic losses, loss of human lives, destruction of environmental and eco-logical systems, and public health threat. Due to uncertain nature of occurrence and severity of failure, it is generally difficult to predict precisely the performance of the project over its anticipated project life.

There are many connotations for the term "risk" in water resource engineering and man-agement ranging from the conventional idea of probability of failure to the notion of con-sidering the consequence of failure. To follow current trend and avoid possible confusion, this chapter adopts the definition of risk as being the product of failure probability and consequence of all undesirable events. In other words, reliability-based analysis repre-sents mathematical assessments of failure and reliability whereas risk-based analysis explicitly accounts for consequences associated with the failures. Note that the conse-quence of failure does not have to be confined only to economic aspects.

The risk-based design and management of an engineering project involves following major steps:

1. *Identification of failure events and consequences.* The step involves qualitative def-inition of undesirable events constituting the failure of the project, recognition of causative factors (or scenarios) and their interactions leading to failure, and attempts to identify the adverse consequences associated with the failure.

2. *Quantification of failure probability.* The step involves quantitative assessment of fail-ure probability for each of the undesirable event. Depending on the nature of the per-formance function describing system performance, appropriate methods can be applied to determine the probability of occurrence of undesirable events (Tung et al., 2006).

3. *Quantification of consequences.* This task is by no means a trivial one. Take flood consequence as an example. It can involve casualties, psychological trauma, damage to buildings and properties, economic losses in business, industry, and agriculture, as well as ecologic/environmental damage. Having information about the failure con-sequence is essential for a good and realistic assessment of the risk. One should also be aware that quantification of failure consequence is subject to uncertainty.

4. *Assessment of risk acceptability.* Acceptable risk is a reference point based on which results from risk analysis are compared. Its establishment is based on various crite-ria from laws, regulations, standards, experience, and scientific knowledge. The level of acceptable risk might change with societal development and should be reviewed periodically. It can be expressed numerically or verbally. Building a water resource engineering project may introduce new types of risk to affected areas. It is expected

that involuntary risks introduced to the public (e.g., dam failure risk) will have to be much lower than those voluntary risks (e.g., skiing and driving). In Europe, the *as low as reasonably possible* (ALARP) criteria stipulate societal acceptable risk on fatalities and casualties of various engineering projects. An application considering acceptable risk on flood defense system in the Netherlands can be found elsewhere (Waarts and Vrouwenvelder, 2004).

5. *Determination of optimal design or course of action.* If failure consequence data is available, reliability analysis methods can be integrated with optimization to determine the optimal scale for project development that satisfies acceptable risk and other relevant constraints.

These steps in risk management can be cast into the risk-based decision-making framework.

Any sustainable development of water resource projects requires consideration of several goals and involves different stakeholders. The decision making is multidimensional in nature which generally involves objectives and constraints arising from political, environmental, social, economical, and engineering aspects. Engineering design is often at the final stage to find technical means to best accomplish the project goals. This chapter presents a risk-based analysis framework for sustainable development of water resource engineering projects. that integrates uncertainties, risks, and multicriteria decision.

RISK-BASED HYDROSYSTEMS ENGINEERING AND MANAGEMENT

Risk-based engineering and management involve assessments of uncertainties, failure probabilities, and consequences of failure. The development of a risk-based concept and its application in hydrosystems engineering are summarized in Tung et al. (2006). The basic concept of risk-based design is shown schematically in Fig. 7.1. The risk function considering uncertainties of various factors is a function of failure probability and undesirable consequences associated with the failure of hydrosystems.

Basic Concept

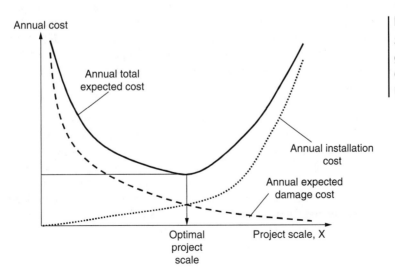

Figure 7.1
Schematic diagram of risk-based engineering and management.

Assume information on tangible costs associated with the failure (referred herein as the risk costs) is available. As risk due to the failure of hydrosystem projects vary from year to year, a practical way for its quantification is to use the expected value on an annual basis. The *total annual expected cost* (TAEC) is the sum of the annual installation cost and annual expected damage cost which can be mathematically expressed as:

$$\text{TAEC}(\mathbf{x}) = \text{FC}(\mathbf{x}) \times \text{CRF} + E(D \mid \mathbf{x}) \tag{7.1}$$

where FC is the first or total installation costs that is the function of decision vector $\mathbf{x}$, defining the level of project development; $E(D \mid \mathbf{x})$ is the annual expected damage cost associated with the project failure; and CRF is the capital recovery factor, which brings the present worth of the installation costs to an annual basis.

Referring to Fig. 7.1, as the scale of project development gets larger, the annual installation cost increases while the annual expected damage cost due to failures decreases. The optimal risk-based project development is the one having the lowest total annual expected cost. Mathematically, the optimal risk-based design problem can be stated as:

Minimize

$$\text{TAEC}(\mathbf{x}) = \text{FC}(\mathbf{x}) \times \text{CRF} + E(D \mid \mathbf{x}) \tag{7.2a}$$

subject to

$$g_j(\mathbf{x}) = 0, j = 1, 2, \ldots, m \tag{7.2b}$$

where $g_j(\mathbf{x})$ is 0, j is 1,2, ..., m which are constraints representing the design specifications that must be satisfied. The solution to Eqs. (7.2a, 7.2b) requires the use of an appropriate optimization algorithm. The application of the solution algorithm is largely problem dependent. In general, the failure probability of a hydrosystem holds a one-to-one correspondence with the scale of the development.

Risk Considerations in Hydrosystems Engineering

The design philosophy in hydrosystems engineering has evolved over several decades starting from judgment- or experience-based design of a particular protection level (say, the largest flood event or fraction of it in record) to return-period design, to optimal risk-based design with consideration given to various uncertainty. Safety factors are frequently employed to implicitly account for the presence of uncertainties. However, the value of safety factor is determined primarily on the basis of judgment or experience of the designer.

For many years, hydrologic risk dealing with the random occurrences of hydrologic extremes (rainstorms, floods, or droughts) has been practiced through frequency analysis. Through frequency analysis the occurrence probability (or return periods) of loads can be quantified based on the level of protection (or project scale) determined. Once the design return period is determined, the corresponding design hydrologic quantities remain fixed throughout the entire design process. Again, the selection of the design return period for a flood defense system is a complex procedure involving the consideration of economic, social, legal, environmental, and other factors. However, the procedure does not account for these factors explicitly.

The risk-based hydrosystems engineering and management evaluates among alternatives by considering the trade-off between the investment cost and the expected damages due to failures. Instead of being a preselected design parameter as with the return period design procedure, in risk-based analysis the design return period is a decision variable. Specifically, the conventional risk-based approach considers the inherent hydrologic uncertainty in the calculation of the annual expected damages. Example applications include the design of highway drainage structures, such as culverts (Young et al., 1974; Corry et al., 1980), bridges (Schneider and Wilson, 1980), and storm surge protection structures (Voortman et al., 2003). The inherent randomness of hydrologic processes is also integrated with reliability analysis in seismic, structural, and geotechnical aspects in the design of new dams (Pate-Cornell and Tagaras, 1986), evaluation of alternatives for rehabilitating existing dams (McCann et al., 1984; Bureau of Reclamation, 1986), and large flood defense systems (Waarts and Vrouwenvelder, 2004). Furthermore, calculations of risk cost have incorporated hydrologic parameters uncertainty in the design of levee systems (Wood et al., 1977; Bodo et al., 1976; Van Gelder, 2000) and the incorporation of hydraulic uncertainty in design of highway drainage structures (Mays, 1979; Tung and Mays, 1982; Tung and Bao, 1990), storm sewer systems (Yen and Ang, 1971; Tang et al., 1975), levee systems (Tung and Mays, 1981), storm surge barrier (Voortman et al., 2003), and river diversion (Afshar et al., 1994).

In the conventional risk-based hydrosystems engineering, risk-costs are calculated considering only the randomness of hydrologic events. In reality, there exist various types of uncertainty in hydrosystem planning and design. Advances have been made to incorporate other aspects of uncertainty in the design of various types of hydrosystem infrastructurer.

Evaluations of Annual Expected Damage Cost Due to Failure

In risk-based analysis of hydrosystems, the thrust of the exercise, after uncertainty and risk analyses are performed, is to evaluate $E(D \mid \mathbf{x})$ as the function of the probability density functions (PDFs) of loading and resistance, damage function, and the types of uncertainty considered.

Conventional Approach

In conventional risk-based design, in which only inherent hydrologic uncertainty is considered, the system scale $\mathbf{x}$ and its corresponding capacity q_c, in general, have a one-to-one, monotonically increasing relationship. The annual expected damage cost, in the conventional risk-based approach, can be computed as:

$$E_1(D|\mathbf{x}) = \int_{q_c^*}^{\infty} D(Q|q_c^*)f(q)dq \tag{7.3}$$

where q_c^* is the deterministic capacity of the hydrosystem subject to random hydrologic load following a PDF, $f(q)$; and $D(q \mid q_c^*)$ is the damage function corresponding to the magnitude of load and the system capacity q_c^*. A graphical representation of the Eq. (7.3) is shown in Fig. 7.2. Note that Eq. (7.3) assumes a perfect knowledge about the probability distribution of hydrologic load and it is generally not the case in reality.

Figure 7.2

Graphical repre-
sentation of annual
expected damage.

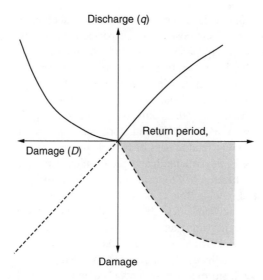

Incorporation of Hydraulic Uncertainty

Uncertainties also exist in hydraulic computations for determining the capacity of the hydrosystem infrastructure. In other words, q_c is subject to uncertainty. To incorporate the uncertainty of q_c in the risk-based design, the annual expected damage is calculated using:

$$E_2(D) = \int_0^\infty \left[\int_{q_c}^\infty D(q \,|\, q_c) f(q) dq \right] g(q_c) dq_c = \int_0^\infty E_1(q \,|\, q_c) g(q_c) dq_c \qquad (7.4)$$

where $g(q_c)$ is the PDF of the random capacity.

Extension of Conventional Approach by Considering Hydrologic Parameter Uncertainty

Because the occurrence of hydrologic load is random by nature, its statistical properties such as the mean, standard deviation, and skew coefficient estimated from a finite sample are also subject to sampling errors.

Consider hydrologic load (flood or rainfall) that could potentially cause failures of the hydrosystem infrastructures. Due to the uncertainty associated with the estimated values of statistical properties, the estimated magnitude of hydrologic load with a specified probability, q_p, is also a random variable associated with its probability distribution (see Fig. 7.3) instead of being a single-valued quantity presented by its "average," as commonly done in hydrologic frequency analysis. Hence, there is an expected damage corresponding to the pth quantile of hydrologic load which can be expressed as:

$$E(D_p \,|\, q_c^*) = \int_{q_c^*}^\infty D(q_p \,|\, q_c^*) h(q_p) dq_p \qquad (7.5)$$

where $E(D_p \,|\, q_c^*)$ is the expected damage corresponding to the pth-order quantile of hydrologic load given a known capacity of the hydrosystem infrastructure, q_c^*; $h(q_p)$ is

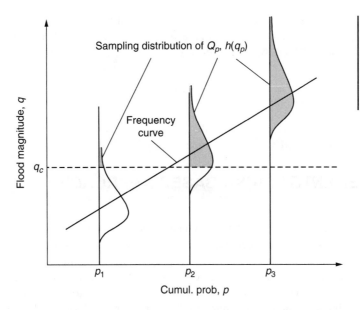

Figure 7.3

Schematic sketch
of sampling
distribution of
flood magnitude
estimator.

the sampling PDF of the estimator of the pth-order quantile of hydrologic load; and q_p is
the dummy variable for the pth-order quantile of hydrologic load. To combine the inherent
hydrologic uncertainty, represented by the PDF of random hydrologic load, $f(q)$, and the
hydrologic parameter uncertainty, represented by the sampling PDF for the pth-order quantile of hydrologic load, $h(q_p)$, the annual expected damage cost can be written as:

$$E_3(D|q_c^*) = \int_{q_c^*}^{\infty} \left[\int_{q_c^*}^{\infty} D(q_p|q_c^*) h(q_p|q) dq_p \right] f(q) dq \qquad (7.6)$$

Incorporation of Both Hydrologic Inherent/Parameter and Hydraulic Uncertainties

To include hydrologic inherent and parameter uncertainties along with the hydraulic
uncertainty associated with the hydrosystem capacity, the annual expected damage cost
can be written as:

$$E_4(D) = \int_0^{\infty} \left[\int_{q_c}^{\infty} \left[\int_{q_c}^{\infty} D(q_p, q_c) h(q_p) dq_p \right] f(q) dq \right] g(q_c) dq_c$$

$$= \int_0^{\infty} E_3(D|q_c) g(q_c) dq_c \qquad (7.7)$$

Based on the preceding formulations for computing annual expected damage in risk-
based design of hydrosystems, one realizes that the mathematical complexity increases as
the more uncertainties are considered. However, to obtain an accurate estimation of annual expected damage associated with the system failure would require the consideration of
all uncertainties, if such could be accomplished practically. Otherwise, the annual expected
damage could be overestimated or underestimated, which potentially leads to inaccurate
optimal design (Bao et al., 1987; Tung, 1987).

To incorporate uncertainty associated with the distribution model of hydrologic loads due to a finite sample size, the annual expected flood damage cost can be obtained by the weighted average (Wood et al., 1977; Van Gelder, 2000):

$$E(D) = \sum_{m=1}^{M} p_m \times E(D \mid f_m) \qquad (7.8)$$

where p_m is the likelihood that PDF, $f_m(q)$ is the true distribution model for the hydrologic loads and M is the total number of plausible distribution models.

DATA REQUIREMENTS IN RISK-BASED APPROACH

Design of a hydrosystem, by nature, is a decision-making problem requiring an analysis of system performance for various aspects of concern and to determine the most economical and feasible design. The objective is to minimize the sum of capital investment cost, the expected damage costs due to failures, and operation and maintenance costs subject to various constraints. Consequences due to failures of water resource projects involve, but are not limited to, economic loss, agricultural damage, loss of human lives, and environmental damage.

Economic loss consists of direct and indirect losses. The former involves loss of properties, damage to lifelines, and cost of recovery. The indirect economic loss includes loss of productivity and income due to interruption of electricity and/or blockage of roads. This type of damage, taking the failure of flood defense system as an example, depends primarily on water depth on the site. Damage to agriculture depends on the economic value of the crops in the field affected by the failure. Water depth and inundation duration affects the extent of the damage to crops.

Potential human casualties due to floods depends on a number of factors, such as water depth, flow velocity, rate of water rise, warning and evacuation, and epidemic aftermath. Other elements such as water temperature, debris, and the collapse of buildings might contribute to the threat.

Failure of a water resource project might contribute to the degradation of ecologic systems in affected areas. For example, flood water could carry large quantities of sediment that alters an otherwise relatively stable channel and destroys fish spawning ground. Polluted water and suspended particulate matters adversely affect the sources of potable water and may alter an existing ecosystem.

The basic data required for risk-based engineering and management of hydrosystems can be categorized into the following types:

a. *Hydrologic/physiographical data* include flow and precipitation data, drainage area, channel bottom slope, drainage basin slope, soil types, vegetation cover, and land use pattern. These data are needed to predict the magnitude of hydrologic events, such as stream flow and rainfall to conduct accurate by at-site and/or regional frequency analyses, and perform rainfall-runoff analysis. Many of the geographical features of a basin can be extracted through a geographical information system (GIS) approach.

b. *Hydraulic data* include floodplain slopes, geometry of channels, roughness coefficients, size and type of drainage structures, and height of embankments. These data are needed to determine the flow carrying capacities of hydraulic structures, and to calculate spatial and temporal distributions of flow characteristics such as depth and velocity.

c. *Infrastructure data* include materials geometry, dimension, type, and layout of the infrastructure

d. *Economic data* include:

1. Type, location, distribution, and economic value of properties, such as crops and buildings
2. Unit costs of structural materials, equipment, operation of vehicles, accidents, occupancy, and labor fee
3. Rate of repair, incident rate of accidents
4. Time of repair, length of traffic detours
5. Economic structure of the region

As an example, Table 7.1 shows the types of data needed to conduct risk-based design of highway drainage structures.

In the design of hydrosystem infrastructure, the installation cost often depends on the location, geomorphic and geological conditions, soil type, type and price of construction material, hydraulic conditions, flow conditions, recovery factor of the capital investment, labor, and transportation costs. In reality, these factors introduce uncertainties in the cost functions used in the analysis. Tung (1994) has touched on some preliminary issues of economic uncertainties in the risk-based design of hydraulic structures. A more rigorous treatment of the subject remains to be developed.

RISK-BASED ANALYSIS CONSIDERING INTANGIBLE FACTORS

The risk-based approach described considers only tangible economic costs consisting of installation costs, operation/maintenance costs, and various failure-related damages. Besides the economic factors that can be quantified in monetary terms, there are other intangible factors that are noncommensurable and cannot be quantified. Examples of intangible factors in the design and planning of hydrosystems are the potential loss of human lives, stability of water course, impacts on local society and ecological environment, health hazards after floods, litigation potential, and others. Plate and Duckstein (1988), list a number of performance measures, called figures of merit, for risk-based design of hydraulic structures and water resource systems. Additional work on the subject of intangible factors in risk management are found elsewhere (Jonkman et al., 2003; Vrijling et al., 1995). A simple risk-based framework considering intangible factors is presented later.

Considering all or some of these factors provides a much more complete picture of the problem than the conventional risk-based design and analysis. As more intangible factors are considered in the risk-based design of hydrosystems, it becomes a *multiobjective or*

Table 7.1

Damage Categories with Related Economic Variables and Site Characteristics in Risk-Based Design of Highway Drainage Structures

Damage Category	Economic Variables	Site Characteristics
Floodplain property damage: • Losses to crops • Losses to buildings	• Types of crops • Economic value of crops • Types of buildings • Economic values of buildings and contents	• Location of crop fields • Location of buildings • Physical layout of drainage structures • Roadway geometry • Flood characteristics • Stream valley cross-section • Slope of channel profile • Channel and flood-plain roughness
Damage to pavement and embankment: • Pavement damage • Embankment damage	• Material cost of pavement • Material cost of embankment • Equipment costs • Labor costs • Repair rate for pavement and embankment	• Flood magnitude • Flood hydrograph • Overtopping duration • Depth of overtopping • Total area of pavement • Total volume of embankment • Types of drainage structure and layout • Roadway geometry
Traffic-related losses: • Increased travel cost due to detour • Lost time of vehicle occupants • Increased risk of accidents on detour • Increased risk of accidents on a flooded highway	• Rate of repair • Operational cost of vehicle • Distribution of income for vehicle occupants • Cost of vehicle accident • Rate of accident • Duration of repair	• Average daily traffic volume • Composition of vehicle types • Length of normal detour paths • Flood hydrograph • Duration and depth of overtopping

Source: Tung et al., 2006.

multicriteria decision-making (MCDM) problem in which economic efficiency is one of many factors to be considered simultaneously. Use of a multiple-criteria approach results in more realistic decision making.

Methods of varying degrees of sophistication have been developed to deal with (MCDM) problems (Zeleny, 1982; Steuer, 1986). A simple yet effective method is the *simple additive weighing* (SAW) technique. This technique analyses an information matrix consisting of the decision maker's subjective ratings to each of the criteria involved in a number of alternatives. A typical information matrix for an MCDM problem by the SAW technique is shown in Fig. 7.4. The relative merit of each alternative is judged on the basis of its final rating computed by:

Design alternatives	Factors (criteria)					
	Factor-1	Factor-2	...	Factor-j	...	Factor-J
Alt-1	R_{11}	R_{12}	...	R_{1j}	...	R_{1J}
Alt-2	R_{21}	R_{22}	...	R_j	...	R_{2J}
.	.	.	...	.	...	.
.	.	.	...	.	...	.
.	.	.	...	.	...	.
Alt-k	R_{k1}	R_{k2}	...	R_{kj}	...	R_{kJ}
.	.	.	...	.	...	.
.	.	.	...	.	...	.
.	.	.	...	.	...	.
Alt-K	R_{K1}	R_{K2}	...	R_{Kj}	...	R_{KJ}
Relative importance	W_1	W_2	...	W_j	...	W_J

Figure 7.4

Information matrix for the SAW technique.

$$F_k = \frac{\sum_{j=1}^{J} R_{kj} W_j}{\sum_{j=1}^{J} W_j}, \text{ for } k = 1, 2, \ldots, K \qquad (7.9)$$

where F_k = final rating for alternative k

R_{kj} = rating for alternative k with respect to criterion j

W_j = weight representing the relative importance of criterion j

K and J = total number of alternatives and criteria, respectively

The conventional risk-based framework can be extended to include intangible factors by using MCDM. In a multiple-criteria risk-based design problem, the criteria would include the economic (tangible) factors and those intangible factors that are relevant to decision makers. Using the SAW technique, the design variables (such as the scale of hydrosystem development) are discretized into alternatives. Further, the decision makers use a numerical rating system to reflect the desirability of different alternatives under various criteria as well as the relative importance of the criteria. In general, the list of alternatives should use the economically efficient risk-based design alternative as the lower bound. Considering additional intangible criteria (factors), such as environmental preservation, will generally lead to adopting a more expensive alternative.

For practical implementation of the MCDM to incorporate intangible factors in risk-based analysis in highway drainage structure design, Tung et al. (1993) adopted a system of verbal ratings from "very desirable" to "less desirable" for design alternatives in terms of return period varying from 2 years to 200 years. Criteria included economic efficiency, litigation potential, maintenance frequency, and public service. To facilitate numerical analysis in the framework of the SAW technique, numerical ratings (0–10) associated with different verbal ratings were established through an extensive survey among engineers in highway departments at the state and federal levels in the United States. To account for the uncertain nature of the numerical ratings associated with verbal ratings a fuzzy membership function was used. The calculation of the final rating for each alternative by Eq. (7.9) was conducted by applying fuzzy arithmetic (Kaufmann and Gupta, 1985). Another example of a practical use of MCDM was presented by Waarts and Vrouwenvelder (2004) reporting a multicriteria analysis of a flood defense system in the Netherlands that considered economic fatalities and environmental factors.

RISK-BASED DESIGN WITHOUT FLOOD DAMAGE INFORMATION

In general, risk-based analysis of hydrosystems requires information with regard to various failure-related damages. Such information requires an extensive survey of the type and value of various properties, economic and social activities, and other demographical information of the affected regions. Hence, failure-related damage information is very much location specific with little that can be generalized from one place to another. For areas where failure-related damage data are not available, the conventional risk-based framework described so far cannot be implemented.

Taking the analysis of a flood defense system as an example, one normally would have to conduct hydraulic simulation to delineate the flood-affected zone and other related flood characteristics, such as water depth, flow velocity, and inundation period. The hydraulic characteristics, combining with property survey data, would allow estimating flood damage for a specified flood event. In the situation where flood-related damage data are unavailable, the risk-based analysis still can be conducted by replacing the flood-related damage functions with relevant physical performance characteristics of the flood defense systems. For example, useful physical performance characteristics in urban drainage systems could be pipe length (or street area) subject to surcharge, volume of surcharged water, maximum (or average) depth, and velocity of overland flow. Although these performance characteristics may not completely reflect the flood damages, they nonetheless indicate the potential seriousness of the flooding situation.

For a given design, the corresponding amortized annual installation cost can be estimated. Also, the system's responses under the different hydrologic loadings can be obtained by a hydraulic simulation model. As an example the hydraulic simulation outputs for an urban drainage system designed for a five-year capacity is shown in Table 7.2 (a). Each row corresponds to a rainfall event of specified return period. The hydraulic responses of the system tabulated in Table 7.2 (a) can be used in Eqs. (7.3–7.8), depending on the types of uncertainties considered, to calculate the annual expected hydraulic responses under a specified system capacity. Table 7.2 (b) shows the costs corresponding to the hydraulic responses given in Table 7.2 (a). The process can be repeated to consider other design alternatives and protection levels.

Based on the annual project cost of the system and the expected hydraulic response of the system, a trade-off analysis can be performed by examining the marginal improvement in the system responses due to a one unit increase in capital investment. Referring to Fig. 7.5, it is observed that the annual expected surcharge volume decreases as the annual capital cost of the system increases due to increasing level of protection. The marginal cost corresponding to one reduction in surcharge volume can be written as:

$$\text{MC} = -\frac{\partial C}{\partial S_v} \tag{7.10}$$

(a) Hydraulic Responses of Urban Drainage System Designed for Five-Year Protection

Level of Protection: Five-Year Storm				
Storm with T(year)	Exc. Prob.	Nonexc. Prob.	Flood Vol. (m³)	Flood Area (m²)
2	0.5	0.5	0	0
5	0.2	0.8	0	0
10	0.1	0.9	141	864
20	0.05	0.95	487	1875
50	0.02	0.98	605	2020
100	0.01	0.99	811	3838
200	0.005	0.995	1313	5046

(b) Annual Cost Needed to Improve Existing System to Different Levels of Protection

Protection Level	Installation Cost ($M)	Annual Cost* ($M)	Annual Expct. Surcharge Vol. (m³)	Annual Expct. Surcharge Area (m²)
Existing	0	0	727	1777
2 years	4.05	0.34	89	372
5 years	5.80	0.49	52	209
10 years	7.74	0.65	10	53
20 years	9.25	0.78	6	36
50 years	16.80	1.41	0.14	2.4
100 years	18.26	1.53	0.05	0.9
200 years	18.37	1.54	0	0

*The interest rate is 8 percent and the service life is assumed to be 40 years.

Table 7.2

Hydraulic Performance of Example Urban Drainage System

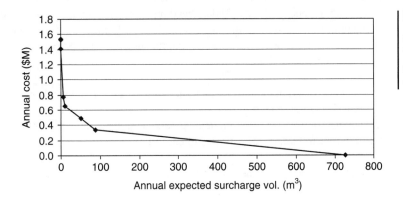

Figure 7.5

Annual project cost versus annual expected surcharge volume.

where S_v = surcharge volume
   C = capital cost
   MC = marginal cost

The example has a low MC for the existing system and an annual capital cost around HK$0.6M for a 10-year protection. For levels of protection higher than 10 years the rate of increase in capital investment per unit reduction in surcharge volume becomes very high. Thus the trend of marginal cost presented in Fig. 7.5 would allow a decision maker to choose a sensible level of protection for project implementation.

UNCERTAINTY AND RELIABILITY ANALYSES

Risk-based analysis and design of hydrosystems require quantifying the probability of occurrence of various failure events. This section offers a brief overview on calculating uncertainty in hydrosystems engineering. Detailed descriptions can be found elsewhere (Tung and Yen, 2005; Tung et al., 2006). Once the uncertainties of the system are understood and statistically quantified, reliability analysis may be performed. An overview of reliability analysis is presented in the next section.

Uncertainty Analysis

In hydrosystems engineering and management, uncertainties arise mainly due to our lack of (1) perfect understanding with regard to the phenomena and processes involved and (2) perfect knowledge of how to determine parameter values for processes that are fairly well understood.

More specifically, in water resources engineering analyses and designs uncertainties can be categorized into natural, model, parameter, data, and operational uncertainties (Yen et al., 1986). Natural uncertainty is associated with the inherent randomness of natural processes such as precipitation and flood events. Model uncertainty reflects the inability of the model or design technique to accurately represent the system's true physical behaviour. Parameter uncertainties result from the inability to quantify accurately the parameters of a model. Parameter uncertainty can be caused by changes in the operational conditions of hydraulic structures and the inherent variability of inputs and parameters in time and in space. Parameter uncertainty is exacerbated by the lack of a sufficient amount of data. Data uncertainties include (1) measurement errors (2) inconsistency and nonhomogeneity of data (3) data handling and transcription errors (4) inadequate representation of data sample due to time and space limitations. Operational uncertainties include those associated with construction, manufacture, equipment and supply, deterioration, maintenance, and human factors.

In general, uncertainty due to inherent randomness of physical processes cannot be eliminated. Other uncertainties such as those associated with the lack of complete knowledge of the process, models, parameters, data, and so forth can be reduced through research, data collection, and careful manufacture.

Several expressions have been used to describe the degree of uncertainty of a parameter, a function, a model, or a system. They range from the most complete information in the form of probability density function, statistical moments, to an expression in terms of a confidence interval.

The main objective of uncertainty analysis is to identify the statistical properties of outputs of a system as functions of its stochastic input parameters. Uncertainty analysis provides a formal and systematic framework to quantify the uncertainty associated with the system outputs. Furthermore, it offers the designer insight regarding the contribution of each input parameter to the overall uncertainty of the system outputs. Such knowledge is essential to identify the parameters to which more attention should be given to better assess their values and, accordingly, to reduce the overall uncertainty of the outputs.

Uncertainty analysis techniques involve different levels of mathematical complexity and data requirements. The appropriate techniques to be used depend on the nature of the problem at hand including the availability of information, model complexity, and type and accuracy of results desired.

Several useful analytical approaches for uncertainty analysis include derived distribution techniques and integral transform techniques. Although these analytical approaches are rather restrictive in practical applications, they are powerful tools for deriving information about a stochastic process in some situations. Also, there exist several techniques that are particularly useful for problems involving complex functions which are analytically difficult. They are primarily developed to estimate the statistical moments about the underlying random processes.

The resistance or strength of a hydrosystem infrastructure is its ability to accomplish the intended mission satisfactorily when subject to a demand load or external stresses. Failure occurs when the resistance is exceeded by the load.

Reliability Analysis

The reliability of a hydrosystem infrastructure is the probability of safety, p_s, that the load L does not exceed the resistance R of the structure:

$$p_s = P_r[L \leq R] \qquad (7.11a)$$

where $P_r[]$ denotes the probability of $L \leq R$. Conversely, the failure probability, p_f, can be expressed as:

$$p_f = P_r[L > R] = 1 - p_s \qquad (7.11b)$$

Reliability analysis can be applied to various types of engineering problems, such as design of engineering facilities; evaluation of the safety of existing structures; inspection, repair and other maintenance operations; routing, distribution, allocation, and other planning and management operations; data sampling and measurement network design; real-time prediction and long-term forecasting; and comparison and assessment of techniques and procedures.

Failure of hydrosystem infrastructures can be classified as structural failure and performance failure (Yen and Ang, 1971; Yen et al., 1986). Structural failure involves damage or change of the structure or facility, preventing its ability to function as desired. On the other hand, performance failure does not necessarily involve structural damage, but the performance limit of the structure is exceeded and undesirable consequences occur. Generally, the two types of failure are related. Some hydraulic structures, such as dams and levees, are designed on the concept of structural failure, whereas other hydraulic structures like sewer and water supply networks are designed on the basis of performance failure.

A common practice for measuring the reliability of a hydrosystem is to consider its return period or recurrence interval. The return period is defined as the long-term average (or expected) time between two successive failure causing events. Flood frequency analysis using annual maximum flow series is a typical example of this kind. The main disadvantage of using the return period is that reliability is measured only in terms of time of occurrence of loads without considering their interactions with resistance (Melchers, 1999).

Two other types of reliability measures are frequently used in engineering practice. One measure is the *safety margin* (SM) defined as the difference between the resistance and the anticipated (or design) load, that is:

$$SM = R - L \tag{7.12}$$

The other measure is the *safety factor* (SF) which is the ratio of resistance to load as:

$$SF = R/L \tag{7.13}$$

Several types of safety factors and their applications to hydraulic engineering design are discussed by Yen (1979).

There are two basic probabilistic approaches to evaluate the reliability of a hydrosystem. The most direct approach is a statistical analysis of data of past failure records for similar systems. The other approach is through reliability analysis which considers and combines the contribution of each potentially influencing factors. The former is a lumped system approach requiring no knowledge about the behaviour of the facility or structure and its load and resistance. However, in many cases this direct approach is impractical because (1) the sample size may be too small to be statistically reliable, especially for low-probability/high-consequence events; (2) the sample may not be representative of the system or of the population; and (3) the physical conditions of a system may vary with respect to time, that is, the system is nonstationary.

There are two major steps in reliability analysis:

1. To identify and analyze the uncertainties of each of the contributing factors
2. To combine the uncertainties of the stochastic factors to determine the overall reliability of the system.

The second step may proceed in two different ways:

1. Directly combining the uncertainties of all the factors
2. Separately combining the uncertainties of the factors belonging to different components or subsystems to evaluate first the respective subsystem reliability and then

combining the reliabilities of the different components or subsystems to yield the overall reliability of the system.

CONCLUDING REMARKS

The concepts and theory of reliability analysis and risk-based design of hydrosystem infrastructures have been around for sometime. Yet, the application of probabilistic design/analysis to real-life hydraulic engineering projects has been largely limited to hydrologic frequency analysis. One major difficulty that hinders practical application of risk-based analysis is the availability of reliable flood-related damage data.

This chapter describes the basic elements of risk-based analysis and demonstrates how the analysis can be modified to deal with problems when damage data due to failure are absent. Furthermore, the risk-based analysis discussed can be integrated into multiple-criteria decision-making framework to consider the multiple objectives which are essential to water resource design and analysis.

REFERENCES

Afshar, A., A. Barkhordary, and M. A. Marino, "Optimzing River Diversion Under Hydraulic and Hydrologic Uncertainties," *Water Resources Planning and Management*, Vol. 120, No. 1, pp. 36–47, 1994.

Bao, Y., Y. K. Tung, and V. R. Hasfurther, "Evaluation of Uncertainty in Flood Magnitude Estimator on Annual Expected Damage Costs of Hydraulic Structures," *Water Resources Research*, Vol. 23, No. 11, pp. 2023–2029, 1987.

Bodo, B., and T. E. Unny, "Model Uncertainty in Flood Frequency Analysis and Frequency Based Design," *Water Resources Research*, Vol. 12, No. 6, pp. 1109–1117, 1976.

Bureau of Reclamation, "Guideline for Decision Analysis," *ACER Technical Memorandum*, No. 7, U.S. Dept. of Interior, 1986.

Corry, M. L., J. S. Jones, and D. L. Thompson, "The Design of Encroachments of Floodplain Using Risk Analysis," *Hydraulics Engineering Circular*, No. 17, U.S. Dept. of Transp. Federal Highway Admin., Washington, DC, 1980.

Jonkman, S. N., P. H. A. J. M. van Gelder, and J. K. Vrijing, "An Overview of Quantitative Risk Measures for Loss of Life and Economic Damage," *Journal of Hazardous Materials*, Vol. A99, pp. 1–30, 2003.

Kaufmann, A., and M. M. Gupta, *Introduction to Fuzzy Arithmetic—Theory and Applications,* Van Nostrand Reinhold, New York, 1985.

McCann, M. M., J. B. Franzini, E. Kavazanjiam, and H. C. Shah, "Preliminary Safety Evaluation of Existing Dams, Vol. 1," Report Prepared for Federal Emergency Mgmt. Agency by Dept. of Civ. Eng., Stanford University, CA, 1984.

Mays, L. W., "Optimal Design of Culverts Under Uncertainties," *Journal of Hydraulics Engineering*, Vol. 105, No. 5, pp. 443–460, 1979.

Melchers, R. E., *Structural Reliability: Analysis and Prediction*, 2d ed., John Wiley and Sons, New York, 1999.

Pate-Cornell, M. E., and G. Tagaras, "Risk Cost for New Dams: Economic Analysis and Effects of Monitoring," *Water Resources Research*, Vol. 22, No. 1, pp. 5–14, 1986.

Plate, E. J., and L. Duckstein, "Reliability Based Design Concepts in Hydraulic Engineering, *Water Resources Bulletin*, Vol. 24, No. 2, pp. 234–245, 1988.

Schneider, V. R., and K. V. Wilson, "Hydraulic Design of Bridges with Risk Analysis," *Report* FHWA-TS-80-226, U.S. Dept. of Transp. Federal Highway Admin., Washington, DC, 1980.

Steuer, R.E., *Multiple Criteria Optimization: Theory, Computation, and Application.* John Wiley and Sons, New York, 1986.

Tang, W. H., L. W. Mays, and B. C. Yen, "Optimal Risk-Based Design of Storm Sewer Networks," *Journal of Environmental Engineering*, Vol. 101, No. 3, pp. 381–398, 1975.

Tang, W. H., and B. C. Yen, "Probabilistic Inspection Scheduling for Dams," in B.C. Yen and Y.K. Tung (eds.), *Reliability and Uncertainty Analyses in Hydraulic Design,* pp. 107–122, ASCE, New York, 1993.

Tung, Y. K., and L. W. Mays, "Optimal Risk-Based Design of Flood Levee Systems," *Water Resources Research*, Vol. 17, No. 4, pp. 843–852, 1981.

Tung, Y. K., and L. W. Mays, "Optimal Risk-Based Hydraulic Design of Bridges," *Journal of Water Resources Planning and Management*, Vol. 108, No. 2, pp.191–202, 1982.

Tung, Y. K., and Y. Bao, "On the Optimal Risk-Based Designs of Highway Drainage Structures," *J. of Stoch. Hydrol. and Hydraul.*, Vol. 4, No. 4, pp. 311–324, 1990.

Tung, Y. K., V. Hasfurther, A. M. Wacker, Y. Bao, and B. Zhao, "Least Total Expected Cost (LTEC) Analysis for Selecting a Defensible Design Flood Frequency for Highway Drainage Structures in Wyoming," *Technical Report*, Wyoming Water Resour. Center, Univ. of Wyoming, Laramie, WY, 444 + xv pp. 1993.

Tung, Y. K., "Effects of Uncertainties on Optimal Risk-Based Design of Hydraulic Structures," *Journal of Water Resources Planning and Management*, Vol. 113, No. 5, pp. 709–722, 1987.

Tung, Y. K., "Probabilistic Hydraulic Design: A Next Step to Experimental Hydraulics," *Journal of Hydraulics Research*, Vol. 32, No. 3, pp. 323–336, 1994.

Tung, Y. K., and B. C. Yen, *Hydrosystems Engineering Uncertainty Analysis*, McGraw-Hill, New York, 2005.

Tung, Y. K., B. C. Yen, and C. S. Melching, *Hydrosystems Engineering Reliability Assessments and Risk Analysis*, McGraw-Hill, New York, 2006.

Van Gelder, P. H. A. J. M., *Statistical Methods for the Risk-Based Design of Civil Structures,* Delft University of Technology, p. 249, 2000.

Voortman, H. G., P. H. A. J. M. van Gelder, and J. K. Vrijling, "Risk-Based Design of Large-Scale Flood Defense Systems," *Proceedings of the 28th International Conference on Coastal Engineering (ICC2002)*, Solving Coastal Conundrums, J.M. Smith (ed.), pp. 2360–2372, World Scientific Publishing, 2003.

Vrijling, J. K., W. van Hengel, and R. J. Houben, "A Framework for Risk Evaluation," *Journal of Hazard Materials*, Vol. 43, No. 3, pp. 245–261, 1995.

Waarts, P. H., and A. C. W. M. Vrouwenvelder, "Risk Management of Large Scale Floodings," *Heron*, Vol. 49, No. 1, pp. 7–32, 2004.

Wood, E. F., "An Analysis of Flood Levee Reliability," *Water Resources Research*, Vol. 13, No. 3, pp. 665–671, 1977.

Yen, B. C., and A. H. S. Ang, "Risk Analysis in Design of Hydraulic Projects," *Stochastic Hydraulics*, 1st International Symposium on Stochastic Hydaulics, pp. 694–709, 1971.

Yen, B.C., "Safety Factor in Hydrologic and Hydraulic Engineering Design," in E. A. McBean, K. W. Hipel, and T. E. Unny (eds.), *Reliability in Water Resources Management*, pp. 389–407, Water Resources Publications, Littleton, CO, 1979.

Yen, B. C., S. T. Cheng, and C. S. Melching, "First-Order Reliability Analysis," in B.C. Yen (ed.), *Stochastic and Risk Analysis in Hydraulic Engineering*, pp. 1–36, Water Resources Publications, Littleton, CO, 1986.

Young, G. K., M. R. Childrey, and R. E. Trent, "Optimal Design of Highway Drainage Culverts," *Journal of Hydraulics Engineering*, ASCE, Vol. 107, No. 7, pp. 971–993, 1974.

Zeleny, M., *Multiple Criteria Decision Making,* McGraw-Hill, New York, 1982.

8

CLIMATE CHANGE EFFECTS AND WATER MANAGEMENT OPTIONS

Larry W. Mays
Department of Civil and Environmental Engineering
Arizona State University
Tempe, Arizona

INTRODUCTION

The Pacific Institute has compiled a comprehensive bibliography of peer-reviewed literature dealing with climate change and its effects on water resources and water systems. As of June 2006 the bibliography included over 3600 citations. The address for this bibliography is http://biblio.paoinst.org/biblio/. The International Panel on Climate Change (IPCC) has been active in compiling information concerning future climate change (IPCC, 1990, 1996a, 1996b, 1996c, 1998, 2001a, 2001b, 2001c). These reports can be obtained from the Web address www.ipcc.ch.

Other reports have been published that contain our state-of-the-art knowledge on the topic of global climate change and water resources. Gleick (1993) edited the book *Water Crisis: A Guide to the World's Fresh Water Resources*, in which he and several authors presented summaries on several freshwater resources topics related to our resources on earth. In particular, this work presented in one place many tables of information and data related to water resources around the world. The Conference on Climate and Water held in Helsinki, Finland in September 1989 and the Second International Conference on Climate and Water held in Espoo, Finland in August 1998 provide other sources of information on this topic, particularly for the European region. The December 1999 and April 2000 issues of the *Journal of the American Water Resources Association* were devoted to the topic of water resources and climate change. Gleick (2000) and several authors of the National Water Assessment Group presented a report for the U.S. Global Change Group.

The Climate System

The *climate system* (refer to Fig. 8.1) as defined by the IPCC (2001a) is "an interactive system consisting of five major components: the atmosphere, the hydrosphere, the cryosphere, the land surface, and the biosphere, forced or influenced by various external forcing mechanisms, the most important of which is the Sun." The effect of human activities on the climate system is considered as external forcing. The sun provides the ultimate source of energy that drives the climate system.

The *atmosphere*, which is the most unstable and rapidly changing part of the system, is composed (earth's dry atmosphere) mainly of nitrogen (78.1 percent), oxygen (20.9 percent), and argon (0.93 percent). The atmospheric distribution of ozone in the lowest part of the atmosphere, the troposphere, and the lower stratosphere acts as greenhouse gas. (There are five basic layers in the atmosphere: troposphere, stratosphere, mesosphere, thermosphere, and exosphere.) In the higher part of the stratosphere there is a natural layer of high ozone concentration. This ozone absorbs ultraviolet radiation, playing an essential role in the stratosphere's radiation balance, and at the same time filters out this potentially damaging form of radiation.

The *hydrosphere* comprises all liquid surface water and subterranean water, both fresh water, including rivers, lakes, and aquifers, and saline water of the oceans and seas. The *cryosphere*, which includes the ice sheets of Greenland and Antarctia, continental glaciers and snow fields, sea ice, and permafrost, derives its importance to the climate system from its reflectivity (*albedo*) for solar radiation, its low thermal conductivity, and its large thermal

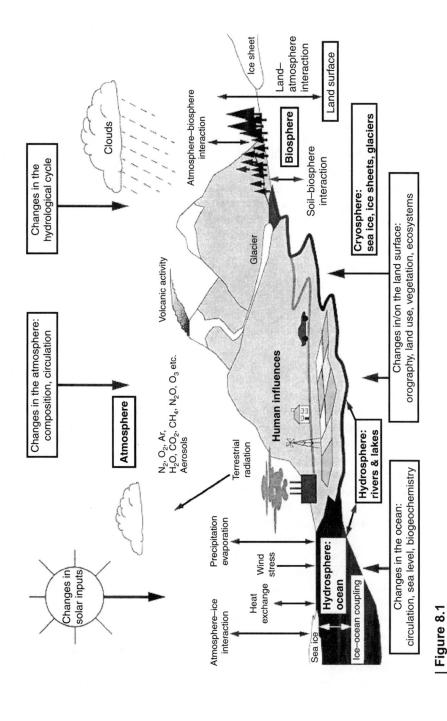

Figure 8.1

Schematic view of the components of the global climate system (bold), showing their processes and inter-actions (thin arrows) and some aspects that may change (bold arrows). (*Source: IPCC, 2001a.*)

inertia. The cryosphere has a critical role in driving the deep-water circulation. Ice sheets store a large volume of water so that variations in their volume are a potential source of sea-level variations. *Marine* and *terrestrial biospheres* have major impacts on the atmosphere's composition through *biota*, influencing the uptake and release of greenhouse gases, and through the *photosynthetic process*, in which both marine and terrestrial plants (especially forests) store significant amounts of carbon from carbon dioxide. Interaction processes (physical, chemical, and biological) occur among the various components of the climate system on a wide range of space and time scales. This makes the climate system, as shown Fig. 8.1, extremely complex.

The mean energy balance for the earth is illustrated in Fig. 8.2, which shows on the left-hand side what happens with the incoming solar radiation and on the right-hand side how the atmosphere emits the outgoing infrared radiation. A *stable climate* must have a balance between the incoming radiation and the outgoing radiation emitted by the climate system. On average the climate system must radiate 235 W/m² back into space. Climate variations may result not only from the radiative forcing but also from internal interactions among components of the climate system; therefore, a distinction is made.

Definition of Climate Change

The earth's temperature is affected by numerous causes, including (1) the incoming solar radiation that is absorbed by the atmosphere and the earth's surface, (2) the characteristics (emissivity) of the matter that absorbs the radiation, and (3) the fact that part of the longwave radiation emitted by the surface is absorbed by the atmosphere, and is then reemitted as longwave radiation either in the upward or downward direction.

The *greenhouse effect* is caused by the net change of the internal radiation balance of the atmosphere due to the continued increased emission of greenhouse gases, resulting in both the atmosphere and the earth's surface being warmer (Mitchell, 1989; Mitchell et al., 1995). Magnitude of the greenhouse effect is dependent on the composition of the atmosphere, the most important factors of which are the concentrations of water vapor and carbon dioxide, and less importantly on certain trace gases such as methane. There is mounting evidence that global warming is under way (Houghton et al., 1990; Levine, 1992; Mitchell et al., 1995; Santer et al., 1996; Tett et al., 1996; Harris and Chapman, 1997; Kaufmann and Stern, 1997). Scientists agree that a global temperature rise of 1 to 2°C is likely by 2050; however, a precise knowledge of the global-average temperature change is of little predictive value for water resources. On a global scale the warming would result in increased evaporation and increased precipitation. What is more important is how the temperature will change regionally and seasonally, affecting precipitation and runoff, which is by no means clear. The articles by Karl et al. (1995) and Karl and Quayle (1988) on climate change may be of interest to the reader.

In general, the hydrologic effects are likely to influence water storage patterns throughout the hydrologic cycle and impact the exchange among aquifers, streams, rivers, and lakes. Chalecki and Gleick (1999) provided a bibliography of the impacts of climate change on water resources in the United States. Climate change can have a significant effect on the organisms living in these aquatic systems (see Fisher and Grimm, 1996; Firth and Fisher, 1992). Other literature concerning lakes, ecosystems, and other related topics includes articles by Burkett and Kusler (2000), Cohon (1987), and Hostetler and Small (1999).

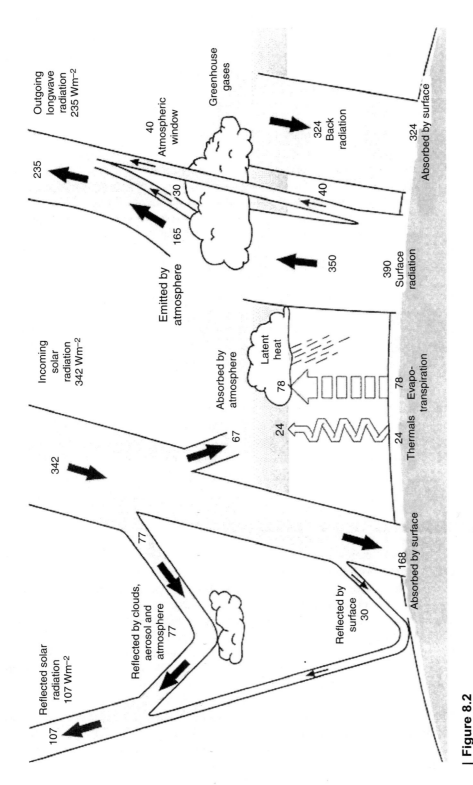

Figure 8.2

The earth's annual and global mean energy balance. Of the incoming solar radiation, 49 percent (168 W/m²) is absorbed by the surface. That heat is returned to the atmosphere as sensible heat, evapotranspiration (latent heat), and thermal infrared radiation. Most of this radiation is absorbed by the atmosphere, which in turn emits radiation both up and down. The radiation lost to space comes from cloud tops and atmospheric regions much colder than the surface. This causes a greenhouse effect. [Source: Kiehl and Trenberth (1997), as presented in IPCC (2001a).]

163

Climate Change Prediction

Climate change predictions are based mostly on computer simulations using *general circulation models* (GCMs) of the atmosphere. These models solve the conservation equations (in discrete space and time) that describe the geophysical fluid dynamics of the atmosphere. GCMs have the same general structure as *numerical weather prediction models* (NWPMs) that are used for weather prediction; however, they have a much larger spatial distribution and are run for much longer time periods. Scenarios of changes in atmospheric variables (such as temperature, precipitation, and evapotranspiration) that effect the surface and subsurface hydrology and determined using methods that range from prescribing hypothetical climate conditions to the use of GCMs. Lettenmaier et al. (1996) discussed the advantages and disadvantages of using three methods: historical and spatial analogs, output from GCMs, and prescribed climate change scenarios. They further discussed the various types of climate change scenarios based on GCM simulations that include scenarios based directly on GCM simulations, ratio and difference methods, scenarios based on GCM atmospheric circulation patterns, stochastic downscaling models, and nested models. They felt that perhaps the most useful scenarios of future climatic conditions are those developed by using a combination of methods.

Figure 8.3 shows the results of models that were used to make projections of atmospheric concentrations of greenhouse gases and aerosols, and future climate, based on emissions from the IPCC Special Report on Emission Scenarios (SRES) as discussed in the IPCC (2001a) report. A great deal of uncertainty lies in the estimate of climate change, which is not focused on in this chapter. Refer to Dickinson (1989) and the IPCC reports for more information on this topic.

Droughts

Droughts can be classified into different types: *meteorological droughts*, which refer to lack of precipitation; *agricultural droughts*, which refer to lack of soil moisture; and *hydrological droughts*, which refer to reduced streamflow or groundwater levels. *Drought analysis* is used to characterize the magnitude, duration, and severity of the respective type of drought in a region of interest. Figure 8.4 illustrates the progression of droughts and their impacts over a standard 30-year climatic period.

CLIMATE CHANGE EFFECTS

Hydrologic Effects

Evaporation and Transpiration

The two main factors influencing evaporation from an open water surface are the supply of energy to provide the latent heat of vaporization and the ability to transport the vapor away from the evaporative source (see Chow et al., 1988). Solar radiation is the main source of heat energy, and the ability to transport vapor away from the evaporative surface depends on the wind velocity over the surface and the specific humidity gradient in the air above it. As temperatures rise, the availability of energy for evaporation increases and the atmospheric demand for water in the hydrologic cycle increases.

Evaporation from the land surface includes evaporation directly from the soil and vegetation surface, and transpiration through plant leaves, in which water is extracted by the plant's roots, transported upward through its stem, and diffused into the atmosphere through tiny openings in the leaves called *stomata* (see Chow et al., 1988). The processes

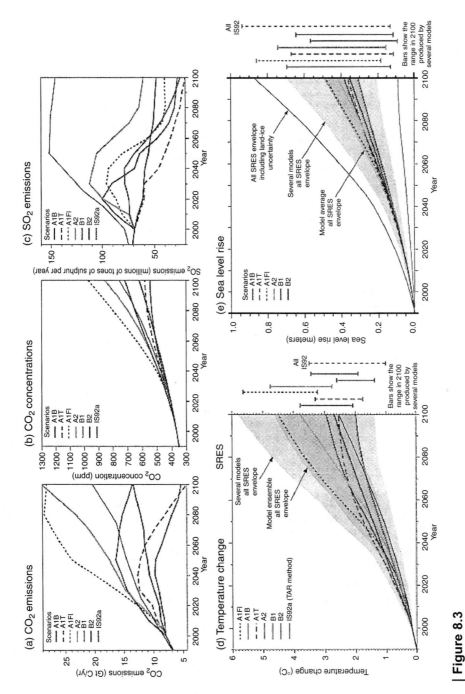

Figure 8.3

The global climate of the twenty-first century will depend on natural changes and the response of the climate system to human activities. (*IPCC, 2001a.*)

Figure 8.4

Progression of droughts and their impacts. (*Source: www.drought.unl. edu/whatis/ concept.htm.*)

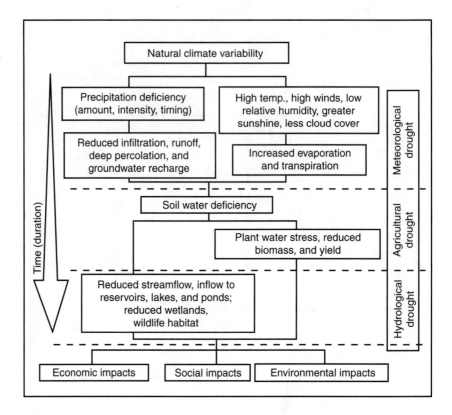

of evaporation from the land surface and transpiration from vegetation collectively are termed *evapotranspiration,* which is influenced by the two factors described previously for open water evaporation, and a third factor, the supply of moisture at the evaporative surface. Potential evapotranspiration is that which occurs from a well-vegetated surface when moisture supply is not limiting. The actual evapotranspiration drops below its potential level as the soil dries out.

Soil Moisture A major part of the hydrologic cycle is subsurface water flow, which can be discussed as three major processes; infiltration of surface water into the soil to become soil moisture, subsurface (unsaturated) flow through the soil, and groundwater (saturated) flow through soil or rock strata. Climate changes that alter the precipitation patterns and the evapotranspiration regime will directly affect soil moisture storage, runoff processes, and groundwater recharge dynamics (see Rind et al., 1990; Vinnikov et al., 1996). Decreases in precipitation may significantly decrease soil moisture. In regions where precipitation increases along with increases in evaporation as a result of higher temperatures, or the timing of the precipitation or runoff changes, the soil moisture on the average or over certain periods may decrease. In regions where precipitation increases significantly, soil moisture will probably increase and in some locations may significantly increase.

Sensitivity studies by Gregory et al. (1997) reported reduced soil-moisture conditions in midlatitude summers in the Northern Hemisphere as temperature and evaporation rise and winter snow cover and spring runoff decline. Wetherald and Manabe (1999) investigated the temporal spatial variation of soil moisture in a coupled ocean-atmosphere model, with results that showed both summer soil moisture dryness and winter wetness in middle and high latitudes of North America and southern Europe. The study results showed a large percentage reduction of soil moisture in summer and a soil moisture decrease for nearly the entire year in response to warming.

Snowfall and Snowmelt   Temperature increases are expected to be greater in the higher-latitude regions because of the dynamics of the atmosphere and feedbacks among ice, albedo, and radiation. Research has established that higher temperatures will lead to dramatic changes in snowfall and snowmelt dynamics in watersheds with substantial snow. Higher temperatures will have several major effects, including (Gleick, 2000):

- Increase in the ratio of rain to snow
- Delay in the onset of the snow season
- Accelerated rate of spring snowmelt
- Shortening of the overall snowfall season
- More rapid and earlier seasonal runoff
- Significant changes in the distribution of permafrost and mass balance of glaciers

One of the most persistent and well-established findings on the impacts of climate change is that higher temperatures will affect the timing and magnitude of runoff in watersheds with substantial snow dynamics. Both climate models and theoretical studies of snow dynamics have long projected that higher temperatures will lead to a decrease in the extent of snow cover in the Northern Hemisphere. More recently, field studies have corroborated these findings.

Storm Frequency and Intensity   In general, global climate change will result in a wetter world. Many model estimates indicate that for much of the northern midlatitude continents, summer rainfall will decrease slightly and winter precipitation will increase. In other studies precipitation changes in midlatitudes are highly variable and uncertain. Unfortunately, general circulation models poorly reproduce detailed precipitation patterns. As pointed out by Gleick (2000), potential changes in rainfall intensity and variability are difficult to evaluate because intense convective storms tend to occur over regions smaller than global models are able to resolve.

Runoff: Floods and Droughts   A major challenge for the hydrologic community in the study of climate change is to establish the linkage between local-scale and global-scale processes.

Wigley and Jones (1985) used a simple water balance model to conclude that

- Changes in runoff are more sensitive to changes in precipitation than to changes in evaporation.
- Relative change in runoff is always greater than that in precipitation.
- Runoff is most sensitive to climate changes in arid and semiarid regions.
- Relative change in runoff exceeds the relative change in evapotranspiration only in regions where the *runoff ratio w* (ratio of long-term average runoff to long-term average precipitation) is less than 0.5.

- Overall it is expected that there will be very large increases in average runoff in response to predicted warming, unless there is a compensating large increase in land evapotranspiration.
- Higher surface temperatures will increase global precipitation in the range of 3 to 11 percent, probably leading to even greater relative increases in runoff.

Loaiciga et al. (1996) reviewed the findings and limitations of predictions of hydrologic responses to global warming.

Streamflow Variability Karl and Riebsame (1989) used historical data from the United States with a water balance model to study the sensitivity of streamflow to changes in temperature and precipitation. Their study concluded that a temperature change of 1 to 2°C typically had little effect on streamflow. On the other hand, a relative change in precipitation magnified to a one- to six-fold change in relative streamflows.

Schaake (1990) and Nemec and Schaake (1982) showed that on an average basis (e.g., using mean annual precipitation, evaporation, and runoff) that the *elasticity of runoff* with respect to precipitation is greater than one, and is larger relative to the *long-term runoff ratio*, which is less in arid than in humid climates.

Because streamflow is climatically determined and consequently highly variable, it is instructive to look at historical data on streamflow variability in order to develop a context for detecting and predicting the effects of climate change on the hydrologic cycle. Dingman (2002) illustrates that there is a much higher relative variability of streamflow in arid regions (upper and lower plains and the south-western United States) than in humid regions (northeastern and northwestern regions). Also evident is that the time series of river discharge shows considerable synchronism over the United States.

Long records of flow in large rivers such as the Nile show evidence of the climate-related persistence of streamflow variability. The Nile River has the longest streamflow record in the world, with information available from 622 to 1520 and from 1700 to the present. Riehl and Meitin (1979) discussed three contrasting patterns in the record: (1) the period from 622 to about 950 had periods of high flow alternated with periods of low flow, with these cycles lasting from 50 to 90 years with moderate amplitude; (2) the period from 950 to 1225 had no major trends; and (3) the period from 1700 to present had alternating periods of high and low flow, having cycles of 100 to 180 years with much higher amplitudes than during the period from 622 to 950. These periods of variability appear to be linked to global climatic fluctuations. Eltahir (1996) found that the Nile River flows were influenced by the El Niño—Southern Oscillation (ENSO) cycle. Richey et al. (1989) studied the Amazon River record (1903–1985) and found no indication of climate or land-use change over that period; however, they did find a 2- to 3-year period of declining flow following the warm phase of the ENSO cycle. They found that periods of high flow were coincident with the ENSO cold phase.

Groundwater The effects of climate change on groundwater sustainability include those mentioned by Alley et al. (1999):

- Changes in groundwater recharge resulting from changes in average precipitation and temperature or in seasonal distribution of precipitation
- More severe and longer-lasting droughts
- Changes in evapotranspiration resulting from changes in vegetation
- Possible increased demands of groundwater as a backup source of water supply

Surficial aquifers are likely to be part of the groundwater system, which is most sensitive to climate. These aquifers supply much the flow to streams, lakes, wetlands, and springs. Because groundwater systems tend to respond more slowly to short-term variability in climate conditions than do surface water systems, the assessment of groundwater resources and related model simulations are based on average conditions, such as annual recharge and/or average annual discharge to streams. The use of average conditions may underestimate the importance of droughts (Alley et al., 1999).

The impacts of climate change on (1) specific groundwater basins, (2) the general groundwater recharge characteristics, and (3) groundwater quality have received little attention in the literature. Vaccaro (1992) addressed the climate sensitivity of groundwater recharge, finding that a warmer climate (doubling CO_2) resulted in a relatively small sensitivity to recharge and depended on land use. Sandstorm (1995) studied a semiarid basin in Africa and concluded that a 15 percent reduction in rainfall could lead to a 45 percent reduction in groundwater recharge. Sharma (1989) and Green et al. (1997) reported similar sensitivities of the effects of climate change on groundwater in Australia. Panagoulia and Dimou (1996) studied the effect of climate change on groundwater–streamflow interactions in a mountainous basin in central Greece. They realized large impacts in the spring and summer months as a result of temperature-induced changes in snowfall and snowmelt patterns. Oberdorfer (1996) looked at the impacts of climate change on groundwater discharge to the ocean using a simple water balance model to study the effect of changes in recharge rates and sea level on groundwater resources and flows in a California coastal watershed. The impacts of sea-level rise on groundwater will include increased intrusion of salt water into coastal aquifers.

Water Resource System Effects Studies that have looked at the impact of climate change on water resource management include those by Lettenmaier et al. (1999), Kirshen and Fennessey (1993), Lettenmaier and Sheer (1991), Nash and Gleick (1991), and Nemec and Schaake (1982).

A broad assessment was performed by Lettenmaier et al. (1999) of the sensitivity to climate change of six major water systems in the United States: Columbia River basin, Missouri River basin, Savannah River basin, Apalachicola-Chattahooche-Flint (ACF) River basin, Boston water supply system, and Tacoma water supply system. The investigators evaluated *system reliability* (percentage of time that the system operates without failure), the *system resiliency* (ability of a system to recover from failure), and *system vulnerability* (average severity of failure). Some thoughts or conclusions from this study were

- The most important factors determining climate sensitivity of system performance were changes in runoff, even when direct effects of climate change on water demands were affected.

- Sensitivities depended on the purpose for which water was needed and the priority given to those uses.

- Higher temperatures increased system use in many basins, but these increases tended to be modest. Effects of higher temperatures on system reliability were similar.

- Influence of long-term demand on system performance had a greater impact than did climate change when long-term withdrawals were projected to substantially grow.

Regional assessments can focus on three major questions (IPCC, 1996b; Miles et al., 2000):

1. How sensitive is the region to climate variability?

2. How adaptable is the region to climate variability and change?

3. How vulnerable is the region to climate variability and change?

Sensitivity is the degree to which a system will respond to a change in climatic conditions (IPCC, 1996b). *Adaptability* refers to the degree to which adjustments in systems' practices, processes, or structures to projected or actual changes of climate are possible (IPCC, 1996b), *Vulnerability* defines the extent to which climate change may damage or harm a system (IPCC, 1996b). Vulnerability depends not only on a system's sensitivity but also on its ability to adapt to new climatic conditions. Major (1998) discussed the role of risk management methods for climate change and water resources.

River Basin/ Regional Runoff Effects

Numerous studies of the effect of climate change on various river basins have been reported in the literature. These include those by Ayers et al. (1993), Cohon (1991), Georgakakos et al. (1998), Hamlet and Lettenmaier (1999), Hay et al. (2000), Hotchkiss et al. (2000), Hurd et al. (1999), Kaczmarek et al. (1996a, 1996b), Kirshen and Fennessey (1993), Lettenmaier and Gan (1990), Leung and Wigmosta (1999), McCabe and Ayers (1989), Miles et al. (2000), Nash and Gleick (1991), Stone et al. (2001), and many others.

Stone et al. (2001) studied the impacts of climate change on the Missouri River basin water yield (Fig. 8.5). They considered the effect of doubling the atmospheric carbon dioxide on water yield using a Regional Climate Model (RegCM) and the Soil and Water Assessment (SWAT) model. Results of this analysis include the following (Stone et al., 2001):

- Precipitation and temperature both increase more in the northern part of the basin, thus resulting in the greatest changes in the predicted water yields.

- Water yields (modeled) for the spring, summer, and fall seasons showed more distinct spatial trends than water yields for the winter months. SWAT simulation results exhibited dramatic increases (70 percent and much more in some subbasins) in water yields across the northern and northwestern portions of the Missouri River basin in the spring, summer, and fall.

- Southeastern portions of the basin displayed an overall decrease (most decreased by less than 20 percent, but some decreased by more than 80 percent) in water yields during these three seasons. During the spring and summer the decreased water yields in the southeastern portions are sufficient to lower the total water yield of the entire basin by 10 to 20 percent.

- Water balance summaries for the principal tributaries indicate the large increase of runoff in the northern subbasins and decreased runoff in the southern subbasins.

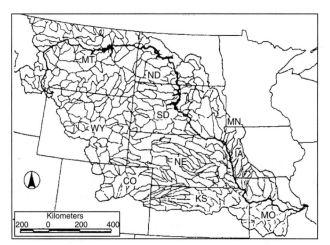

Figure 8.5
Missouri River basin. (*Stone et al., 2001.*)

(a) The Missouri river basin separated into the 310 USGS eight-digit subbasin

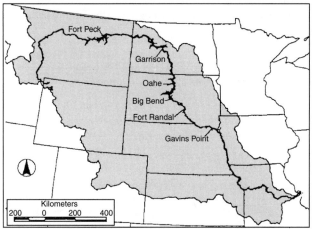

(b) The six Missouri river main stem reservoirs.

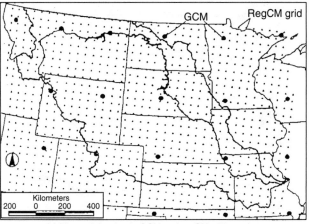

(c) Missouri river basin GCM and regCM grid spacing.

- Six main stem reservoirs are operated in the upper river basin and control 55 percent of the area upstream of the reservoirs with the storage capacity of three times the mean annual runoff from the area. Total main stem reservoir storage changed dramatically from the baseline conditions, doubling the average annual releases, enough additional water under doubling CO_2 conditions to provide 24 cm of additional water annually for all irrigated areas in the Missouri River basin.

Hamlet and Lettenmaier (1999) and Miles et al. (2000) studied the impact of climate change on the hydrology and water resources of the Columbia River basin. The Columbia River system and the major dam locations are shown in Fig. 8.6. Figure 8.7 illustrates the dominant impact pathway through which changes in regional climate are manifested in the basin. Some of the major findings from Hamlet and Lettenmaier (1999) are outlined below:

- Hydrologic response to climate change will tend to shift flow volumes from the summer to the winter.
- Management of the system to compensate for this effect would require that water be stored in the winter and released in the summer, which is basically the opposite to what is done now.

Figure 8.6

The Columbia River basin and major dams. The gray shaded area defines the drainage basin upstream of the Dalles Dam in Oregon. Triangles represent major storage reservoirs (size of icon represents relative storage capacity), and circles are major run-of-river projects. (*Hamlet and Lettenmaier, 1999.*)

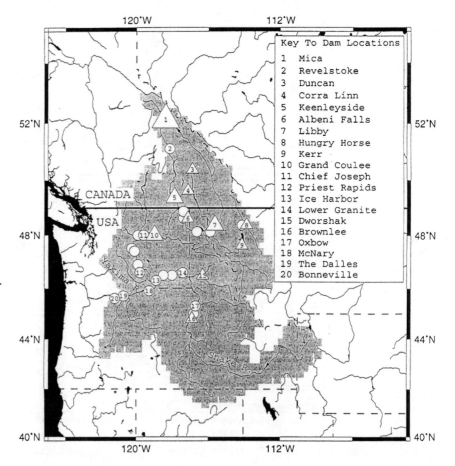

Key To Dam Locations

1. Mica
2. Revelstoke
3. Duncan
4. Corra Linn
5. Keenleyside
6. Albeni Falls
7. Libby
8. Hungry Horse
9. Kerr
10. Grand Coulee
11. Chief Joseph
12. Priest Rapids
13. Ice Harbor
14. Lower Granite
15. Dworshak
16. Brownlee
17. Oxbow
18. McNary
19. The Dalles
20. Bonneville

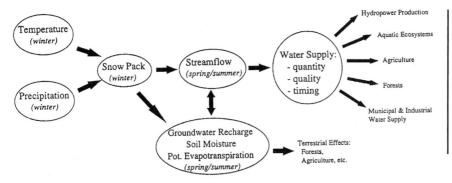

Figure 8.7

The dominant impact of the pathway through which changes in regional climate are manifested in the Pacific Northwest. (*Miles et al., 2000.*)

• Energy production, however, has its greatest economic value during the winter months. The result would be potential reductions in the required spring/summer flood storage and a shift in primary hydropower production to the summer.

Summary of Regional Runoff Effects Found in the United States

Arid and Semiarid Western Regions   Relatively modest changes in precipitation can have proportionately larger impacts on runoff, and higher temperatures result in higher evaporation rates, reduced streamflows, and increased frequency of droughts.

Cold and Cool Temperate Zones   Where a large portion of annual runoff comes from snowmelt, the major effect of warming is the change in timing of streamflow, both timing and intensity of the peaks. A declining proportion of the total precipitation occurs as snow with rising temperatures, and the remaining snow melts sooner and faster in the spring, resulting in more winter runoff. In some basins there may be increases in spring peak runoff, and in other basins runoff volumes may significantly shift to winter months.

Warmer Climates   In the southern region of the United States, where seasonal cycles of rainfall and evaporation are predominant, runoff is affected much more significantly by total precipitation with seasonal cycles of rainfall and precipitation.

WATER MANAGEMENT OPTIONS

Table 8.1 summarizes some supply-side and demand-side adaptive options for various water-use sectors.

UNCERTAINTIES

As pointed out by many investigators, the limitations of state-of-the-art climate models are the primary sources of uncertainty in the experiments that study the hydrologic and water resources impact of climate change. Future improvement to the climate models, hopefully resulting in more accurate regional predictions, should greatly improve the types of experiments to more accurately define the hydrologic and hydraulic impacts of climate change.

Table 8.1

Supply-Side and Demand-Side Adaptive Options for Municipal Water Supply

Supply-Side		Demand-Side	
Option	Comments	Option	Comments
Increase reservoir capacity	Expensive; potential environmental impact	Incentives to use less (e.g., through pricing)	Possibly limited opportunity; needs institutional framework
Extract more from rivers or groundwater	Potential environmental impact	Legally enforceable water standards (e.g., for appliances)	Potential political impact; usually cost-inefficient
Alter system operating rules	Possibly limited opportunity	Increase use of gray water turbines; encourage energy efficiency	Potentially expensive
Interbasin transfer	Expensive; potential environmental impact; may not be feasible	Reduce leakage	Potentially expensive to reduce to very low levels, especially in old systems
Desalination	Expensive; potential environmental impact	Development of non-water-based sanitation systems	Possibly too technically advanced for wide application

Source: IPCC (2001b).

Future precipitation and temperature are the primary drivers for determining future hydrologic response. Because of the uncertainties of the predictions of the future precipitation and temperatures, the hydrologic responses of various river basins are uncertain, resulting in uncertainties of our future hydraulic resources.

REFERENCES

Alley, W. M., T. E. Reilly, and O. L. Franke, *Sustainability of Ground-Water Resources,* U.S. Geological Survey Circular 1186, U.S. Geological Survey, Denver Colorado, 1999.

Ayers, M. A., D. M. Wolock, G. J. McCabe, L. E. Hay, and G. D. Tasker, *Sensitivity of Water Resources in the Delaware River Basin to Climate Variability and Change,* U.S. Geological Survey Open-File Report 02-52, 1993.

Burkett, V., and J. Kusler, "Climate Change: Potential Impacts and Interactions in Wetlands of the United States," *Journal of the American Water Resources Association,* 33:313–320, 2000.

Chalecki, L. H., and P. H. Gleick, "A Comprehensive Bibliography of the Impacts of Climate Change and Variability on Water Resources of the United States," *Journal of the American Water Resources Association,* 35:1657–1665, 1999.

Chow, V. T., D. R. Maidment, and L. W. Mays, *Applied Hydrology,* McGraw-Hill, New York, 1988.

Cohon, S. J., "Influences of Past and Future Climates on the Great Lakes Region of North America," *Water International,* 12:163–169, 1987.

Cohon, S. J., "Possible Impacts of Climatic Warming Scenarios on Water Resources in the Saskatchewan River Sub-Basin, Canada," *Climatic Change,* 19:291–317, 1991.

Dickinson, R. E., "Uncertainties of Estimates of Climate Change: A Review," *Climate Change,* 15:5–14, 1989.

Dingman, S. L., *Physical Hydrology,* 2d ed., Prentice-Hall, Upper Saddle River, N.J., 2002.

Eltahir, E. A. B., "El Niño and the Natural Variability of the Flow of the Nile River," *Water Resources Research,* 32:131–137, 1996.

Falkenmark, M., and G. Lindh, "Water and Economic Development," in P. H. Gleick (ed.), *Water in Crisis: A Guide to the World's Fresh Water Resources,* Oxford University Press, Oxford, U.K., 1993.

Firth, P., and S. G. Fisher (eds.), *Climate Change and Freshwater Ecosystems,* Springer-Verlag, New York, 1992.

Fisher, S. G., and N. B. Grimm, "Ecological Effects of Global Climate Change on Freshwater Ecosystems with Emphasis on Streams and Rivers," in L. W. Mays (ed.), *Water Resources Handbook,* McGraw-Hill, New York, 1996.

Georgakakos, A., H. Yao, M. Mullusky, and K. Georgakakos, "Impacts of Climate Variability on the Operational Forecast and Management of the Upper Des Moines River Basin," *Water Resources Research,* 34:799–821, 1998.

Gleick, P. H. (ed.), *Water in Crisis: A Guide to the World's Fresh Water Resources,* Oxford University Press, Oxford, U.K., 1993.

Gleick, P. H. (lead author), *Water: The Potential Consequences of Climate Variability and Change for the Water Resources of the United States,* a report of the National Water Assessment Group, for the U.S. Global Change Research Program, Pacific Institute for Studies in Development, Environment, and Security, Oakland, Calif., September 2000.

Green, T. R., B. C. Bates, P. M. Fleming, S. P. Charles, and M. Taniguchi, "Simulated Impacts of Climate Change on Groundwater Recharge in the Subtropics of Queensland, Australia," *Subsurface Hydrological Responses to Land Cover and Land Use Changes,* Kluwer Academic Publishers, Norwell, Mass., pp. 187–204, 1997.

Gregory, J. M., J. F. B. Mitchell, and A. J. Brady, "Summer Drought in Northern Midlatitudes in a Time-Dependent CO_2 Climate Experiment," *Journal of Climate,* 10:662–686, 1997.

Hamlet, A. F., and D. P. Lettenmaier, "Effects of Climate Change on Hydrology and Water Resources Objectives in the Columbia River Basin," *Journal of the American Water Resources Association,* 35:1597–1624, 1999.

Harris, R. N., and D. S. Chapman, "Borehole Temperatures and a Baseline for 20th-Century Global Warming Estimates," *Science,* 275:1618–1621, 1997.

Hay, L. E., R. L. Wilby, and G. H. Leavesley, "A Comparison of Delta Change and Downscaled GCM Scenarios for Three Mountainous Basins in the United States," *Journal of the American Water Resources Association,* 36:387–398, 2000.

Hostetler, S. W., and E. E. Small, "Response of North American Freshwater Lakes to Simulated Future Climates," *Journal of the American Water Resources Association,* 35:1625–1638, 1999.

Hotchkiss, R. H., S. F. Jorgensen, M. C. Stone, and T. A. Fontaines, "Regulated River Modeling for Climate Impacts Assessments: The Missouri River," *Journal of the American Water Resources Association,* 36:375–386, 2000.

Houghton, J. T., G. J. Jenkins, and J. J. Ephraums (eds.), *Climate Change: The IPCC Scientific Assessment,* Cambridge University Press, Cambridge, U.K., 1990.

Hurd, B., N. Leary, R. Jones, and J. Smith, "Relative Regional Vulnerability of Water Resources to Climate Change," *Journal of the American Water Resources Association,* 35:1399–1410, 1999.

Intergovernmental Panel on Climate Change (IPCC), *Climate Change: The IPCC Scientific Assessment,* J. T. Houghton, G. J. Jenkins, and J. J. Ephrauns (eds.), Cambridge University Press, Cambridge, U.K., 1990.

Intergovernmental Panel on *Climate Change (IPCC), Climate Change 1995: The Science of Climate Change: Contribution of Working Group I to the Second Assessment Report of the Intergovernmental Panel on Climate Change,* Cambridge University Press, New York, 1996a.

Intergovernmental Panel on Climate Change (IPCC), *Climate Change 1995: Impacts, Adaptations, and Mitigation of Climate Change: Scientific-Technical Analysis: Contribution of Working Group II to the Second Assessment Report of the Intergovernmental Panel on Climate Change,* Cambridge University Press, New York, 1996b.

Intergovernmental Panel on Climate Change (IPCC), "Hydrology and Freshwater Ecology," in *Climate Change 1995: Impacts, Adaptations, and Mitigation of Climate Change: Contribution of Working Group II to the Second Assessment Report of the Intergovernmental Panel on Climate Change,* Cambridge University Press, New York, 1996c.

Intergovernmental Panel on Climate Change (IPCC), *Regional Impacts of Climate Change: An Assessment of Vulnerability,* Cambridge University Press, New York, 1998.

Intergovernmental Panel on Climate Change (IPCC), *Climate Change 2001: The Scientific Basis,* www.ipcc.ch. 2001a.

Intergovernmental Panel on Climate Change (IPCC), *Climate Change 2001: Impacts, Adaptation, and Vulnerability,* www.ipcc.ch, 2001b.

Intergovernmental Panel on Climate Change (IPCC), *Climate Change 2001: Mitigation,* www.ipcc.ch, 2001c.

Kaczmarek, Z., N. W. Arnell, and E. Z. Stakhiv, "Water Resources Management," Chap. 10 in Intergovernmental Panel on Climate Change (IPCC), *Second Assessment Report,* Cambridge University Press, 1996a.

Kaczmarek, Z., J. J. Napiorkowski, and K. M. Strzepek, "Climate Change Impact on the Water Supply System in the Warta River Catchment," *Water Resources Development,* 12(2):165–180, 1996b.

Karl, T. R., and R. G. Quayle, "Climate Change in Fact and Theory: Are We Collecting the Facts?" *Climate Change,* 13:5–17, 1988.

Karl, T. R., and W. E. Riebsame, "The Impact of Decadal Fluctuations in Mean Precipitation and Temperature on Runoff: A Sensitivity Study over the United States," *Climate Change,* 15:423–447, 1989.

Karl, T. R., R. W. Knight, and N. Plummer, "Trends in High Frequency Climate Variability in the Twentieth Century," *Nature,* 377:217–220, 1995.

Kaufmann, R. K., and D. I. Stern, "Evidence for Human Influence on Climate from Hemispheric Temperature Relations," *Nature,* 388:39–44, 1997.

Kiehl, J. T., and K. E. Trenberth, "Earth's Annual Global mean Energy Budget," *Bulletin American Meteorological Society,* 78:197–208, 1997.

Kirshen, P. H., and N. M. Fennessy, *Potential Impacts of Climate Change upon the Water Supply of the Boston Metropolitan Area,* U.S. Environmental Protection Agency, 1993.

Lettenmaier, D. P., and T. Y. Gan, "Hydrologic Sensitivities of the Sacramento–San Joaquin River Basin, California, to Global Warming," *Water Resources Research,* 26:69–86, 1990.

Lettenmaier, D. P., and D. P. Sheer, "Climatic Sensitivity of California Water Resources," *Journal of Water Resources Planning and Management,* 117(1):108–125, 1991.

Lettenmaier, D. P., G. McCabe, and E. Z. Stakhiv, "Global Climate Change: Effect on Hydrologic Cycle," in L. W. Mays (ed.), *Water Resources Handbook,* McGraw-Hill, New York, 1996.

Lettenmaier, D. P., A. W. Wood, R. N. Palmer, E. F. Wood, and E. Z. Stakhiv, "Water Resources Implications of Global Warming: A U.S. Regional Perspective," *Climate Change,* 43(3): 537–579, 1999.

Leung, L. R., and M. S. Wigmosta, "Potential Climate Change Impacts on Mountain Watersheds in the Pacific Northwest," *Journal of the American Water Resources Association,* 36(6): 1463–1471, 1999.

Levine, J. S., "Global Climate Change," in P. Firth and S. Fisher (eds.), *Climate Change and Freshwater Ecosystems,* Springer-Verlag, New York, 1992.

Loaiciga, H. A. et al., "Global Warming and the Hydrologic Cycle," *Journal of Hydrology,* 174:83–127, 1996.

Major, D. C., "Climate Change and Water Resources: The Role of Risk Management Methods," *Water Resources Update,* 112:47–50, 1998.

McCabe, G. J., and M. A. Ayers, "Hydrologic Effects of Climate Change in the Delaware River Basin," *Water Resources Bulletin,* 25(6):1231–1242, 1989.

Miles, E. L., A. K. Snover, A. F. Hamlet, B. Callahan, and D. Fluharty, "Pacific Northwest Regional Assessment: Impacts of the Climate Variability and Climate Change on the Water Resources of the Columbia River Basin," *Journal of the American Water Resources Association,* 36(2):399–420, April 2000.

Mitchell, J. F. B., "The 'Greenhouse' Effect and Climate Change," *Reviews of Geophysics,* 27:115–139, 1989.

Mitchell, J. F. B., T. C. Johns, J. M. Gregory, and S. F. B. Tett, "Climate Response to Increasing Levels of Greenhouse Gases and Sulphate Aerosols," *Nature,* 376:501–504, 1995.

Nash, L. L., and P. H. Gleick, "Sensitivity of Streamflow in the Colorado Basin to Climatic Changes," *Journal of Hydrology,* 125:221–241, 1991.

Nemec, J., and J. C. Schaake, "Sensitivity of Water Resource Systems to Climate Variation," *Journal of Hydrological Sciences,* 27:327–343, 1982.

Oberdorfer, J. A., "Numerical Modeling of Coastal Discharge: Predicting the Effect of Climate Change," in R. W. Buddemeier (ed.), *Groundwater Discharge in the Coastal Zone: Proceedings of an International Symposium,* Moscow, Russia, pp. 85–91, July 1996.

Panagoulia, D., and G. Dimou, Sensitivities of Groundwater-Streamflow Interactions to Global Climate Changes, *Hydrological Sciences Journal,* 41:781–796, 1996.

Richey, J. E., C. Nobre, and C. Deser, "Amazon River Discharge and Climate Variability: 1903 to 1985," *Science,* 246:101–103, 1989.

Riehl, H., and J. Meitin, "Discharge of the Nile River; A Barometer of Short-Period Climatic Fluction," *Science,* 206:1178–1179, 1979.

Rind, D. R., R. Goldberg, J. Hansen, C. Rosensweig, and R. Ruedy, "Potential Evapotranspiration and the Likelihood of Future Drought," *Journal of Geophysical Review,* 95:9983–10,004, 1990.

Sandstrom, K., "Modeling the Effects of Rainfall Variability on Groundwater Recharge in Semi-Arid Tanzanian," *Nordic Hydrology,* 26:313–320, 1995.

Santer, B. et al., "A Search for Human Influences on the Thermal Structure of the Atmosphere," *Nature,* 382:39–46, 1996.

Schaake, J. C., "From Climate to Flow," in P. E. Wagner (ed.), *Climate Change and U.S. Water Resources,* Wiley, New York, 1990.

Sharma, M. L., "Impact of Climate Change on Groundwater Recharge," *Proceedings of the Conference on Climate and Water,* Valton Painatuskeskus, Helsinki, Finland, vol. I:511–520, 1989.

Stone, M. C., R. H. Hotchkiss, C. M. Hubbard, T. A. Fontaine, L. O. Mearns, and J. G. Arnold, "Impacts of Climate Change on Missouri River Basin Water Yield," *Journal of the American Water Resources Association,* 37(5):1119–1129, 2001.

Tett, F. B., J. F. B. Mitchell, D. E. Parker, and M. R. Allen, "Human Influence on the Atmospheric Vertical Structure: Detection and Observations," *Science,* 274:1170–1173, 1996.

Vaccaro, J. J., "Sensitivity of Groundwater Recharge Estimates to Climate Variability and Change, Columbia Plateau, Washington," *Journal of Geophysical Research,* 97(D3):2821–2833, 1992.

Vinnikov, K. Y., A. Robock, N. A. Speranskaya, and C. A. Schlosser, "Scales of Temporal and Spatial Variability of Midlatitude Soil Moisture," *Journal of Geophysical Research,* 101:7163–7174, 1996.

Wetherald, R. T., and S. Manabe, "Detectability of Summer Dryness Caused by Greenhouse Warming," *Climate Change,* 43(3):495–522, 1999.

Wigley, T. M. L., and P. D. Jones, "Influence of Precipitation Changes and Direct CO_2 Effects on Streamflow," *Nature,* 314:149–151, 1985.

FURTHER READING

Alexandrov, V., "A Strategy Evaluation of Irrigation Management of Maize Crop under Climate Change in Bulgaria," *Proceedings of the Second International Conference on Climate and Water,* R. Lemmela and N. Helenius (eds.), Espoo, Finland, August 17–20, vol. 3, pp. 1545–1555, 1998.

American Water Works Association (WWA), "Climate Change and Water Resources, Committee Report of the AWWA Public Advisory Forum," *Journal of the American Water Works Association,* 89(11): 107–110, 1997.

Arnell, N. W., *Global Warming, River Flows and Water Resources,* Wiley, Chichester, United Kingdom, 1996.

Arnell, N. W., "The Effect of Climate Change on Hydrological Regimes in Europe: A Continental Perspective," *Global Environmental Change,* 9:5–23, 1999.

Arnold, J. G., R. Srinivasan, R. S. Muttiah, and J. R. Williams, "Large Area Hydrologic Modeling and Assessment, Part I: Model Development," *Journal of American Water Resources Association,* 34(1):73–89, 1998.

Arnold, J. G., R. Srinivasan, R. S. Muttiah, and P. M. Allen, "Continental Scale Simulation of the Hydrologic Balance," *Journal of American Water Resources Association,* 35:1037–1051, 1999.

Dvorak, V., J. Hladny, and L. Kasparek, "Climate Change Hydrology and Water Resources Impact and Adaptation for Selected River Basins in the Czech Republic," *Climate Change,* 36:93–106, 1997.

Eagleson, P. S., "The Emergence of Global-Scale Hydrology," *Water Resources Research,* 22:6S–14S, 1986.

Gleick, P. H., "Climate Change, Hydrology, and Water Resources," *Reviews of Geophysics,* 7(3):329–344, 1989.

Gore, A., *An Inconvenient Truth,* Rodale, Emmaus, PA, 2006.

Grabs, W., *Impact of Climate Change on Hydrological Regimes and Water Resources Management in the Rhine,* Cologne, Germany, 1997.

Hayden, B. P., "Climate Change and Extratropical Storminess in the United States: An Assessment," *Journal of the American Water Resources Association,* 35:1387–1398, 1999.

Herrington, P., *Climate Change and the Demand for Water,* Her Majesty's Stationary Office, London, United Kingdom, 1996.

Jacques, G. and H. Le Trent, *Climate Change,* UNESCO Publishing, Paris, 2005.

Kandel, R., *Water from Heaven,* Columbia University Press, New York, 2003.

Klein, R. J. T., J. Aston, E. N. Buckley, M. Capobianco, N. Mizutani, R. J. Nicholls, P. D. Nunn, and S. Ragoonaden, in Metz et al. (eds.), *Coastal Adaptation Technologies: IPCC Special Report on Methodological and Technological Issues in Technology Transfer,* Cambridge University Press, New York, 2000.

Kos, Z., "Sensitivity of Irrigation and water Resources Systems to Climate Change," *Journal of Water Management,* 41(4–5):247–269, 1993.

Mander, U., and A. Kull, "Impacts of Climatic Fluctuations and Land Use Change on Water Budget and Nutrient Runoff: The Porijogi," *Proceedings of the Second International Conference on Climate and Water,* R. Lemmela and N. Helenius (eds.), Espoo, Finland, August 17–20, 2:884–896, 1998.

Saelthun, N. R. et al., "Climate Change Impacts on Runoff and Hydropower in the Nordic Countries," *TemaNord,* 552:170, 1998.

Vorosmarty, C. V., C. A. Federer, and A. L. Schloss, "Potential Evaporation Functions Compared on U.S. Watersheds: Possible Implications for Global-Scale Water Balance and Terrestrial Ecosystem Modeling," *Journal of Hydrology,* 207:147–169, 1998.

Vorosmarty, C. V., P. Green, J. Salisbury, and R. B. Lammers, "The Vulnerability of Global Water Resources: Major Impacts from Climate Change or Human Development?" *Science,* 289:284–288, 2000.

Vorosmarty, C., L. Hinzman, B. Peterson, D. Bromwich, L. Hamilton, J. Morison, V. Romanovsky, M. Sturm, and R. Webb, "Artic-CHAMP: A Program to Study Artic Hydrology and Its Role in Global Change," EOS, *Transactions, American Geophysical Union,* 83(22), May 28, 2002.

Walsh, J. et al., *Enhancing NASA's Contributions to Polar Science: A Review of Polar Geophysical Data Sets,* Commission on Geosciences, Environment, and Resources, National Research Council, National Academy of Science Press, Washington, D.C., 2001.

9 WATER SUSTAINABLITY: THE POTENTIAL IMPACT OF CLIMATE CHANGE

John A. Dracup
Professor of the Graduate School
Department of Civil & Environmental Engineering
University of California
Berkeley, California 94720
USA

INTRODUCTION

The future sustainability of water becomes more complex and critical when climate change and its potential impacts are considered. Water systems in the western United States are especially vulnerable to these impacts due to their dependence on snow accumulation and the snowmelt process. These systems could experience more precipitation falling as rain rather than snow during the winter months and a snowmelt runoff that occurs earlier in the year. The result would especially impact irrigation systems, which accounts for 80 percent of water withdrawal in the western United States. These systems would experience more water earlier in the year, when it is not needed, and less water during the summer months when it is needed.

The latest *Intergovernmental Panel on Climate Change* (IPCC) report reaffirms that the climate is changing in ways that cannot be accounted for by natural variability, and that "global warming" is occurring (IPCC, 2001). This global warming is likely to have significant impacts on the hydrologic cycle, affecting water resources systems (IPCC, 2001; Arnell, 1999). These impacts will be different for different regions of the world. Regions that have a large fraction of runoff driven by snowmelt would be especially susceptible to changes in temperature, because temperature determines the fraction of precipitation that falls as snow or as rain and determines the timing of snowmelt process.

The impact of climate change on water resource systems can be analyzed in three different approaches:

1. The first approach is to consider what is currently observed in the geophysical record. Here we look at the historical trend of temperature, precipitation, streamflows, and other geophysical variables in order to determine if there is any evidence of climate change in the historical record.
2. The second approach is to predict future temperatures, precipitation and streamflow using *global circulation models* (GCM).
3. Then using the predicted temperature, precipitation, and streamflow, changes in the natural runoff caused by this climate change can be modeled to determine the economic, ecologic, or institutional impacts.

In this chapter we look at each of these three approaches, the methodological aspects involved, and their significant conclusions.

IS THERE EVIDENCE OF THE EFFECTS OF CLIMATE CHANGE ON HYDROLOGY IN THE HISTORICAL RECORD?

It has been almost two decades since Roos (1987, 1991, 2003) first brought attention to changes that are occurring in California's streamflow patterns. He looked specifically at the Sacramento basin and determined that the seasonal fraction of runoff during the snowmelt/spring season (from April to July) was decreasing throughout the twentieth century. This same behavior was confirmed by other studies using more complex statistical measures and extended to other basins (Fox et al., 1990; Wahl, 1991; Aguado et al., 1992; Pupacko, 1993; Dettinger and Cayan, 1995; Shelton and Fridirici, 1997; Shelton, 1998; Freeman, 2002). Those studies that compared different basins show that the trend has been most pronounced for midaltitude river basins (1000 to 2000 m). This seems intuitively correct as high altitude basins may be too cold to be affected by a few degree change in temperature and low altitude basins may already be receiving rainfall rather than snow.

Using the fractional runoff as a measure of changes in streamflow patterns could be misleading because it could imply one or both of changes in spring runoff or changes in other seasons (Wahl, 1991). To avoid this uncertainty Cayan et al. (2001) and Stewart et al. (2004) used a different approach to measure changes in streamflow pattern. Cayan et al. (2001) considered what they called the "spring pulse" defined as the day when cumulative departure of daily streamflows from mean is most negative. Their study, covering several basins in the Sierra Nevada, correlated the spring pulse with other measures of spring onset (e.g., flower blooming), showing an organized earlier spring onset. Stewart et al. (2004) obtained a similar result but used the flow-weighted timing, or "center of mass" of streamflow, as the metric to determine runoff timing.

Several factors have been postulated as the cause for the decline in fractional spring streamflow such as increases in winter precipitation and increases in spring temperatures. This latter factor explains earlier spring timing of runoff as studied by Cayan et al. (2001) and Stewart et al. (2004). However, whether these trends are a signature of a changing climate in California remains uncertain. Although some authors have suggested that these trends are due to climate change (e.g., Pupacko, 1993; Shelton et al., 1997; Shelton, 1998), other have been more cautious, suggesting the possibility that the length of the historical record only shows one realization of a long-term variability process such as the Pacific Decadal Oscillation (e.g., Aguado et al., 1992; Dettinger and Cayan, 1995; Cayan et al., 2001).

FUTURE POTENTIAL IMPACTS ON NATURAL STREAMFLOW DUE TO CLIMATE CHANGE

The methodologies used to address climate change impacts on water resources have been addressed by Gleick (1989) and Wood et al. (1997). There are two major steps involved in this process:

Methodology

1. Determining changes in temperature, precipitation, and other climatologic variables such as evapotranspiration
2. Determining changes in runoff using the outputs from the previous step

Determining Changes in Temperature and Precipitation
To determine the changes in temperature and precipitation associated with climate change there are two alternative approaches. The most simple and direct is to consider hypothetical scenarios of changes in temperature and precipitation. The second approach considers the use of climate output data obtained from a "general circulation model" (GCM).

Hypothetical climate scenarios include changes in temperature covering the plausible range for the twenty-first century according to climate change impact assessments (e.g., +2 to +5°C). There is less consistency on the direction of changes in precipitation, so the hypothetical scenarios considered cover both increase and decrease (e.g., from ±15 percent). The "hypothetical scenario" approach was used in the earliest studies (Revelle and Waggoner, 1983; Gleick, 1987; Jeton et al., 1996) and a recent study by Miller et al. (2003). It is important to mention that of all these studies, only Revelle and Wagonner (1983), relied solely on the "hypothetical scenarios" approach. All of the other studies also consider output from GCMs. The advantage of the hypothetical scenario approach is its simplicity in representing a wide range of alternative scenarios, to easily gauge the sensibility of a particular basin to changes in climate conditions. However, the approach is not realistic enough to be used as a tool for making policy decisions in the water resources management.

GCMs are state-of-the-art representations of the coupled atmosphere-land-ocean-ice systems and interactions. These models provide information on the response of the atmosphere to different scenarios of greenhouse gas concentrations. Most climate change studies considered using GCMs' output data as the "climate change perturbed" climatic conditions (Lettenmaier et al., 1988; Lettenmaier and Gan, 1990; Tsuang and Dracup, 1991; Leung and Ghan, 1999; Miller et al., 1999; Miller and Kim, 2000; Hay et al., 2000; Wilby and Dettinger, 2000; Carpenter and Georgakakos, 2001; Kim, 2001; Kim et al., 2002; Snyder et al., 2002; Knowles and Cayan, 2002; Vanrheenen et al., 2003; Miller et al., 2003; Dettinger, 2004; Dettinger et al., 2004; Vanrheenen et al., 2004; Stewart et al., 2004; Knowles and Cayan, 2004; Leung et al., 2004; Hayhoe et al., 2004; Maurer and Duffy, 2005). Several factors distinguish all of these studies. The most important are the choice of GCM used, the downscaling methodology, and the method used to characterize uncertainty.

With respect to the choice of GCM output, it is interesting to note how the most prominent GCMs used in earlier studies [the models of the Geophysical Fluid Dynamics Laboratory (GFDL), the Goddard Institute for Space Studies (GISS), and the Oregon State University (OSU)] were replaced by a different set of models for later studies. These are the U.K.'s Hadley Center (HadCM2 and HadCM3) and the NCAR's (CCM3 and PCM) models.

One major limitation on using GCM output climatologic data is that their spatial and temporal resolution does not match the resolution needed for hydrologic models. For example, the spatial resolution of GCMs (about 200 km) is too coarse to resolve complex orography and subgrid scale processes such as convective precipitation, which are of major relevance for mountainous terrains like the California Sierra Nevada (Wilby and Dettinger, 2000). Several methods have been developed to "downscale" or transfer GCM output to surface variables at the river basin scale. The most recurrent are (a) ratio and difference (perturbation) methods (b) stochastic/statistic downscaling (c) dynamic downscaling or nested models (Wood et al., 1997). From the Californian experience it is interesting to note how the evolution of this field of study has followed the evolution of the downscaling methodologies. Earlier studies did not consider downscaling GCM output but used the raw GCM output instead. The delta/ratio method then became the preferred method but as time progressed, the more complex downscaling methods (either statistic or dynamic) have been preferred. Improved downscaling methodologies have provided means of expanding the temporal aspect of the analysis. Earlier studies relied solely on monthly perturbations of historical time series but later studies have explored the derivation of totally "new" (not based on historical data) time series of climate variables (either daily or monthly), allowing the analysis of changes in the frequency of extreme events (e.g., flood or droughts) or changes in interannual variability (e.g., Miller et al., 2003; Dettinger et al., 2004; Vanrheenen et al., 2004; Stewart et al., 2004; Hayhoe et al., 2004; Maurer and Duffy, 2005).

The output of different GCMs for California tend to be consistent in terms of temperature predictions, all indicating an increase for the twenty-first century of between 2 and 6°C. However, the GCMs are less consistent in their precipitation predictions, indicating a great variability. This inconsistency among GCM predictions brings a great degree of uncertainty at the moment for making water resources management decisions. Later studies have tried to tackle this uncertainty problem using different approaches. Some have considered using multiple (and differing) GCMs' outputs, under multiple greenhouse gas emission scenarios (e.g., Miller et al., 2003; Hayhoe et al., 2004; Leung et al., 2004; Maurer and Duffy, 2005). With that approach they intend to get a bracket of equally plausible results but they still do not account explicitly for the uncertainty in predictions. Dettinger (2004) was the first to attempt to focus explicitly on the uncertainty related to GCM predictions in California. In his study Dettinger determines *projection distribution functions* (pdfs) based on a resampling technique of 18 available projection scenarios (6 models*3 emission scenarios each). The explicit consideration of the uncertainty embedded in climate change predictions is the preferred approach because it will improve the tools available for the decision-making process of the water resource community in California.

Determining Changes in Natural Runoff

The perturbed series of climatological data (mainly temperature and precipitation) is used to drive a hydrologic model to predict changes in streamflow runoff. There are two alternative approaches, using either statistically or physically based hydrologic models.

Statistical hydrologic models determine future potential runoff values according to the characteristics of the data embedded in the historical record. Examples of such statistical models used for climate change impact assessment are models that determine (through regression or observational analysis) the relation between streamflow runoff and climate variables such as temperature and precipitation. The first study of the impacts of climate change on Californian water resources was a statistical model (Revelle and Wagoner, 1983). The relations obtained in that study considered annual values of streamflow, temperature, and precipitation. Most recent studies use a monthly basis to determine not only changes in the total annual volume but also changes in the seasonal pattern (Cayan et al., 1993; Duell, 1994; Risbey and Entekhabi, 1996; Stewart et al., 2004).[1] One major limitation of these statistical approaches is that they

---

[1] Some of these studies (Cayan et al., 1993; Risbey and Entekhabi, 1996) did not actually perform a climate change assessment but were intended to be used for that purpose. So although, they do not provide actual results in terms of climate change impacts they still prove to be worth including in this review of the literature.

do not incorporate the physical mechanisms and processes that determine basin response to climate forcing. Also, this approach is bound to react to only those levels of forcing that have historically occurred.

Physically based hydrologic models have been the preferred tools to assess the impacts of climate change in the California hydrology. Some researchers developed their own models for their analysis (e.g., Gleick, 1987; Tsuang and Dracup, 1991). However, most studies used previously developed models. Examples of the models used are the USGS Precipitation-Runoff Modeling System (PRMS); the U.S. National Weather Service River Forecast System Sacramento Soil Moisture Accounting and Anderson Snow Models (SAC-SMA); and the Variable Infiltration Capacity (VIC) model. The spatial parameterization of these models range from distributed to lumped. Their spatial resolution ranges from a regional scale (West Coast) to the subbasin level. Although some of the models include state of the art levels of complexities there are still some important factors that have not been taken into account. Probably the most important of these missing factors is the dynamic interaction between the vegetated land cover and the climatic variables. So far such factors that affect watershed responses have been treated as static parameters in the models, not changing with time or future climate.

Potential Climate Change Impacts on Streamflows: Most Significant Results

Although the studies that have assessed the impacts of climate change in hydrology have differed in their methodological approach, their results tend to agree. Results consistently show that increasing temperatures associated with climate change will impact hydrology by changing the seasonal streamflow pattern to an earlier (and shorter) spring snowmelt and an increased winter runoff as a fraction of total runoff. These impacts vary by basin, with the key parameter being the basin elevation relative to the snow line during snow accumulation and melt periods. Basins located at medium altitudes will be affected more by climate change.

Changes in the total runoff volume depend on the precipitation prediction (scenario), which depends on the GCM chosen as previously discussed. Assessments using GCMs that predict wetter conditions (e.g., U.K.'s HadCM2) tend to predict higher overall streamflow runoff as compared to "drier" GCMs (e.g., NCAR's PCM). An example of such diverging results is shown by Miller et al. (2003). Improvements in the characterization of uncertainty related to climate impact assessment would improve the ability to use these diverging results in a manner useful for water resources management.

FUTURE POTENTIAL IMPACTS ON CALIFORNIAN WATER RESOURCES SYSTEMS DUE TO CLIMATE CHANGE

Most of the streamflow in California and in the western United States is regulated by large reservoirs. Significant changes in the timing of streamflow that feeds these reservoirs will change their ability to serve their design functions: flood control, water supply, hydropower generation, environmental services, navigation, and recreation. For example, an earlier and shorter snowmelt spring runoff will make it more difficult to refill reservoir flood space (determined considering historical hydrologic conditions) during the late spring and early summer, thus reducing the amount of water supply that can be delivered (Roos, 2003). The ultimate impact on California water resources and their associated functions will depend on the ability of the man-made infrastructure to cope with these changes. The analysis of the performance of the California water system under hypothetical hydrologic scenarios such as the ones associated with climate change requires the aid of water resources systems models, also called *reservoir system analysis models.*

The performance of the California water system under climate change scenarios was first studied by Lettenmaier and Sheer (1991), and by Sandberg and Manza (1991). Both these groups examined the implications of climate change scenarios on the performance of the state water project (SWP) and the central valley project (CVP) using simulation models. Most of the later studies (Dracup et al., 1993; Yao and Georgakakos, 2001; Vanrheenen et al., 2003; Knowles and Cayan, 2002; Brekke et al., 2004; Quinn et al., 2004; Knowles and Cayan, 2004; and Vanrheenen et al., 2004) on the impact of climate change on Californian water resources have also relied on simulation models. An exception to this is the use of the optimization model (CALVIN) developed by Lund et al. (2003). The results of these studies in terms of water deliveries and reservoir storage also reflect the climatological predictions used to derive streamflow runoffs (see previous section). In this regard, predicted drier conditions reduce the ability of the system to perform at historical levels. On the other hand, predicted wetter conditions would tend to overcome changes in the seasonality of streamflows, producing an overall improvement of system performance (see Brekke et al., 2004).

Although water storage and deliveries from reservoirs have been the preferred measure used to assess the final impacts of climate change on water resources, it is worthwhile mentioning that some studies have also included other means of measuring impacts. For example, Knowles and Cayan (2002, 2004) studied the impacts on San Francisco Bay salinity levels. Dracup et al. (1993) performed a series of studies on several water resources functions such as hydropower generation (for CVP/SWP system), agricultural economic costs, and Chinook salmon population.

SUGGESTED FUTURE STUDIES

There are potential climate change impacts that need to be addressed such as:

1. The impact on hydropower generation for high altitude facilities.
2. The impact associated with changes in reliability (and therefore economic costs) among different water users (with different water rights and water sources) in the system.
3. How the different conflicting reservoir objectives (e.g., hydropower vs. water supply or flood control vs. water supply) could be adjusted under changes in seasonal patterns predicted by climate change assessments.

A further step in the analysis of impacts to water resources associated with climate change is to derive changes in the management practices of the water resources systems to cope with the predicted changes in streamflows. This approach was pursued by Vanrheenen et al. (2004) who developed a series of mitigation strategies such as changing the flood control rule curves of reservoir releases to lessen the impacts of climate change. Vanrheenen et al. (2004) concluded that even with the most comprehensive approaches "achieving and maintaining status quo (control scenario climate) system performance in the future would be nearly impossible, given the altered climate (change) scenario hydrologies." In a different approach, Yao and Georgakakos (2001) developed an integrated forecast-decision system used to assess the sensitivity of reservoir performance to various forecast-management schemes under historical and future climate scenarios. Their assessments are based on various combinations of inflow forecasting models, decision rules, and climate scenarios. We believe that although these two studies reflect advances in this issue, there are still some areas of work that have not been properly addressed. These include studying the mitigation opportunities brought by a coupled surface and ground water management system or using optimization techniques (e.g., stochastic dynamic programming) to determine whether reservoir release policies should be modified according to changes in streamflow patterns.

CONCLUSIONS

Presented here is an analysis on the impact of climate change on water sustainability, using the western United States as an example. Major conclusions that can be derived from this analysis are the following:

1. In the last few decades there has been a consistent trend of changes in the timing of streamflow particularly in the California Sierra Nevada. These trends correlate well with increases in temperatures, although it is premature to say that they show an evidence of climate change.

2. Several different approaches have been used to assess potential future changes in streamflow related to climate change. These approaches consider either using GCMs or hypothetical scenarios as predictors of changes in climatic variables. These changes of climatic variables have been used to assess changes in natural runoff using different types of hydrologic models (e.g., statistical or physically based). Results derived from these studies consistently show that there will be a change in timing in streamflow runoff due to a consistent increase in temperature. However, changes in the total volume of runoff are still uncertain, mainly due to uncertainties in precipitation predictions. The methodology used to assess changes in hydrology has improved both in terms of downscaling outputs from GCMs and in the way uncertainty is characterized. These improvements will improve the ability to use this information in policy decisions.

3. The hydrologic changes associated with climate change will potentially affect the performance of the infrastructure and thereby will affect the different uses of water in California. There have been some attempts to address these issues using water resource system models (either optimization or simulation) to distribute the hydrologic changes throughout the California system. There is a large potential for more research on the impacts of climate change on California water resource systems.

REFERENCES

Aguado, E., D. R. Cayan, L. G. Riddle, and M. Roos, "Climatic Fluctuations and the Timing of West Coast Streamflow," *Journal Of Climate*, Vol. 5, pp. 1468–1483, 1992.

Arnell, N.W., "Climate Change and Global Water Resources," *Global Environmental Change*, Vol. 9, S31–S49 Suppl., 1999.

Brekke, L. D., N. L. Miller, K. E. Bashford, N. W. T. Quinn, and J. A. Dracup, "Climate Change Impacts Uncertainty for Water Resources in the San Joaquin River Basin, California," *Journal of the American Water Resources Association*, Vol. 40, No. 1, pp. 149–164, 2004.

Carpenter, T. M., and K. P. Georgakakos, "Assessment of Folsom Lake Response to Historical and Potential Future Climate Scenarios: 1. Forecasting," *Journal of Hydrology*, Vol. 249, No. 1–4, pp. 148–175, 2001.

Cayan, D. R., S. Kammerdiener, M. D. Dettinger, J. Caprio, and D. H. Peterson, "Changes in the Onset of Spring in the Western United States," *Bulletin Of American Meteorological Society*, Vol. 82, pp. 399–415, 2001.

Cayan, D. R., and L. G. Riddle, "The Influence of Precipitation and Temperature on Seasonal Streamflow in California," *Water Resources Research*, Vol. 29, No. 4, pp. 1127–1140, 1993.

Dettinger, M. D., "From Climate Change Spaghetti to Climate Change Distribution," Prepared for California Energy Commission, *Public Interest Research Program,* Sacramento, CA, 2004.

Dettinger, M. D., and D. R. Cayan, "Large-Scale Atmospheric Forcing of Recent Trends toward Early Snowmelt Runoff in California," *Journal Of Climate,* Vol. 8, No. 3, pp. 606–623, 1995.

Dettinger, M. D., D. R. Cayan, M. Meyer, and A. E. Jeton, "Simulated Hydrologic Responses to Climate Variations and Change in the Merced, Carson, and American River basins, Sierra Nevada, California, 1900-2099," *Climatic Change,* Vol. 62, No. 1–3, pp. 283–317, 2004.

Dracup, J. A., S. D. Pelmulder, R. Howitt, G. Horner, W. M. Hanemann, C. F. Dumas, R. McCann, J. Loomis, and S. Ise, "Integrated Modeling of Drought and Global Warming: Impacts on Selected California Resources," *Report by the National Institute for Global Environmental Change,* University of California, Davis, CA, P. 112, 1993.

Duel, L. F. W. Jr. "The Sensitivity of Northern Sierra Nevada Streamflow to Climate Change," *Water Resources Bulletin*, Vol. 30, No. 5, pp. 841–859, 1994.

Fox, J. P., T. R. Mongan, and W. J. Miller, "Trends in Freshwater Inflow to San Francisco Bay from the Sacramento-San Joaquin Delta," *Water Resources Bulletin*, Vol. 26, No. 1, pp. 101–116, 1990.

Freeman, G. J., "Looking for Recent Climatic Trends and Patterns in California's Central Sierra." *Proceedings of the 19th Annual Pacific Climate (PACLIM) Workshop*, Pacific Grove, CA, pp. 35–48, 2002.

Gleick, P. H., "Regional Hydrologic Consequences of Increases in Atmospheric CO_2 and other Trace Gases," *Climatic Change*, Vol. 10, pp. 137–161, 1987.

Gleick, P. H., "Climate Change, Hydrology, and Water Resources," *Review Of Geophysics*, Vol. 27, No. 3, pp. 329–344, 1989.

Gleick, P. H., and E. L. Chalecki, "The Impacts of Climatic Change for Water Resources of the Colorado and Sacramento-San Joaquin River Basins," *Journal of the American Water Resources Association,* Vol. 35, No. 6, pp. 1429–1441, 1999.

Hay, L. E., R. L. Wilby, and G. H. Leavesley, "A Comparison of Delta Change and Downscaled GCM Scenarios for Three Mountainous Basins in the United States," *Journal of American Water Resources Association*, Vol. 36, No. 2, pp. 387–397, 2000.

Hayhoe, K., D. R. Cayan, C. Field, P. Frumhoff, E. Maurer, N. Miller, S. Moser, S. Schneider, K. Cahill, E. Cleland, L. Dale, R. Drapek, R. M. Hanemann, L. Kalkstein, J. Lenihan, C. Lunch, R. Neilson, S. Sheridan , and J. Verville, "Emissions Pathways, Climate Change, and Impacts on California,." *Proceedings of the National Academy of Sciences* (PNAS), Vol. 101, No. 34, pp. 12422–12427, 2004.

IPCC, "Climate Change 2001: Scientific Basis," Contribution of Working Group III to the Third Assessment Report of the Intergovernmental Panel on Climate Change, Bert Metz, et al. (eds.), Cambridge, New York, Published for the Intergovernmental Panel on Climate Change [by] Cambridge University Press, 2001.

Jeton, A. E., M. D. Dettinger, and J. L. Smith, "Potential Effects of Climate Change on Streamflow: Eastern and Western Slopes of the Sierra Nevada, California and Nevada," U.S. Geological Survey, *Water Resources Investigations Report 95-4260*, P. 44, 1996.

Kim, J., "A Nested Modeling Study of Elevation-Dependent Climate Change Signals in California Induced by Increased Atmospheric CO2," *Geophysics Research Letters*, Vol. 28, No. 15, pp. 2951–2954, 2001.

Kim J., T. K. Kim, R. W. Arritt, and N. L. Miller, "Impacts of Increased Atmospheric CO2 on the Hydroclimate of the Western United States," *Journal of Climate*, Vol. 15, pp. 1926–1942, 2002.

Knowles, N., and D. Cayan, "Potential Effects of Global Warming on the Sacramento/San Joaquin Watershed and the San Francisco Estuary," *Geophysics Research Letters*, Vol. 29, No. 18, P. 1891, 2002.

Knowles, N., and D. R. Cayan, "Elevational Dependence of Projected Hydrologic Changes in the San Francisco Estuary and Watershed," *Climatic Change,* Vol. 62, pp. 319–336, 2004.

Lettenmaier, D. P., and T. Y. Gan, "Hydrologic Sensitivities of the Sacramento-San Joaquin River Basin, California, to Global Warming," *Water Resources Research*, Vol. 26, pp. 69–86, 1990.

Lettenmaier, D. P., T. Y. Gan, and D. R. Dawdy, "Interpretation of Hydrologic Effects of Climate Change in the Sacramento-San Joaquin River Basin, California," Prepared for U.S. Environmental Protection Agency, Department of Civil Engineering, University of Washington, Seattle, WA, 1988.

Lettenmaier, D. P., and D. P. Sheer, "Climate Sensitivity of California Water Resources," *Journal of Water Resources Planning and Management*, Vol. 117, No. 1, pp. 108–125, 1991.

Leung, L. R., and S. Ghan, "Pacific Northwest Climate Sensitivity Simulated by a Regional Climate Model Driven by a GCM. Part II: $2 \times CO\,2$ Simulations," *Journal of Climate,* Vol. 12, pp. 2031–2053, 1999.

Leung, L. R., Y. Qian, W. M. Washington, J. Han, and J. O. Roads, "Mid-Century Ensemble Regional Climate Change Scenarios for the Western United States," *Climatic Change,* Vol. 62, pp. 75–113, 2004.

Lund, J. R., R. E. Howitt, M. W. Jenkins, T. Zhu, S. K. Tanaka, M. Pulido, M. Tauber, R. Ritzema, and I. Ferriera, "Climate Warming & California's Water Future," A Report for the California Energy Commission, Center for Environmental and Water Resource Engineering, University of California, Davis. Sacramento, CA, 2003.

Maurer, E. P., and P. B. Duffy, "Uncertainty in Projections of Streamflow Changes due to Climate Change in California," *Geophysics Research Letters,* Vol. 32, No. 3, 2005.

Miller, N. L., K. E. Bashford, and E. Strem, "Potential Impacts of Climate Change on California Hydrology," *Journal of the American Water Resources Association,* Vol. 39, No. 4, pp. 771–784, 2003.

Miller, N. L., and J. Kim, "Climate Change Sensitivity Analysis for Two California Waterhseds: Addendum to Downscaled Climate and Streamflow Study of the Southwestern United States," *Journal of American Water Resources Association,* Vol. 36, No. 3, pp. 657–661, 2000.

Miller, N. L., J. Kim, R K. Hartman, and J. Farrara, "Downscaled Climate and Streamflow Study of the Southwestern United States," *Journal American Water Resources Association,* Vol. 35, pp. 1525–1537, 1999.

Pupacko, A., "Variations in Northern Sierra Nevada Streamflow: Implications of Climate Change," *Water Resources Bulletin*, Vol. 29, No. 2, pp. 283–290, 1993.

Quinn, N. W. T., L. D. Brekke, N. L. Miller, T. Heinzer, H. Hidalgo, and J. A. Dracup, "Model Integration for Assessing Future Hydroclimate Impacts on Water Resources, Agricultural Production and Environmental Quality in the San Joaquin Basin, California," *Environmental Modelling & Software,* Vol. 19, pp. 305–316, 2004.

Revelle, R. R., and P. E. Waggoner, "Effects of a Carbon Dioxide Induced Climatic Change on Water Supplies in the Western United States," *Changing Climate,* National Academy of Sciences Press, Washington, DC, 1983.

Risbey, J. S., and D. Entekhabi, "Observed Sacramento Basin Streamflow Response to Precipitation and Temperature Changes and its Relevance to Climate Change Impact Studies," *Journal of Hydrology,* Vol. 184, pp. 209–223, 1996.

Roos, M., "Possible Changers in California Snowmelt Patterns," *Proceedings of the Fourth Annual Pacific Climate (PACLIM) Workshop*, Pacific Grove, CA, pp. 22–31,1987.

Roos, M., "A Trend of Decreasing Snowmelt Runoff in Northern California," *Proceedings of the 59th Western Snow Conference*, Juneau, AK, pp. 29–36, 1991.

Roos, M., "The Effects of Global Climate Change on California Water Resources," A Report for the Energy California Commission, *Public Interest Energy Research Program,* Research Development and Demonstration Plan, Sacramento, CA, 2003.

Sandberg, J., and P. Manza P., "Evaluation of Central Valley Project Water Supply and Delivery System," Global Change Response Program, U.S. Department of Interior, Bureau of Reclamation, Sacramento, California, 1991.

Shelton, M. L., "Seasonal Hydroclimate Change in the Sacramento River Basin, California," *Physical Geography*, Vol. 19, No. 3, pp. 239–255, 1998.

Shelton, M. L., and R. M. Fridirici, "Decadal Changes of Inflow to the Sacramento-San Joaquin Delta, California," *Physical Geography*, Vol. 18, No. 3, pp. 215–231, 1997.

Snyder, M. A., J. B. Bell, L. C. Sloan, P. B. Duffy, and B. Govindasamy, "Climate Responses to a Doubling of Atmospheric Carbon Dioxide for a Climatically Vulnerable Region," *Geophysics Research Letters,* Vol. 29, No. 11, pp. 1514–1517, 2002.

Stewart, I. T., D. R. Cayan, M. D. Dettinger, "Changes in Snowmelt Runoff Timing in Western North America Under a 'Business as Usual' Climate Change Scenario," *Climatic Change,* Vol. 62, No. 1–3, pp. 217–232, 2004.

Tsuang, B. J., and J. A. Dracup, "Effect of Global Warming on Sierra Nevada Mountain Snow Storage," *Proceedings of the 59th Western Snow Conference*, Juneau, AK, pp. 17–28, 1991.

Vanrheenen, N. T., R. M. Palmer, and M. Hahn, "Evaluating Potential Climate Change Impacts on Water Resource System Operations: Case Studies of Portland, Oregon and Central Valley, California," *Water Resources Update,* Vol. 124, pp. 35–50, 2003.

Vanrheenen, N.T., A. W. Wood, R. N. Palmer, and D. P. Lettenmaier, "Potential Implications of PCM Climate Change Scenarios for Sacramento-San Joaquin River Basin Hydrology and Water Resources," *Climatic Change,* Vol. 62, No. 1–3, pp. 257–281, 2004.

Wahl, K. L., "Is April to June Runoff Really Decreasing in the Western United States?" *Proceedings of the 59th Western Snow Conference*, Juneau, AK, pp. 67–78, 1991.

Wilby, R. L., and M. D. Dettinger, "Streamflow Changes in the Sierra Nevada, California, Simulated Using Statistically Downscaled General Circulation Model Output," in S. McLaren and D. Kniven (eds.), *Linking Climate Change to Land Surface Change*, Kluwer Academic Pub, pp. 99–121, 2000.

Wilkinson, R., K. Clarke, M. Goodchild, J. Reichman, and J. Dozier, "The Potential Consequences of Climate Variability and Change for California: The California Regional Assessment," U.S. Global Change Research Program, Washington, DC, 2002.

Wood, A. W., D. P. Lettenmaier, and R. N. Palmer, "Assessing Climate Change Implications for Water Resources Planning," *Climatic Change*, Vol. 37, pp. 203–228, 1997.

Yao, H., and A. Georgakakos, "Assessment of Folsom Lake Response to Historical and Potential Future Climate Scenarios 2. Reservoir Management," *Journal of Hydrology*, Vol. 249, No. 1–4, pp. 176–196, 2001.

10 WATER SUPPLY SECURITY: AN INTRODUCTION

Larry W. Mays
Department of Civil and Environmental Engineering
Arizona State University
Tempe, Arizona

HISTORY

A long history, in fact since the dawn of history, of threats to drinking water systems during conflicts has plagued humans. Water has been a strategic objective in armed conflicts throughout history. Gleick (1994, 1998, 2000) has developed a water conflict chronology in which he categorizes the conflicts as the following: control of water resources, military tool, political tool, terrorism, military target, and development disputes. Terrorism is defined as, "water resources, or water systems, are either targets or tools of violence or coercion by nonstate actors." There are many historical conflicts that caused flooding by diversion or eliminated water supplies by building dams or other structures, whereas in the following only a few examples of some of the water conflicts that included water supply systems are summarized.

During the time of King Hezekiah—the period of the First Temple (the latter part of the eighth century B.C.), Jerusalem was under military threat from Assyria (2 Kings 20:20; Isaiah 22:11; 2 Chronicles 32:2-4,30). The Gihon spring, located just outside the city walls, was the main water source for the ancient city of Jerusalem (Bruins, 2002), requiring strategic planning on King Hezekiah's part. He had a water tunnel (533 m) dug to channel the water underground into the city, with the outlet at a reservoir known as the Pool of Siloam. Two crews of miners dug through solid limestone from both ends of the tunnel, meeting at the same spot (Bruins, 2002).

During the second Samnite War, ca. 310 B.C., the Romans realized the need for alternate water sources for Rome due to the insufficient and unreliable local supplies. The Roman Senate procured and distributed water rights from estates surrounding Rome in order to develop the supply and security needed for Rome.

In 1503, Leonardo da Vinci and Machiavelli planned to divert the Arno River away from Pisa during the conflict between Pisa and Florence.

During the Civil War (1863) in the United States, General U.S. Grant cut levees in the battle against the Confederates during the campaign against Vicksburg.

In 1948, during the first Arab-Israeli War, Arab forces cut off the West Jerusalem water supply.

In 1982, Israel cut off the water supply of Beirut during the siege.

In 1990 in South Africa, the pro-apartheid council cut off water to the Wesselton township of 50,000 blacks following protests over miserable sanitation and living conditions.

During the 1991 Gulf War, the Allied coalition targeted Baghdad's water supply and sanitation system. Discussions were held about using the Attaturk Dam to cut off flows to the Euphrates to Iraq. Also during the Gulf War, Iraq destroyed much of Kuwait's desalination capacity during retreat. In 1993, Saddam Hussein reportedly poisoned and drained water supplies of southern Shiite Muslims.

In Kosovo (1999), water supplies/wells were contaminated by Serbs who disposed of the bodies of Kosovar Albanians in local wells. Serbian engineers shut down the water system in Pristina prior to occupation by NATO. Also during that same year in Yugoslavia, NATO targeted utilities and shut down water supplies in Belgrade.

Gleick (2000) developed a water conflict chronology (1503 to 2000) that can be found at the following site: http://www.worldwater.org/conflict.htm.

THE WATER SUPPLY SYSTEM: A BRIEF DESCRIPTION

The events of September 11, 2001 have significantly changed the approach to management of water utilities. Previously, the consideration of the terrorist threat to the U.S. drinking water supply was minimal. Now we have an intensified approach to the consideration of terrorist threat. The objective of this chapter is to provide an introduction to the very costly process of developing water security measures for U.S. water utilities.

Figure 1.1 illustrates a typical municipal water utility showing the water distribution system as a part of this overall water utility. In some locations, where excellent quality groundwater is available, water treatment may include only chlorination. Other handbooks on the subject of water supply/water distribution systems include Mays (1989, 2000, 2002, 2003).

Water distribution systems are composed of three major components: pumping stations, distribution storage, and distribution piping. These components may be further divided into subcomponents, which in turn can be divided into sub-subcomponents. For example, the pumping station component consists of structural, electrical, piping, and pumping unit subcomponents. The pumping unit can be divided further into sub-subcomponents: pump, driver, controls, power transmission. The exact definition of components, subcomponents, and sub-subcomponents depends on the level of detail of the required analysis and, to a somewhat greater extent, the level of detail of available data. In fact, the concept of component-subcomponent-sub-subcomponent merely defines a hierarchy of building blocks used to construct the water distribution system. Figure 1.2 shows the hierarchical relationship of system, components, subcomponents, and sub-subcomponents for a water distribution system.

A water distribution system operates as a system of independent components. The hydraulics of each component is relatively straightforward; however, these components depend directly upon each other and as a result effect the performance of one another. The purpose of design and analysis is to determine how the systems perform hydraulically under various demands and operation conditions. These analyses are used for the following situations:

- Design of a new distribution system
- Modification and expansion of an existing system
- Analysis of system malfunction such as pipe breaks, leakage, valve failure, pump failure
- Evaluation of system reliability

Figure 10–1
A typical water distribution system.

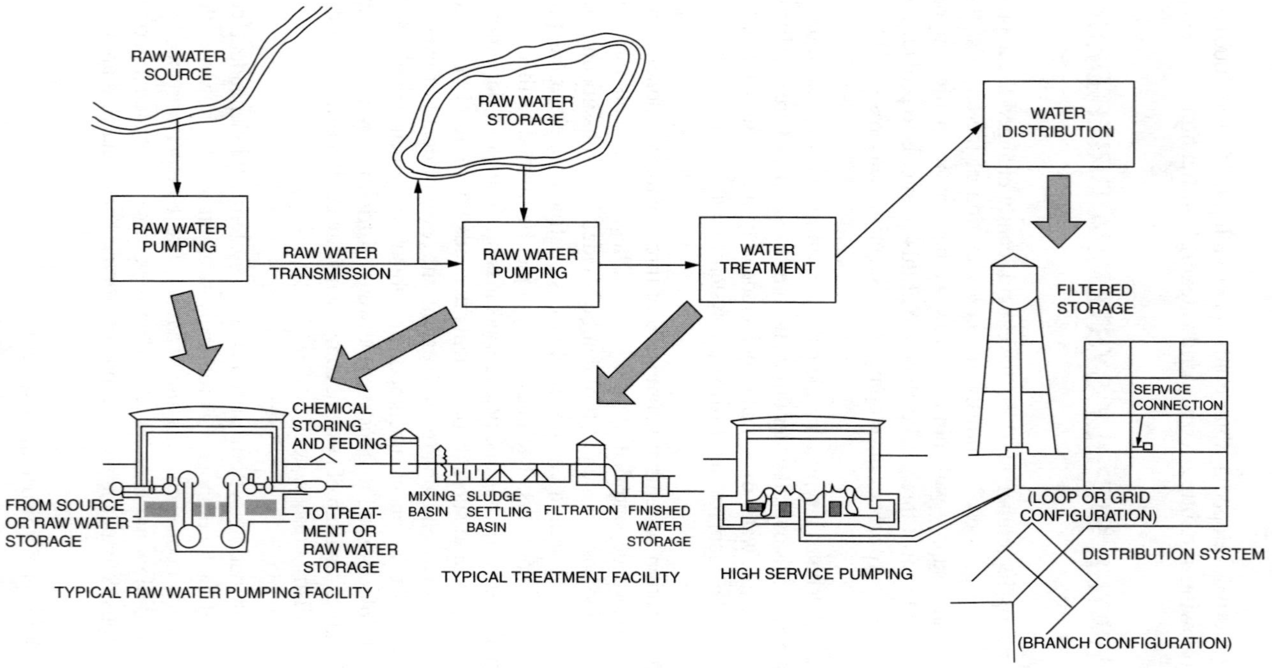

Figure 10–2
Hierarchy of building blocks in water distribution systems.

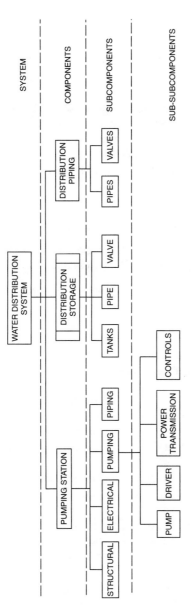

197

• Preparation for maintenance
• System performance and operation optimization

WHY WATER SUPPLY SYSTEMS?

A distribution system of pipelines, pipes, pumps, storage tanks, and the appurtenances such as various types of valves, meters, etc. offers the greatest opportunity for terrorism because it is extensive, relatively unprotected and accessible, and often isolated. The physical destruction of a water distribution system's assets or the disruption of water supply could be more likely than contamination. A likely avenue for such an act of terrorism is a bomb, carried by car or truck, similar to the recent events listed in Table 10.1. Truck or car bombs require less preparation, skill, or manpower than complex attacks such as those of September 11, 2001. However, we must consider all the possible threats no matter how remote we may think that they could be.

Table 10–1

Recent Terrorist Attacks Against American Targets Using Car-Bomb Technologies

Date	Target/location	Delivery/material	TNT equivalent (lb)	Reference
Apr. 1983	U.S. Embassy Beirut, Lebanon	Van	2,000	www.beirut-memorial.org
Oct. 1983	U.S. Marine Barracks Beirut, Lebanon	Truck, TNT with gas enhancement	12,000	www.usmc.mil
Feb. 1993	World Trade Center New York, U.S.A.	Van, urea nitrate and hydrogen gas	2,000	www.interpol.int
Apr. 1995	Murrah Federal Bldg Oklahoma City, U.S.A.	Truck, ammonium nitrate fuel oil	5,000	U.S. Senate documents
June 1996	Khobar Towers Dhahran, Saudi Arabia	Tanker truck, plastic explosive	20,000	www.fbi.gov
Aug. 1998	U.S. Embassy Nairobi, Kenya	Truck, TNT, possibly Semtex	1,000	news reports, U.S. Senate documents
Aug. 1998	U.S. Embassy Dar es Salaam, Tanzania	Truck	1,000	U.S. Senate documents
Oct. 2000	Destroyer USS Cole	Small watercraft,	440	www.al-bab.com
	Aden Harbor, Yemen	possibly C-4		news.bbc.co.uk

Source: Peplow et al. (2003).

THE THREATS

The probability of a terrorist threat to drinking water is probably very low; however, the consequences could be extremely severe for exposed populations. Various types of threats may have higher probabilities than others. The following four major types of threats are discussed further throughout this book. The term *weaponized* when referring to chemical and biological agents means that it can be produced and disseminated in large enough quantities to cause the desired effect (Hickman, 1999).

Cyber Threats

- Physical disruption of a supervisory control and data acquisition (SCADA) network
- Attacks on central control system to create simultaneous failures
- Electronic attacks using worms/viruses
- Network flooding
- Jamming
- Disguise data to neutralize chlorine or add no disinfectant allowing addition of microbes

Physical Threats

- Physical destruction of system's assets or disruption of water supply could be more likely than contamination. A single terrorist or a small group of terrorists could easily cripple an entire city by destroying the right equipment.
- Loss of water pressure compromises firefighting capabilities and could lead to possible bacterial buildup in the system.
- Potential for creating a water hammer effect by opening and closing major control valves and turning pumps on and off too quickly that could result in simultaneous main breaks.

Chemical Threats

Table 10.2 lists some chemicals that are effective in drinking water. The list includes both chemical warfare agents and industrial chemical poisons. There are five types of chemical warfare (CW) agents: nerve agents, blister agents, choking agents, blood agents, and hallucinogens. The list includes some of the chemical warfare agents and some of the industrial chemical poisons along with their acute concentrations.

Biological Threats

Several pathogens and biotoxins (see Chap. 2) exist that have been weaponized, are potentially resistant to disinfection by chlorination, and are stable for relatively long periods in water (Burrows and Renner, 1998, 1999). The pathogens include, *Clostridium perfringens*, plague and others, and biotoxins that include botulinum, aflatoxin, ricin, and others. Even though water provides dilution potential, a neutrally buoyant particle of any size could be used to disperse pathogens into drinking water systems. Other more sophisticated systems such as microcapsules also could be used to disperse pathogens in drinking water systems. Because of dilution effects, the effectiveness of a bioattack would be enhanced by introduction of the bioweapon near the tap.

Water storage and distribution systems can facilitate the delivery of an effective dose of toxicant to a potentially very large population. These systems also can facilitate a lower-level of chronic dose (for chemicals) with longer-term effects and lower-detection thresholds (Foran and Brosnan, 2000).

Table 10–2

Summary of
Chemicals
Effective in
Drinking Water

Chemical agents ((mg/L) unless otherwise noted)	Acute concentration* (0.5 L)	Recommended guidelines[†]	
		5 L/day	15 L/day
Chemical warfare agents			
Hydrogen cyanide	25	6.0	2.0
Tabun (GA, μg/L)	50	70.0	22.5
Sarin (GB, μg/L)	50	13.8	4.6
Soman (GD, μg/L)	50	6.0	2.0
VX (μg/L)	50	7.5	2.5
Lewisite (arsenic fraction)	100–130	80.0	27.0
Sulfur mustard (μg/L)		140.0	47.0
3-Quinucli dinyl benzilate (BZ, μg/L)		7.0	2.3
Lysergic acid diethylamide (LSD)	0.050		
Industrial chemical poisons			
Cyanides	25	6.0	2.0
Arsenic	100–130	80.0	27.0
Fluoride	3000		
Cadmium	15		
Mercury	75–300		
Dieldrin	5000		
Sodium fluoroacetate[‡]		Not provided	
Parathion[‡]		Not provided	

*Major John Garland, Water Vulnerability Assessments, (Armstrong Laboratory, AL-TR-1991-0049), April 8–9, 1991. The author assumes acute effects (death or debilitation) after consumption of 0.5 L.

[†]National Research Council, Committee on Toxicology, Guidelines for Chemical Warfare Agents in Military Field Drinking Water, 1995, 10. Listed doses are safe.

[‡]W. Dickinson Burrows, J. A. Valcik, and Alan Seitzinger, "Natural and Terrorist Threats to Drinking Water Systems," presented at the American Defense Preparedness Association 23rd Environmental Symposium and Exhibition, 7–10 April 1997, New Orleans, LA, 2. The authors consider the organophospate nerve agent VX, the two hallucinogens BZ and LSD, sodium cyanide, fluoroacetate and parathion as potential threat agents. They do not provide acute concentrations or lethal doses.

Hydrogen cyanide (blood agent), the nerve agents Tabun, Sarin, Soman, and VX, the blistering agents Lewisite and sulfur mustard, and the hallucinogen BZ are potential drinking water poisons. Garland focuses on LSD (a hallucinogen), nerve agents (VX is listed as most toxic), arsenic (Lewisite) and cyanide (hydrogen cyanide). Burrows, et al., list BZ, LSD, and VX. These agents, however, are not the only chemicals a saboteur might use in drinking water.

Source: As presented in Hickman (1999).

PRIOR TO SEPTEMBER 11, 2001

Prior to September 11, 2001 the literature contained numerous articles concerning the threat of terrorist attacks to our water supply infrastructure. A few of these included: Burrows and Renner (1998, 1999), DeNileon (2001), Dickey (2000), Foran and Brosnan (2000), Grayman et al. (2001), Haimes et al. (1998), Hickman (1999), and many others. The topic was receiving a little attention but basically the water utility industry was not implementing mitigation measures to such threats (Mays, 2004).

The following news article summarizes this:

> Washington, MSNBC, Jan 14. 2002—The vulnerability of the nation's water supply isn't in the headlines, it's in the details of the country's 54,065 public and private water systems. For years, experts have warned about the need to upgrade, repair and thoroughly assess the risk of terrorists targeting the nation's water supply and distribution channels. Yet most of those warnings have been ignored, under-funded or relegated to the back burner as policy-makers addressed "more important" projects. (By Brock N. Meeks, Washington, D.C. correspondent, MSNBC).

Before the events of September 11, 2001, there was a growing concern, by some, with the potential for terrorist use of biological weapons (bioweapons) to cause civilian harm (Lederberg, 1997; Simon, 1997; Burrows and Renner, 1998; Ableson, 1999; Waeckerle, 2000; Foran and Brosnan, 2000; and many others). These assumptions were focused around two assumptions (Foran and Brosnan, 2000): that a terrorist is most likely to effectively disperse bioweapons through air (Simon, 1997), and that we must be prepared to address terrorists use of bioweapons through treatment of affected individuals, with emphasis on strengthening the response of the health-care community (Simon, 1997; Waeckerle, 2000; Macintyre et al., 2000). For the most part, concern was not focused on the use of bioweapons in drinking systems (Ableson, 1999; Burrows and Renner, 1999), and much less attention was given to preattack detection than to postattack treatment (Foran and Brosnan, 2000).

Conferences such as the one *Early Warning Monitoring to Detect Hazardous Events in Water Supplies* (held May 1999 in Reston, Virginia) concluded that terrorist use of bioweapons poses a significant threat to drinking water. Other experts have agreed that introducing a toxin into a raw water reservoir would have little impact considering the dilution effect that several millions of gallons of water would have on a biohazard. However, the effectiveness of an attack could be enhanced by introducing the bioweapon near the tap, such as in the distribution system after postdisinfection (Foran and Brosnan, 2000).

The President's Commission on Critical Infrastructure Protection (PCCIP) was established by President Clinton in 1996. The PCCIP determined that the water infrastructure is highly vulnerable to a range of potential attacks and convened a public-private partnership called the Water Sector Critical Infrastructure Advisory Group. According to the PCCIP (1997), three attributes (which are obvious) are crucial to water supply users:

- There must be adequate quantities of water on demand.
- It must be delivered at sufficient pressure.
- It must be safe to use.

The first two are influenced by physical damage and the third attribute (water quality) is susceptible to physical damage as well as the introduction of microorganisms, toxins, chemicals, or radioactive materials. Actions (terrorist activities) that affect any one of these three attributes can be debilitating for the water supply system.

RESPONSE TO SEPTEMBER 11, 2001

Within a very short time after September 11, 2001 we began to see a concerted effort at all levels of government to begin addressing issues related to the threat of terrorist activities to U.S. water supply. Articles began appearing including: Bailey (2001), Blomgren (2002), Copeland and Cody (2002), Haas (2002), and many others. We saw a number of Acts passed such as the Security and Bioterrorism Preparedness and Response Act and the Homeland Security Act that addressed the U.S. water supply. These acts resulted in agencies such as the U.S. Environmental Protection Agency (EPA) developing new protocols to address their new responsibilities under these acts.

Public Health, Security and Bioterrorism Preparedness and Response Act ("Bioterrorism Act") (PL 107-188), June, 2002

This act requires every community water system that serves a population of more than 3,300 persons to

- Conduct a vulnerability assessment,
- Certify and submit a copy of the assessment to the EPA Administer,
- Prepare or revise an emergency response plan that incorporates the results of the vulnerability assessment, and
- Certify to the EPA Administer, within 6 months of completing the vulnerability assessment, that the system has completed or updated their emergency response plan.

Table 10.3 lists the key provisions of the security-related amendments.

Homeland Security Act (PL 107-296), November 25, 2002

This act directs the greatest reorganization of the federal government in decades by consolidating a host of security-related agencies into a single cabinet-level department to be headed by Tom Ridge, head of the White House Office of Homeland Security. It creates four major directorates to be led by White House-appointed undersecretaries. These are the Directorates of Information Analysis and Infrastructure Protection (IAIP)—most directly affects USEPA and the drinking water community; Science of Technology; Border and Transportation Security; and Emergency Preparedness and Response.

The law grants IAIP access to all pertinent information, including infrastructure vulnerabilities, and directs all federal agencies to *promptly provide* IAIP with all information they have on terrorism threats and infrastructure vulnerabilities. IAIP is responsible for overseeing transferred functions of the NIPC, the CIAO, the Energy Department's National Infrastructure Simulation and Analysis Center and Energy

1. Requires community water systems serving populations more than 3300 to conduct vulnerability assessments and submit them to U.S. EPA.
2. Requires specific elements to be included in a vulnerability assessment.
3. Requires each system that completes a vulnerability assessment to revise an emergency response plan and coordinate (to the extent possible) with local emergency planning committees.
4. Identifies specific completion dates for both vulnerability assessments and emergency response plans.
5. U.S. EPA is to develop security protocols as may be necessary to protect the copies of vulnerability assessments in its possession.
6. U.S. EPA is to provide guidance to community water systems serving populations of 3300 or less on how to conduct vulnerability assessments, prepare emergency response plans, and address threats.
7. U.S. EPA is to provide baseline information to community water systems regarding types of probable terrorist or other intentional threats.
8. U.S. EPA is to review current and future methods to prevent, detect, and respond to the intentional introduction of chemical, biological, or radiological contaminants into community water systems and their respective source waters.
9. U.S. EPA is to review methods and means by which terrorists or other individuals or groups could disrupt the supply of safe drinking water.
10. Authorizes funds to support these activities.

Table 10–3
Security-Related Amendments to Bioterrorism Act*

*In June 2002, the President signed PL 107-108, the Public Health, Security, and Bioterrorism Preparedness and Response Act (Bioterrorism Act) that includes provisions to help safeguard the nation's public drinking water systems against terrorist and other intentional acts. Key provisions of the new security-related amendments are summarized in this Table

Assurance Office, and the General Services Administration's Federal Computer Incident Response Center. IAIP will administer the Homeland Security Advisory System, which is the government's voice for public advisories about homeland threats as well as specific warnings and counterterrorism advice to state and local governments, the private sector and the public.

On October 2, 2002 the U.S. EPA announced their Strategic Plan for Homeland Security (www.epa.gov/epahome/headline_100202.htm). The goals of the plan are separated into four distinct mission areas: critical infrastructure protection; preparedness, response, and recovery; communication and information; and protection of EPA personnel and infrastructure. EPA's strategic plan lays out goals, tactics, and results for each of these areas.

USEPA's Protocol

The U.S. EPA has developed a compilation of water infrastructure security website links and tools located at www.epa.gov/safewater/security/index.html. Table 10.4 lists the U.S. EPA's strategic objectives to address drinking water system and wastewater utility security needs to meet the requirements of the Bioterrorism Act for public drinking water security.

Table 10–4

U.S. EPA
Objectives to
Ensure Safe
Drinking Water*

1. Providing tools and guidance to drinking water systems and wastewater utilities.
2. Providing training and technical assistance including "Train-the-Trainer" programs.
3. Providing financial assistance to undertake vulnerability assessments and emergency response plans as funds are made available.
4. Build and maintain reliable communication processes.
5. Build and maintain reliable information systems.
6. Improve knowledge of potential threats, methods to detect attacks, and effectiveness of security enhancements in the water sector.
7. Improve networking among groups involved in security-related matters—water, emergency response, laboratory, environmental, intelligence, and law enforcement communities.

*U.S. EPA has developed several strategic objectives to address drinking water system and wastewater utility security needs and also meet requirements set forth in the Bioterrorism Act for public drinking water security. These strategic objectives are as summarized in this table.

REFERENCES

Ableson, P. H., "Biological Warfare," *Science* 286: 1677, 1999.

Bailey, K. C., *The Biological and Toxin Weapons Threat to the United States,* National Institute for Public Policy, Fairfax, VA, October 2001.

Blomgren, P., "Utility Managers Need to Protect Water Systems from Cyberterrorism," *U.S. News,* 19: 10, October 2002.

Bruins, H. J., "Israel: Urban Water Infrastructure in the Desert," in L. W. Mays (ed.), *Urban Water Supply Handbook,* McGraw-Hill, New York, 2002.

Burns, N. L., C. A. Cooper, D. A. Dobbins, J. C. Edwards, and L. K. Lampe, "Security Analysis and Response for Water Utilities," in L. W. Mays (ed.), *Urban Water Supply Handbook,* McGraw-Hill, New York, 2002.

Burrows, W. D., and S. E. Renner, "Biological Warfare Agents as Potable Water Threats," U.S. Army Combined Arms Support Command, Fort Lee, VA, 1998.

Burrows, W. D. and S. E. Renner, "Biological Warfare Agents as Threats to Potable Water," *Environmental Health Perspectives.* 107(12): 975–984, December 1999.

Cheng, S.-T., B. C. Yen, and W. H. Tang, "Stochastic Risk Modeling of Dam Overtopping," in B. C. Yen and Y.-K. Tung, (eds.), *Reliability and Uncertainty Analyses in Hydraulic Design,* American Society of Civil Engineers, New York, pp. 123–132, 1993.

Clark, R. M., and R. A. Deininger, "Protecting the Nations Critical Infrastructure: The Vulnerability of U.S. Water Supply Systems," in L. W. Mays (ed.), *Urban Water Supply Handbook,* McGraw-Hill, New York, 2002.

Copeland, C., and B. Cody, "Terrorism and Security Issues Facing the Water Infrastructure Sector," Order Code RS21026, CRS Report for Congress, *Congressional Research Service,* The Library of Congress, Washington, DC, June 18, 2002.

DeNileon, G. P., "The Who, Why, and How of Counterterrorism Issues," *J. Am. Water Works Assoc.,* 93(5): 78–85, May 2001.

Dickey, M. E., "Biocruise: A Contemporary Threat," Counterproliferation Paper No. 7, Future Warfare Series No. 7 available at: www.au.af.mil/au/awc/awcgate/cpc-pubs/dickey.htm, USAF Counterproliferation Center, Air War College, Air University, Maxwell Air Force Base, Alabama, September 2000.

Foran, J. A., and T. M. Brosnan, "Early Warning Systems for Hazardous Biological Agents in Potable Water," *Environ. Health Perspect.*, 108(10): 993–996, October 2000.

Gleick, P. H., "Water, War, and Peace in the Middle East," *Environment*, vol. 36, no. 3, Heldref Publishers, Washington, DC, p. 6, 1994.

Gleick, P. H., "Water and Conflict," in: P. H. Gleick (ed.), *The World's Water 1998–1999*, Island Press, Washington, DC, pp. 105–135, 1998.

Gleick, P. H., "Water Conflict Chronology," available at: http://www.worldwater.org/conflict.htm, 2000.

Grayman, W. M., R. A. Deininger, and R. M. Males, "Design of Early Warning and Predictive Source-water Monitoring Systems," *AWWA Research Foundation and AWWA*, 2001.

Haas, C. N., "The Role of Risk Analysis in Understanding Bioterrorism," *Risk Anal.*, 22(2): 671–677, 2002.

Haimes, Y. Y., et al., "Reducing Vulnerability of Water Supply Systems to Attack," *J. Infrastruc. Syst., ASCE* 4(4): December 1998.

Hickman, D. C., "A Chemical and Biological Warfare Threat: USAF Water Systems at Risk," Counterproliferation Paper No. 3, Future Warfare Series No. 3, available at: www.au.af.mil/au/awc/awcgate/cpc-pubs/hickman.htm, USAF Counterproliferation Center, Air War College, Air University, Maxwell Air Force Base, Alabama, September 1999.

Lederberg, J., "Infectious Disease and Biological Weapons: Prophylaxis and Mitigation," *JAMA* 278: 435–438, 1997.

Macintyre, A. J., G. W. Christopher, E. Eitzen, R. Gum, S. Weir, C. DeAtley, K. Tonat, and J. A. Barbera, "Weapons of Mass Destruction Events with Contaminated Casualties," *JAMA* 283(2): 242–249, 2000.

Mays, L. W. (ed.), *Reliability Analysis of Water Distribution Systems*, American Society of Civil Engineers, New York, 1989.

Mays, L. W. (ed.), *Water Distribution Systems Handbook*, McGraw-Hill, New York, 2000.

Mays, L. W. (ed.), *Urban Water Supply Handbook*, McGraw-Hill, New York, 2002.

Mays, L. W. (ed.), *Urban Water Supply Management Tools*, McGraw-Hill, New York, 2003.

Mays, L. W. *Water Supply System Security*, McGraw-Hill, New York, 2004.

Peplow, D.E., C.D. Sulforedge, R.L. Saunders R.H. Morris, and T.A. Hann, "Calculating Nuclear Power Plant Vulnerability Using Integrated Gemometry and Event/Fault Tree models, "Oat Ridge National Lnboratory

President's Commission on Critical Infrastructure Protection, Appendix A, Sector Summary Reports, *Critical Foundations: Protecting America's Infrastructure*: A-45, available at: http:// www.ciao.gov/PCCIP/PCCIP_Report.pdf.

Simon, J. D., "Biological Terrorism: Preparing to Meet the Threat," *JAMA* 278: 428–430, 1997.

U.S. Army Medical Research Institute of Infectious Disease, *USAMRID's Medical Management of Biological Causalities Handbook*, available at: www.usamriid.army.mil/education/bluebook.html, 2001.

U.S. EPA, Guidance for Water Utility Response, Recovery, and Remediation Actions for Man-Made and/or Technological Emergencies, available at: http://www.epa.gov/safewater/security/er-guidance.pdf.

U.S. EPA, Guidance for Water Utility Response, Recovery & Remediation Actions for Man-Made and/or Technological Emergencies, EPA 810-R-02-001, Office of Water (4601), available at: www. epa.gov/safewater April 2002.

U.S. EPA, "Water Security Strategy for Systems Serving Populations Less than 100,000/15 MGD or Less," July 9, 2002.

U.S. EPA, "Instructions to Assist Community Water Systems in Complying with the Public Health Security and Bioterrorism Preparedness and Response Act of 2002," EPA 810-R-02-001, Office of Water, available at: www.epa.gov/safewater/security, January 2003.

U.S. EPA, "Vulnerability Assessment Fact Sheet 12-19," EPA 816-F-02-025, Office of Water, available at: www.epa.gov/safewater/security/va fact sheet 12-19.pdf, also at www.epa.gov/ogwdw/ index.html, November 2002.

U.S. EPA, available at: http://www.epa.gov/swercepp/cntr-ter.html.

Waeckerle, J. F., "Domestic Preparedness for Events Involving Weapons of Mass Destruction," *JAMA* 283(2): 252–254, 2000.

11
WATER RESOURCES SUSTAINABILITY ISSUES IN SOUTH KOREA

Joong-Hoon Kim and Hwan-Don Jun
Department of Civil and Environmental Engineering
Korea University
Seoul, S. Korea

Sung Kim
Sustainable Water Resources Research Center (SWRRC)
Koyang, S. Korea

Jae-Soo Lee
Department of Civil and Environmental Engineering
Jeonju University, Jeonju
Jeollabukdo, S. Korea

INTRODUCTION

Sustainability has been an issue in the management of water resources in South Korea for the last decade. The Korean government estimates the amount of water shortage will be 0.1 billion m^3 in 2006, and 1.2 billion m^3 in 2011. In order to resolve water-related problems a new comprehensive water resources plan ("Water Vision 2020") was established in 2001 with the target year as 2020.

After two consecutive disastrous typhoons in 2002 and 2003, costing 392 human lives and property damages amounting to more than 10 billion U.S. dollars, the Korean government set a plan to reduce the property damages due to flooding below 70 percent of current damages by year 2020 by implementing various countermeasures such as improvement of flood forecasting systems, assignment of flood discharges to each upstream subbasin, and mediation of land-use plans. A specific effort was made by the government to develop urban flood disaster management techniques by establishing "Urban Flood Disaster Management Research Center" for a three-phase project—each phase lasting for five years.

River restoration has been an issue in river management for the last decade along with flood protection. Following a success in the restoration of Yangjae-cheon, the city of Seoul recently finished the Cheonggye-cheon restoration project, which is considered as one of the city's most ambitious ecological projects ever. The water resources sustainability project is a good example of harmony that can exist between development and the environment.

Water security has been a worldwide issue recently, especially after the attack on 9/11. Every country has its own areas of concern regarding the water security issue. The issue in Korea is discussed in the last section focusing on specific parameters, such as climate.

The limitations of the water resources policy used in Korea since the 1960s are now surfacing. Although water demand continues to increase, the development of water resources by traditional methods has become difficult to sustain environmentally, politically, and socially. As a result, there is a wide gap between supply and demand and disputes concerning preferential allocation of limited water resources are occurring. The Korean government estimates that the amount of water shortage will be 0.1 billion m^3 in 2006, and 1.2 billion m^3 in 2011. To add to the problems, Korea has endured frequent climatic extremes recently: five years of drought and three years of floods in the ten-year period of 1993 through 2002.

In order to resolve water-related problems the country expects to face in the future, a new comprehensive water resources plan called "Water Vision 2020" (Korean Ministry of Construction and Transportation, 2001) has been established with the target year being 2020. "Water Vision 2020" addresses the changes in the situation around water resources as well as the problems concerning the people and the land. It presents the principal strategies of water resources policy with long-term and comprehensive viewpoints.

To support the "Water Vision 2020" technically, "sustainable water resources research project," a ten-year research project for new water resources policy and technology has been launched in 2001 as one of twenty-first century frontier R&D programs supported by the Ministry of Science and Technology. One of the main objectives of the research project is to solve the water shortage problem for the next 10-year period by adapting sustainable water resources management. To achieve the objective, policies and technologies needed in planning and operation for integrated water management including surface/ground water and alternative water resources are investigated.

This section describes the current state of water resource planning in Korea and a new direction of sustainable water resources management.

Since the 1960s, the Korean government has established the National Water Resources Plan for addressing water demand and supply prospect, water resources development and its use, flood and river environmental management, and has developed guidelines for each of these issues. The first National Comprehensive Water Resources Plan in 1965 made a basic strategy to promote the development of irrigation reservoirs for increasing crop products and hydropower generation dams to cope with electric power demand. The second National Comprehensive Water Resources Plan in 1980 was executed for the development of large multipurpose dams, estuary barrages, and river improvement for reduction of flood damage. In 1990, the third National Comprehensive Water Resources Plan was established for stabilization of nationwide water supply using wide-area water supply networks, reduction of flood damages, and formation of healthy river environments. Recently, in 2001, the fourth National Comprehensive Water Resources Plan, settling 2020 as its target year, was presented. This section describes the difficulties experienced in the past National Water Resources Planning in Korea.

Difficulties of National Water Resources Planning

Large Rainfall Fluctuation

Large differences exist in the annual rainfall in Korea. When one of the world's longest annual rainfall records from Seoul (213 years, from 1770 to 1990) is examined, the large fluctuations are apparent (Kim et al., 1993). Because the standard deviation of the annual precipitation accounts for 31 percent of the mean, 95 percent confidence lower limit is 1.65 times the 31 percent, or 51 percent less than the mean. To satisfy water usage demand with 95 percent confidence level, the supply volume must be guaranteed with approximately 50 percent of an average annual rainfall. In Korea, approximately half of the rainfall is lost to *evapotranspiration*. If the amount of water lost to evapotranspiration is not affected by rainfall volume, then, if rainfall decreases by half, the amount of water available for use would be decreased by one-fourth. As a result, in Korea, a reservoir capable of storing rainfall over a long period of time is needed. In other words, to ensure water supply, Korea needs to maintain at least one year's supply of usable water.

Insufficient Standards of Water Supply System

The representative water resource supply, the multipurpose dams of Korea, currently operate under the standards established during the drought from 1967 to 1968. Most of the large-scale water resource projects developed in the 1970s used the criteria set from the 1967–1968 drought because the data were readily available and the experience of the drought was fresh in the mind of the developers. However, historical data clearly show that the drought of 1967–68 is far from an extreme drought. Korea experienced an extreme drought, from 1884 to 1910. For the 27 years, the average rainfall did not exceed 70 percent of the average annual rainfall for the 213 years recorded. As a result of the drought, insurrections and extreme social unrest occurred and even the Korean monarchy crumbled.

The 1967–68 drought condition standard used to establish the water resource policy for usable water supply does not sufficiently reflect Korea's drought characteristics. This policy was established considering the insufficient financial conditions of the 1970s. In addition, this policy, created for a population of 20 million Seoul metropolitan people, is insufficient to provide usable water supply for the current population of 23 million. This fact has been one of the most serious problems facing Korean water resource policy.

Although the problem of insufficient design level still remains, the "Water Vision 2020" established in 2001 adopted an equivalent drought recurrence interval instead of the fixed 1967–68 drought condition. Thirty years of drought recurrence interval for urban water supply and ten years for agricultural water supply were used as a planning standard throughout the country.

Rapid Increase of Water Demands

Water demand has rapidly increased since 1965. Domestic water use is increasing due to population growth, the improvement of the life standard, and urbanization. Industrial water use had been extensive up to the 1980s, while the water use is only slightly increasing since the 1990s due to the increased recycling ratio and the change in the structure of industries. As for agriculture water use, it is increasing due to the wide utilization of paddy fields and progress in the irrigation of dry fields. Since 1990, in-stream water use (environmental water use) is rapidly increasing due to the rising awareness in river environments (Table 11.1).

It is expected that total water demand is going to increase steadily from 2001 to 2020. In the estimate of future domestic water use in "Water Vision 2020," we consolidate demand management policy, targeting water pricing, leakage control, gray water reuse, and the spread of water-saving equipment. In addition, the estimate takes population growth and economic conditions into account, considering the achievements in domestic water use in the past. In regard the domestic water use is anticipated to reach 348 L/capita/day by 2011, as compared to 280 L/capita/day in 1998. The amount shows a steady increase by about 2 percent per year, reaching 8.6 billion m^3 by 2011, as compared to 7.2 billion m^3 in 1998 (Table 11.2).

A steady increase in industrial water use is also anticipated, although the recycling ratio will be rising. The amount of demanded industrial water use on the national average is likely to reach 4.0 billion m^3, which was 2.9 billion m^3 in 1998.

Table 11.1

Changes in the Water Demand (Unit: Billion Cubic Meters)

Water Use	1965	1980	1990	1998	2001
Total	5.1	15.3	24.9	32.8	33.6
Domestic water	0.2	1.9	4.2	7.2	7.2
Industrial water	0.4	0.7	2.4	2.9	3.3
Agricultural water	4.5	10.2	14.7	15.6	15.6
In-stream water	–	2.5	3.6	7.1	7.5

Water Use	2006	2011	2016	2020
Total	34.7	36.9	37.4	37.8
Domestic water	7.6	8.6	8.8	8.9
Industrial water	3.7	4.0	4.3	4.6
Agricultural water	15.7	15.9	15.9	15.9
In-stream water	7.7	8.4	8.4	8.4

Table 11.2
Prospects of the National Average Water Demand (Unit: Billion Cubic Meters)

In the future, the decrease of acreage of farmland will be expected up to 2011 due to the urbanization and regional development. However, it is expected that agricultural water use will not immediately decrease in proportion to the decrease of acreage of farmland. Rather than that, owing to the change of farming form, irrigation water for winter seasons such as greenhouse farming, increase of irrigation of dry fields and stock-farming water, the agriculture water use will slightly increase. The use will reach 15.7 billion m^3 in 2011, which was 15.6 billion m^3 in 1998.

The needs for conservation and improvement of the water environment are expected to intensify more to maintain and create a natural and living environment, and to improve water quality. Therefore, it is anticipated that the environmental water use will be steadily increased, reaching 8.4 billion m^3 in 2011, which was 7.1 billion m^3 in 1998.

High Usage of Renewable Water Resources

The water available for use is the quantity that remains after evapotranspiration. If more than the remaining amount is used, whether from groundwater or surface water, the ultimate result is the drying up of water resources. As a result, the usable water quantity is referred to as renewable water resources and is used as an index to represent water scarcity of a country. In Korea, the national water resources plan states the renewable water sources to be 73.1 billion m^3/year. When the volume is divided by the population, the total annual volume available for use per person is 1550 m^3. United Nations Economic and Social Commission for Asia and the Pacific (1992) estimated for 26 countries in Asia the average available annual water to be 4143 m^3. The water available to Koreans is only 37 percent of the average for Asia. Among the 26 surveyed countries in Asia, the only two nations below 1700 m^3 are Korea and Singapore.

Korea's heavy water use contributes to the water shortage problems. According to the national water resources plan, Korea uses 33.1 billion m^3 of the 73.1 billion m^3 available for use annually. Because 7.1 billion m^3 is used for in-stream flow, the withdrawal water use becomes 26.0 billion m^3. Accordingly, approximately 35.6 percent of the renewable water resources are being used currently.

United Nations Department for Policy Coordination and Sustainable Development (1997) used this percentage to determine the water stress of a nation. Water usage under 10 percent does not result in water stress. Water usage between 10 and 20 percent is considered normal. Water usage between 20 and 40 percent results in water stress above normal and requires intensive management of supply and demand. If water usage percentage rises above 40 percent, a serious water shortage problem may result. In this scenario, water resources depend more and more on depleting groundwater and desalinization. Accordingly, to manage supply and demand, a special plan is needed. Water usage above 40 percent could not be sustainable and the water shortage may work to limit essential economic progress. The average water use is 35.6 percent in Korea and as a result, an asserted effort to regulate supply and demand is needed. It is important to keep in mind that 35.6 percent is only an average and in some watersheds, the water usage percentage far exceeds the 40 percent limit. In most of these watersheds, the environmental quality continues to degrade. Without a decrease in water use, such a high percentage of water use makes the management of a sustainable water resource very difficult.

Increase of Frequency and Peaks of Floods

In the last 15 years, Korea has experienced large-scale floods never before recorded. In 1987, a large flood in the Keum river watershed caused 1 trillion won (about 2 billion U.S. dollars) worth of property damage. In 1990, an intense rainfall recording 400 mm in the southern Han river basin caused break of levee near Seoul and resulted in a large flood. In 1996, at the Hantan river basin, a concentrated rainfall measuring over 600 mm overwhelmed the Yonchun Dam's flood control volume and flooded the downstream region. Shortly thereafter, the dam collapsed. In August 1998, a surprising record rainfall fell across Korea. As a result, flash floods occurred in the Chiri mountain range, a reservoir in Kangwha-do collapsed, and damage as well as loss of lives occurred in Chungrang stream and mid-and-small river basins property. In 1999, at the Imjin river basin, a rainfall event over 700 mm destroyed Yunchun Dam once again, and caused severe flood damage in Moonsan and Yunchun.

The recent flood damage can be attributed to the change in rainfall characteristics. It is worth noting that rain in 1998 broke the old records, such as the 118.6 mm rainfall in one hour in Seoul. In the last 20-year period, most of the rainfall records were broken throughout the country.

As stated before, the serious problem that Korean water resources policy faces today is how to counteract the extreme floods. The recent flood events are new conditions never dealt with before. Accordingly, the construction methods established during different circumstances are no longer effective. Extreme rainfall events have increased compared to before and as a result, flood occurrences have also increased. In contrast, the flood control ability of multipurpose dams has declined. Questions about the effectiveness of flood control facility properties such as flood frequency, safety factors, and clearance are beginning to arise. Questions are also arising about the flood control ability of multipurpose dams. To maintain the flood prevention level like before, the flood control volume must be increased and the water supply volume must be decreased.

New Directions

To resolve the difficulties of water resources planning, Korea's water resource policy needs to be revised. The current policy focused on water supply needs to be changed to a policy that focuses on the management of water resources as a priority. This shift in policy does not mean the abandonment of water resource development. The conversion to a policy of management should consider water supply guarantee, flood safety guarantee, and the improvement of the environment as priorities.

Sound and Stable Water Utilization

Increase of continuous water demand is expected in the future due to growth of population and economics, urbanization, and extension of domestic water supply. However, implementation of structural measures like building of dams to secure water resources is getting to be more difficult since there is a scarcity of suitable dam building sites and rising awareness in river environments. To solve these problems, we need to promote the sound development of water resources through multiple water sources. At the same time, we should build up societal awareness of water conservation, develop strategies to augment available water resources and develop advanced water management technology. We aim at acquiring balance in water demand and supply in most of the areas on the basis of the existing project standard—a drought with recurrence interval of about 30 years.

Formation of Safe Land Against Flooding

A variety of flood control projects have been implemented to protect life and property against floods, but flood damages are still substantial. Frequent occurrences of rainfall over design flood standards have been the immediate cause. Moreover, rapid growth in population and its concentration in flood prone areas have made flood damages a serious issue since the 1980s. At the same time, "damage potential" of floods is increasing. Principal measures to solve these problems are channel improvements such as channel widening, levee construction and reinforcement, and riverbed dredging so that flooding of less than the design flood can be discharged without inundating land along the river. Besides channel improvement, there is a plan for basin improvement such that the natural flood behavior of drainage is largely retained or improved. In addition, measures of predisaster mitigation and the development of accurate flood forecasting technology will be positively reinforced.

Creation of River Environment in Harmony with Nature

To protect flood damages, we have implemented channel improvement works through river channel straightening and construction of banks covered with concrete so that the fruits of flood loss prevention are obtained. However, these works deprive creatures living in and around the rivers of their living spaces. From a different perspective, the environment of rivers has begun to be reevaluated as being important, and rivers are seen as providing free open spaces for the general public in the cities and for use in outdoor sports. We have been given great responsibility for the care of our natural environment while at the same time are entrusted with devising new ways of protecting and preserving diverse river functions.

Research for Sustainable Water Resources Management

The difficult water situation made the government (Ministry of Science & Technology and Ministry of Construction) launch a huge and long range (10-year long) R&D program called Sustainable Water Resources Research Project in the name of 21C New Frontier Project from the year 2001 (http://www.water21.re.kr/en/index.html).

The program is divided into four research areas:

1. Integrated water resources management technology
2. Surface water management technology
3. Ground water management technology
4. Alternative water resources management technology

Experts in water and wastewater including scientists and engineers are, wholly or partially, involved in this national program. It is expected that Korea can solve the problems of water shortage by securing extra 3 billion m^3 of water in 2011 as the program proceeds.

In the areas of alternative water resource security, development, and application of water reuse, technology is regarded as one of the major approaches to satisfy two aspects, that is, pollution reduction and development of water new sources. It is known that 66 percent of wastewater produced a day ends up with a 10 million m^3/day discharge of treated effluent from municipal wastewater treatment plants. In other words 0.36 billion m^3 of new water resource per year can be available if only 10 percent of wastewater could be reused. Technologies mainly focused on include *membrane bioreactor* (MBR), *nanofiltration* (NF), *soil aquifer treatment* (SAT), and *advanced oxidation processes* (AOPs)/toxicity monitoring.

Summary and Conclusions

According to the precipitation records observed in Seoul from 1777, the Korean peninsula has experienced extreme fluctuation in precipitation. Annual precipitation varies from below 500 mm to 2500 mm. There is a record that extreme drought continued for 26 years from 1884 to 1910 at the end of Joseon Dynasty. Since such extreme precipitation phenomenon has not occurred in recent years and the national water resources plan has been established based on the data collected after 1966, the weaknesses of water resources system in Korea has not been exposed yet. Besides, considering the extensive drought problem occurring all over the globe due to the impact of global climate change, relevant preparation is urgently required.

Korea's water resource policy needs to be revised. The current policy focused on water supply needs to be changed to a policy that focuses on the management of water resources as a priority. This shift in policy does not mean the abandonment of water resource development. The conversion to a policy of management should consider the guaranteed water supply and, flood safety guarantee, and the improvement of the environment as priorities.

The new direction of water management will be guided effectively by the comprehensive and long-range (10-year long) R&D program called sustainable water resources research project in the name of 21C New Frontier Project from year 2001. Most of the scientific and engineering experts in water and wastewater profession are involved in this national program. It is expected that Korea can solve the problems of water shortage by securing an extra three billion m^3 of water in 2011 as the program proceeds.

RESEARCH ON URBAN FLOOD DISASTER MANAGEMENT

Introduction

About 70 percent of the population in Korea live in urban areas, many of which are located in riverine lowlands, which are inundated during floods, so insufficient drainage repeatedly causes loss of human lives and properties (Table 11.3). The increase of discharge and the shortening of flood travel time due to the change of meteorological environment and the extension of urbanization bring about difficulties in river design and management of urban areas.

Water use	Han River	Nakdong River	Gum River	The Other Rivers	Total
Death (person)	400	237	74	656	**1,367**
Flood sufferers (person)	64,065	55,559	26,611	120,138	**266,373**
Inundation area (ha)	85,827	117,947	84,602	244,744	**533,120**
The damage (billion)	3,814.6	4,371.6	1,267.4	7,721.0	**17,174.6**

Table 11.3

Water-Related Damages for the 10 Year Period of 1994—2003

For the purpose of protection of lives and properties from urban floods, the Ministry of Construction and Transportation organized the Urban Flood Disaster Management Research Center (http://www.urbanflood.or.kr) as a part of the construction of essential technical research development, and five year's concentrated research started the development of urban flood disaster management techniques.

The purpose of the Urban Flood Disaster Management Research Center is to make flood-free cities by developing and providing practical techniques that can be used directly for urban river design and management through research, the research including analysis of urban flood disasters, development of techniques for urban flood forecasting and warning, inundation prediction, urban flood disaster mitigation, and that of planning and management of urban flood defenses.

The Goals

Research and Development

The ultimate aim of the research is to prevent loss of life and to mitigate property damage to more than 30 percent through a systematic development of research, to support the theoretical and technical base for establishment and enforcement of comprehensive flood control countermeasures in urban river basins by the Ministry of Construction and Transportation, and to put to practical use research products, such as analytical, forecasting, warning, mitigation, and management techniques. See Fig. 11.1.

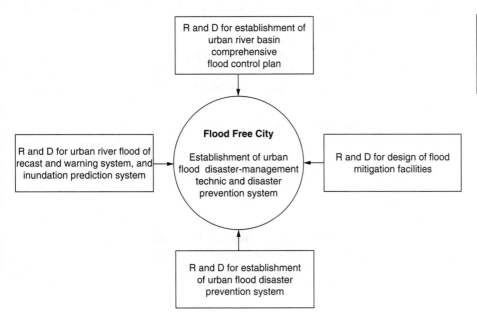

Figure 11.1

Goal of urban flood disaster management research.

Expected Effects

Technological Effects   This research is expected to present an ideal alternative for urban flood defense by providing practical techniques which can be applied to the "urban river basin comprehensive flood control plan" carried out currently by the Ministry of Construction and Transportation.

The techniques which can be obtained by this research development are as follows:

1. Hydraulic and hydrologic analytical techniques for urban river basin
2. Design techniques for drainage network system
3. Techniques to improve the flood control safety for urban river structures
4. Hydraulic restoration techniques for urban river
5. Techniques of operation and data management for the representative basin
6. Techniques to improve rainfall forecast, and flood forecast and warning system
7. Techniques for protecting lowland
8. Design techniques for structural countermeasures such as bank, outlet channel, detention, and infiltration pond
9. Techniques for optimum rainfall management and for the restoration of basin ecological environment
10. Techniques to set up comprehensive basin flood control plan
11. Techniques for the operation and maintenance of urban drainage system
12. Techniques to develop water resisting facility
13. Techniques for the economic evaluation of urban flood
14. Techniques for disaster-free urban design

Through the research, improvement of laws and regulations related to flood disaster, revision of design criterion and development of guidelines, improvements of flood control systems, can be achieved. In addition to these, minimization of sewer contamination by the development of combined sewer overflow(CSO) control techniques, maintenance of river environments by implementing flood control, and river environment evaluation required to the river restoration projects connected with flood control are also possible.

Economical and Industrial Effects   In recent times, damages by overflow and inundation of urban river areas have increased astronomically and the amount of damages has also increased. Compared to the problems in rural areas, damage in the urban areas have been much more serious including loss of lives.

In the case of inundation or washing away of social infrastructure by urban floods, social confusion and economic loss due to the loss of life and property, traffic paralysis, and water supply distribution disturbance can be minimized by applying the results of the research. Considering the fact that damage from floods and recovery expenditure in 2002 were 5.2 trillion and 9.1 trillion hwan respectively, savings of more than three trillion hwan can be expected if recovery expenditure is reduced by 30 percent (1 US dollar ≈ 1000 hwans).

The drainage pumping pond or storm sewer, which plays an important role in the urban drainage system is very expensive to install and operate, so appropriate design and operation are important to achieve overall reduction of urban inundation damage. Technical investigation for implementing natural river maintenance techniques is also important for river restoration and flood control.

The Urban Flood Disaster Management Research comprises four major research groups as follows:

Research Groups

- Group 1: Techniques for disaster analysis of urban flood
  - Comparison and application of urban river discharge models
  - Development of discharge analysis techniques for urban drainage basins
  - Ecological and hydraulic analyses for urban river systems
  - Hydraulic analysis techniques for urban river structures
  - Selection of representative drainage basin and construction of measurement network and operation system
- Group 2: Techniques for flood forecast and warning systems and for the prediction of inundated area
  - Rainfall analysis and forecasting for urban river basin
  - Practical use of urban flood forecast and warning system
  - Analysis of bank failure characteristics and hydraulic analysis
  - Forecast of inundation at urban area
- Group 3: Design techniques to mitigate urban flood disasters
  - Design for urban river bank
  - Design for underground outlet and diversion channel
  - Design for underground multipurpose detention pond
  - Design for mitigation facilities of urban flood discharge
- Group 4: Techniques for defense plans and urban flood management
  - Comprehensive urban river basin flood control plan
  - Operation and maintenance of urban internal drainage system
  - Evaluation of potential risk and damage due to urban flood
  - Urban disaster prevention plan and nonstructural flood response

The different techniques mentioned can be coordinated as in Fig. 11.2 to maximize the effect of Urban Flood Disaster Management Research.

Analysis of Urban Flood

Need for Research Recent rapid industrialization and urbanization cause a heavy load on urban river systems leading to changes in the ecosystem. Inundation happens frequently in riverine lowland areas due to rapid urbanization. Bridges and highways cause rise of flood levels and result in flood damage due to reverse flow' and overbank flooding. Large amounts have been devoted in the budget to strengthen drainage capacity at these areas. However, good design techniques of drainage systems and facilities are lacking.

Figure 11.2

Schematic diagram of urban flood disaster management research.

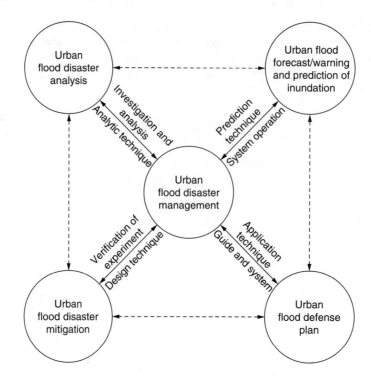

New design techniques are required to address flood control, create favorable water, and environment ecosystems through proper riverine vegetation in urban river basins. In addition, we need scientific and systematic operation of representative urban drainage basins to add to the basic hydraulic and hydrologic data, such as runoff coefficients and travel times.

Research Goals

- Development of techniques for runoff analysis and river design in urban river drainage systems
  - Analysis of existing theories for urban flood modeling and applicability
  - Development of a practical model for domestic urban drainage and water quality analyses
- Development of design techniques for drainage systems
  - Development of techniques to induce and intercept rainwater based on observation and experiment
  - Techniques to improve drainage capability in habitual inundated areas
- Grasp of present state of urban river structures and analysis of hydraulic characteristics
  - Techniques to analyze the hydraulic effects of urban river structures
  - Techniques for riverbed scour and countermeasures
- Development of ecologic and hydraulic analysis of urban rivers during floods, and restoration techniques

- Development of forecasting techniques for the variation of flood level by fluvial vegetation, and for verification using hydraulic model experiments
- Development of techniques for evaluation of ecological effects, and for restoration based on runoff
- Operation of representative drainage basins
  - Collection and distribution of hydraulic, water quality, and hydrologic data
  - Offer of basic data for urban flood forecast, warning, and analysis

Flood Forecast and Warning Systems and Prediction Systems for Inundated Area

Need for Research   Flood forecasting and warning systems, and the prediction of inundated bank areas to mitigate flood damage should be analyzed in conjunction with parameters such as rainfall, urban discharge, drainage system, and inundation, and be operated in real time.

The Korean Meteorological Administration forecasts long-term rainfall with reliable accuracy using weather radars, however they need more accurate techniques for short-term rainfall forecasting, flash floods and concentrated rainfall. In addition, flood forecasting and warning systems, and prediction systems for inundated area should be established.

Research is needed to calibrate, verify, and put to practical use the data collected through the process of rainfall forecast and analysis. Prediction systems to forecast inundated area are also needed through the analysis of hydraulic models and numerical experiments for the failure of urban bank and overflow considering the conditions of river flow, protected lowland area and bank materials.

Research Goals   Research goals are as follows:

Through the establishment of real-time flood forecasting and warning systems for urban floods, improve the accuracy of the system from 50 to 88 percent in coordination with the rainfall forecast system by the Meteorological Administration and the Ministry of Construction and Transportation.

Prepare inundated area maps according to the internal and external overflow in urban areas, and develop practical techniques to mitigate urban flood damage in conjunction with comprehensive urban flood control countermeasures.

- Rainfall analysis for urban basins and practical use of forecasting techniques
- Urban flood forecasting and warning systems
- Analysis of characteristics of dike failure and hydraulic analysis of urban river systems
- Practical application of forecasting inundation in urban area

Design Techniques to Mitigate Urban Flood Disasters

Need for Research   Mid- and long-term investigation, analysis, evaluation, simulation, development and application of techniques for structural countermeasures, and preparation of enforcement schemes are urgent. Theoretical techniques, for simulation, design, and mainframe are needed in the field. In addition, model experiments for proper design of various facilities, application criteria, field application, and operation techniques are also needed.

Efforts in this direction include (1) experiments and evaluation of bank failure connected with flood defense facilities, (2) improvement of design techniques for structural countermeasures, such as outlet channels and underground channels to discharge distributed floods, (3) design and operation of facilities to reduce flood discharge in urban basins, and (4) improvement of urban flood defense capabilities through underground outlet channels and multipurpose underground detention ponds.

Research Goals

- Development of design techniques for urban river banks
  - Development and verification of flood defense facilities through failure experiment considering the material and connection of urban river banks
- Development of design techniques for underground outlet and diversion channels
  - Experimental verification of underground outlet and diversion channels to increase safety of urban flood control structures
- Development of design techniques for multipurpose underground storage ponds
  - Experimental verification to increase hydraulic and hydrologic safety of multipurpose underground detention ponds storing flood discharge
- Development of design techniques for urban flood mitigation facilities
  - Experimental verification for detention and retention ponds to reduce urban flood discharge by detention, retention, and infiltration

Defense Plans and Management Techniques for Urban Flood

Need for Research   Establishment and enforcement of comprehensive basin flood control plans are important. Operation and maintenance of internal drainage facilities, economic evaluation and establishment of countermeasures, establishment of linked techniques between system, administration and urban development law for flood mitigation and prevention of inundation are also important.

There are no concrete procedures or basic laws related to design, utilization and management of national land, and urban plan for disaster prevention connected with comprehensive national land development plan.

Research Goals

- Establishment of techniques to set up comprehensive urban river basin flood control plan
  - Optimal distribution of flood discharge and mitigation of flood damage by responding effectively to the reduction of safety due to increase of flood discharge
- Development of operation and maintenance techniques for urban drainage system
  - Integrated operation and maintenance of urban internal drainage system
- Development of evaluation techniques for potential risk and damage due to urban flood

- Establish flood control safety connected with the overflow simulation technique, and develop evaluation technique for potential risk and damage and efficacy of countermeasures
- Development of disaster-free urban design and nonstructural flood response
  - Establish a systematic enforcement scheme for flood damage reduction in urban planning
- Develop and enforce nonstructural flood response techniques like Flood Insurance

Summary and Conclusions

Through the four major researches on urban flood disaster mitigation, theoretical and practical techniques will be established, and reduction of flood damages and appropriate management for urban flood defense are expected.

This research will contribute to the public welfare through the presentation of standard guidelines for the acquisition and propagation of disaster information, for the development and application of software, and for the establishment of comprehensive flood control plans.

RESTORATION OF CHEONGGYE-CHEON

Introduction

The original name of the Cheonggye-cheon(river) is "Gaecheon" meaning "open stream" and its sources are Inwang-san(Mt.) located northwest of Seoul, the south foot of Bugak-san (Mt.), and the north foot of Nam-san(Mt.). It is an urban river flowing from west to east converging in the center of Seoul. Its total length spans 10.92 km. The total drainage area of the Cheonggye-cheon is 50.96 km^2 and is located at the center of Seoul. Ever since Seoul was designated as the capital, during the Joseon Dynasty in 1394, the Cheonggye-cheon has not only divided the capital geographically but also plays a symbolic role as the boundary in politics, society, and culture.

The Cheonggye-cheon restoration project was not just a planning effort but one that the entire nation was interested in as a symbolic project to revive an important part of Korea's historical and natural heritage. The project was completed successfully in October of 2005, and the capital city became friendly to both the environment and the people. The project not only contributes to renewing the image of Seoul, but also sets a new paradigm for urban management.

With its historical site restored Seoul regained its 600-year history as the capital of Korea becoming a city where the modern era is amalgamated with tradition. The restored Cheonggye-cheon area has become a major tourist attraction for both Korean and overseas tourists. The project focused on improving the living environment and business.

History

Because it is within the capital, the Cheonggycheon has served many important functions. However, flooding has been a prime problem. When the river in Seoul was inundated in 1407, the water management authority called "Gaegeodogam" (later, changed to "Gaecheondogam") was established in November 1411 and large-scale construction on the urban streams was undertaken using 52,800 people. The construction was carried out by digging out the bottom of the stream and expanding the stream width as well as building an embankment with stone and wood. After the construction of the stream, the "Gaecheondogam" continued to exist with the name changed to "Haengrangjoseongdogam." Then during the reign of King Sejong, "Suseonggeumhwadogam," a government organization was dedicated to the repair, maintenance, flood prevention, and fire prevention of the capital. And during the reign of King Seongjong, willows were planted at the edge of the urban river in order to prevent floods.

At that time, urban streams served as a natural sewer system. Though there had been controversy about the use of urban streams in the early days of the Joseon Dynasty, King Sejong decided on its use as a sewer. As the population gradually increased, the amount of sewage also expanded. As the trees in upstream mountains were deforested recklessly for fuel and part of the area was cultivated as arable land, the aggradation of earth and sand continuously increased. During the reign of King Yeongjo, the situation deteriorated to its worst possible level and dredging became the only and unavoidable option. In February 1760, King Yeongjo undertook large-scale dredging by mobilizing 200,000 people for 57 days. The dredging operation progressed by removing sand and mud at the bottom of a stream, changing the water route into a straight line and building a reinforcing stone wall at both sides of the stream (see Fig. 11.3). Ever since, despite the financial difficulties of the government, dredging was conducted persistently every two or three years until 1908.

The most serious problem in the Cheonggye-cheon area was sanitation. During the rainy season, many houses were flooded and contagious diseases swept the entire city. In times of a localized torrential downpour, the sewer of the stream directly flowed backward into the overpopulated area. The mortality rate of Seoul was the highest in the Cheonggye-cheon area.

A simple solution to the Cheonggye-cheon in downtown Seoul that had serious negative influence both on landscape and sanitation was to "cover" it. Covering the stream and utilizing the covered area was planned from the early twentieth century. But the proposals had been turned down by the government authority on the grounds that drainage would be difficult in times of flood.

Figure 11.3

Cheonggye-cheon dredging in 1760 (attended by King Yeongjo).

In 1955, the closed conduit construction was conducted upstream of Gwanggyo spanning 135.8 m. This represented the first covering construction since some areas between Gye-dong, Jongro-gu and Gwanggyo had been covered during the Japanese colonial period. Then the full-scale covering construction to the stream began in May 1958 and was completed in December 1961. Cheonggye-cheon's area of length 2,358.5 m and width 16.54 m between Gwanggyo and Ogansu Bridge (near Pyeonghwa Arcade) in the central downtown was covered and paved with concrete. About 242,000 people were mobilized. Remaining sections were covered as needed until 1978.

Ever since the covering construction was completed, many shops and commercial establishments have been clustered around both sides of the covering road and the amount of traffic has skyrocketed. Therefore, arose the necessity for building new roads that detour the central downtown. Then, a new overpass over the Cheonggye road was built. The construction of the Cheonggye Expressway commenced on August 15, 1967 and ended on August 15, 1971. Its length spanned 5,650 m and the road width was 16 m. See Fig. 11.4.

Safety

Reasons for Restoration

The Cheonggye-cheon covered road and expressway whose construction commenced in 1958 were found to be in need of immediate repair after a safety inspection by the city of Seoul. A comprehensive safety overhaul of the expressway in 1992 by the Korean Society of Civil Engineers recommended it be repaired and reinforced. Upstream portion of the expressway had undergone total repair and reinforcement during 1994–1999. An inspection executed in 2001 recommended a total repair for the rest of the expressway.

A comprehensive safety overhaul for the covered road structure was performed, and repair and reinforcement works were urged due to deteriorated conditions of the structure. After repair, comprehensive maintenance work is still required. See Fig. 11.5.

Figure 11.4
Cheonggye-cheon area before restoration.

Figure 11.5

Deterioration problems in the structure.

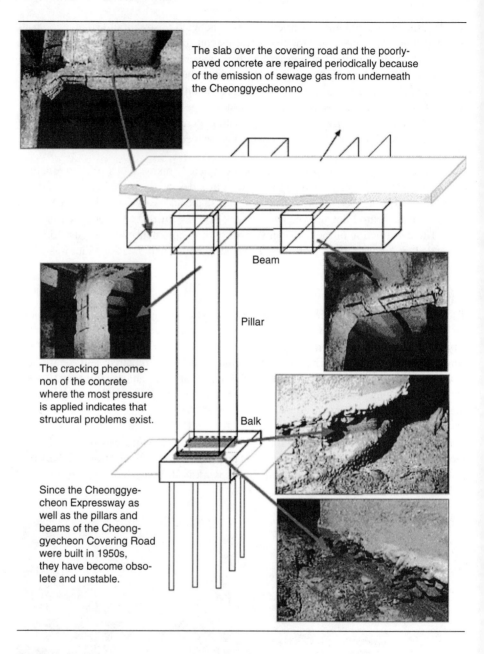

The slab over the covering road and the poorly-paved concrete are repaired periodically because of the emission of sewage gas from underneath the Cheonggyecheonno

Beam

Pillar

The cracking phenomenon of the concrete where the most pressure is applied indicates that structural problems exist.

Balk

Since the Cheonggyecheon Expressway as well as the pillars and beams of the Cheong-gyecheon Covering Road were built in 1950s, they have become obsolete and unstable.

Environment

It has long been a hope for the city that a nature- and human-centered environment-friendly city space be created. The project was a provision for citizen welfare by restoring the stream as a natural river without pollution and by developing the nearby area as an ecological park.

Culture and Industry

The Cheonggye-cheon area has been the central venue of Seoul, Korea's 600-year old capital city. Through the restoration project, the Gwanggyo (bridge) which was located under the covering structure has been recovered and the Supyogyo (bridge) which was relocated to the Jangchungdan Park has been reborn as an imitation. These bridges are stone construction cultural properties. And the Cheonggye-cheon looms as a new historic and cultural symbol of the downtown Seoul.

It was necessary to induce the industrial structure around the underdeveloped Cheonggye-cheon area to reorganize in order to invigorate the downtown economy depressed due to lagging development for the last 50 years.

Construction Procedure

The restoration project was performed by the Seoul Metropolitan Government, research institutes, and citizens. The Seoul Metropolitan Government established the "Cheonggye-cheon Restoration Center" dedicated to restoration operations. The center set up the basic plan and conducted such operations as cultural properties restoration, city planning, construction, structure dismantlement, and stream recovery. Also, the *Seoul Development Institute* (SDI) quickly formed the "Cheonggye-cheon Restoration Research Support Group" to investigate traffic-related affairs.

Adequate accommodation for consensus through harmonized efforts of the Seoul Metropolitan Government, research institutes, and citizens has minimized damage to residents.

Citizens launched the "Cheonggye-cheon Restoration Citizen's Committee" composed of many civilian professionals as well as ordinary citizens. This Committee collected public opinion and provided counseling services such as holding a public hearing on the restoration. The procedure of the restoration construction is as follows. At first, dismantle the Cheonggye Expressway and remove the covering road. Then, using the underground water produced from subway stations and the treated water pulled from the Jungrang Sewage Treatment Plant, clean and clear the Cheonggyecheon (Stream) as it used to be. Minimize the damage to residents by securing roads on both sides of Cheonggyecheon-no and installing screens during construction. After restoration, minimize the damage to neighboring merchants by securing two-lane roads on both sides of Cheonggyecheon. See Fig. 11.6.

Outcome of the Project

The significance of the Cheonggye-cheon restoration was to restore the historical and cultural environment in Seoul, creating a balanced city development between Gangbuk (northern part) and Gangnam (southern part). In the past, Seoul's 600-year old history as the capital was marred due to the disappearance of the water axis which was well-blended with Bukak-san (Mt.) and Nam-san (Mt.). By restoring the covering area in the stream, the historic relics, such as Joseon Dynasty's stone bridges including Gwanggyo and Supyogyo, were restored to their original condition. And the reborn water axis in downtown Seoul made a beautiful riverside.

The project solved problems such as decreasing air pollution caused by the noxious gas emitted from the covering road's underground and the large-scale accidents arising from the obsolescence of the Cheonggye Expressway. Through the restoration project, the environment of the Cheonggye-cheon area was greatly improved and its industrial structure was reorganized, contributing to the booming of the downtown economy.

Figure 11.6

Methods and process of Cheonggye-cheon restoration project.

1. Setting up facilities for transportation, safety, and construction
   erection of scaffolding and demolition chutes under the overpass

2. Dismantling of decks, crossbeams, and the covering
   dismantling of overpass decks by the segment cutting method-dismantling of metal beams by using crane-dismantling of the covering by the segment cutting method

3. Dismantling of piers,
   cutting of piers, and disposal-construction of road for temporary use

4. Construction of intercept seers and a road for temporary use, and dismantling of the coverings in the commerical area
   construction of intercept sewers-dismantling of the covering by sectors and road construction

5. Landscaping for the recovered area including the river
   restoration of the river-landscaping-lighting design

The newly restored Cheonggye-cheon has brought about a drastic change in the ecological environment in downtown Seoul. As a result of the water flowing in the stream, the decrease in the number of vehicles passing the area, and the restored wind passage following the removal of the elevated highway over the stream, the temperature measured in front of Sungin building located near the stream was measured to be 3.5°C lower than that measured at a place in Sinseol-dong, which is 400 m away, on a summer day. The average wind speed at the stream in July 2005 was measured to be about 50 percent faster than the previous year. Thus, the stream is expected to play the role of a dissipater of air pollutants. See Figs. 11.7 and 11.8.

The restoration made Cheonggye-cheon a natural stream where strolling citizens can enjoy the pleasant scenery, take a cozy rest, do shopping, while children swim at the riverside (see Fig. 11.9). It provided resting spaces for citizens in the center of the city by transforming a huge sewer into a natural river. Considering the enormous repair and

Figure 11.7
Cheonggye-cheon
before the restora-
tion (Dongdaemun
area).

Figure 11.8
Cheonggye-cheon
after the restoration
(Dongdaemun
area).

Figure 11.9
Newly opened
channel of
Cheonggye-cheon.

reinforcement cost of the Cheonggye Expressway, covered roads, covered interval, and underground space on an annual basis, the burden of citizens was alleviated and the convenience they use considerably increased in the long-term perspective.

Summary

In one of the city's most ambitious ecological projects ever, the Seoul Metropolitan Government spent more than 380 billion hwans ($380 million) to restore parts of Cheonggye-cheon's original waterways. The stream was opened to the public on October 1, 2005, 44 years after it was covered by concrete amid urban development. The project has been considered a great success by most citizens. People are enjoying the refreshing scenes of nature. Owners of the stores and shopping malls located along the stream are busy welcoming customers.

The restoration project of Cheonggye-cheon is not just a part of Seoul's urban planning but a greater task that the entire nation is interested in as a symbolic project to revive an important part of Korea's historical and natural heritage. The project was completed successfully, creating a friendly environment for the people (http://english.seoul.go.kr/cheonggye/). The project not only contributed to renewing the image of Seoul, but also set a new paradigm for urban management.

With all the positive factors related to restoration of the stream, the Korean government plans to restore other concrete structure-covered streams across the nation in similar way as the Cheonggye-cheon was restored.

WATER SECURITY ISSUES IN KOREA

Introduction

Because water is a critical factor for human life, a crucial question is how water can be acquired and used properly in a secure manner. Many countries including developed ones confront water deficiency, contamination, and transmission problems such as high water loss rate and service interruption by pipe and apparatus failures. Besides those unavoidable problems, a new threat from terrorism to contaminate water sources and to sabotage drinking water supply systems is drawing attention for the secure use of water, especially after 9/11 (Mays, 2004). In response to the new threat, systematic approaches for improving the security of water supply and for protecting the public are being researched and prepared. For example, based on the Public Health Security and Bioterrorism Preparedness and Response Act signed by President Bush in June 2002, United States Environmental Protection Agency (USEPA) prepares the Water System Emergency Response Plan and helps an individual water utility supplying water to more than 3300 customers establish the plan. Providing sufficient and safe water to the public involves not only acquiring sufficient water but also supplying it to the public securely.

The goal of water security will be accomplished when sufficient water is available and it reaches the public in desirable quality by a reliable way (Stockholm Environmental Institute, 2000). Water can be contaminated while being transmitted through water distribution pipes particularly in most of the old water distribution systems. In Korea, most major cities experience drinking water contamination, a considerable amount of water loss, and frequent service interruptions caused by system failures. However, compared to the United States, Korea faces different conditions. Korea falls in the category of water deficient country since the early 90s. The economic loss due to water deficiency is greater than the loss by the heavy traffic volume. Protecting water distribution systems from terrorism may be the focus in United States; reducing water loss in the water distribution system, and replacing deteriorated pipes are main concerns in Korea for achieving better water security. In this section, the unique conditions of Korea that impact water security are explained.

Climate

Korea falls in the continental temperate monsoon climatic zone with four distinct seasons. Currently the average annual temperature is 13.3°C and 24.2°C in summer. However, the global warming has caused the annual average temperature change to go up at 1.5°C for the last 100 years. Because of the change, in summer, there is frequent heavy rainfall, as a result, flood damages are increasing (Kim, 2004).

The annual rainfall is 1,274 mm but it is not evenly distributed over a year. About two-thirds of the annual rainfall is during the rainy season which is from June to September. In the rainy season, flood prevention is the most critical issue, so the dam operation is focused on how to control flood runoff and how to release it with less damage on watersheds. During the dry season, acquisition of water for agriculture, industry, and domestic use becomes a critical problem in most regions. For this reason, many multipurpose and flood control dams have been built in the four major watersheds.

Social and Environmental Changes

In the last three decades, Korea has experienced rapid social and environmental changes. Two of the major changes are an increase of population ensuing urbanization and an increase of income resulting from economic development. In 1970, the population of Korea was 31 million and in 2004 it increased to 46 million. For the same period, GNI per capita increased from $255 to $14,162 (Korea National Statistical Office). Most of the changes resulted from the economic development. Before the 70s, agriculture was the major industry in Korea having now been replaced by manufacturing industries, such as machinery and IT (*information technology*). With those changes the quality of life has improved, whereas environmental contamination has become severe, especially water quality in rivers and reservoirs. Two factors which result in water quality deterioration are growing pollution loads and increasing water consumption. As in other developing countries, industrialization of Korea contributes considerable pollution loads to the natural system. Many water sources including rivers and reservoirs are contaminated and the amount of water which can be used as drinking water is reduced. To make it worse, the pursuit of better quality of life requires more water consumption for domestic and industrial purposes. After the Korean government realized the significance of the problem, efforts have been made to reduce the natural system contamination and provide sufficient water. An example of the efforts is a project to replace deteriorated pipes which is explained later.

As mentioned, in Korea, the daily water consumption per capita has increased while the GNI per capita has risen since 1970s. However there is an interesting pattern found as shown in Fig. 11.10 (Ministry of Environment, 2004). From 1970 to 1997, the daily water consumption per capita had increased and reached the highest point in 1997, it has decreased after that. Although many facts account for this phenomenon, one of the reasons may be the fact that the deteriorated pipe replacement project commenced from the mid-1980s; by the late 1990s most of the project had been completed and then water

Figure 11.10

Changes in daily water consumption (liter/day/capita).

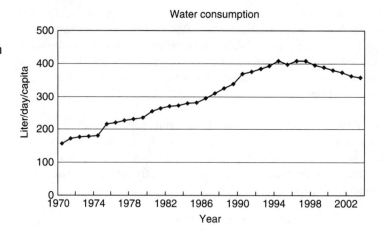

loss through the distribution system reduced drastically. It demonstrates the fact that the amount of water loss is considerable and pipe replacement reduces water loss effectively.

Potential Water Resources

Groundwater

Groundwater is one of the important water resources in Korea and its importance is increasing. In 2003, the portion of groundwater was about 11 percent (37.4 billion tons) of the total water resources used, which is not high compared to developed countries. However, there are 1.2 million wells (with a density of about 11.0 per km^2), respectively. The density of wells of Korea is the highest in the world. In case of the United States, it is 1.56 per km^2. If the portion of groundwater reaches about 20 percent as some other countries, the number and the density of wells will increase and it causes many problems such as groundwater contamination and overuse. For this reason, the government and the *Korea Water Resources Corporation* (KOWACO) established guidelines for governing nationwide and local groundwater use and conservation. Details of groundwater use in Korea are presented in Fig. 11.11.

Figure 11.11

Trend in groundwater usage in Korea.

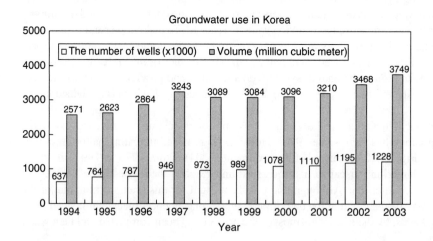

Rainfall Harvesting

Recently, many researchers in Korea have been studying how to use rainfall as a secondary water source for urban and rural areas. These research efforts are supported by the Korean government and government-owned institutes. Important research results are as follows:

- *For urban areas.* Outline and economic valuation of rainwater utilization facility in Seoul National University dormitory by Han and Kim (2004).
- *For rural areas.* The development of rainwater utilizing system for domestic water at agricultural and fishing district by Youngnam University (Gee, 2003).
- *General purpose.* A study on monitoring, operation, and management for rainwater utilization systems by Korean Society of Water and Wastewater (Yoon, 2005).

In general, harvested rainwater can be used for wastewater reclamation and reuse systems. Rainfall harvesting is more useful when it is installed in water scarcity regions such as islands and seaside towns. In urban regions, the rainfall harvesting systems can reduce the direct runoff if it is used widely. The first practical rainfall utilization facility in Korea was built at Seoul National University dormitory, designed to use rainwater for their gardening system. And other rainfall harvesting systems are installed in houses and on a military base (Han and Kim, 2004).

However, as shown earlier, the use of rainwater is in an early stage in Korea and still requires extensive studies to develop a proper system. Even so, it is obvious that rainfall harvesting has many advantages as a potential water source for urban and rural areas.

Deteriorated Pipe Replacement

Water Supply System Improvement

With time, pipes are subject to deterioration and the rate is dependent on the condition surrounding them. Once pipes are deteriorated and scaled, water flow is reduced and they are vulnerable to stress such as pressure, traffic, or earth load. Additional pumping head and cost is necessary to supply water at demand nodes with the design head. Under this condition, pipes are easily cracked. Leaking through cracks is another problem found with deteriorated pipes. The more water required, the more frequently the pipes fail and leaking occurs. As a consequence, water loss rate soars and frequent service interruptions follow. In terms of water quality, the inside surface of deteriorated pipes on which biofilms are formed, contaminates water before it reaches a faucet. In brief, deteriorated pipes result in high water loss, frequent service interruptions, and water quality problem.

Many cities having an old water distribution system have to deal with the same problem like Seoul. The condition of the water distribution system became worse as the population of Seoul rapidly increased. To resolve the problem, Seoul started the pipe replacement project in 1984 and completed its first phase in 2005. The purposes of the project were:

- To prevent water contamination in pipes so that water quality will be maintained from treatment plants to faucets.
- To reduce water loss so that the water distribution system's efficiency will be improved.
- To reduce service interruption due to pipe or apparatus failures.

Cast iron and PVC pipes, which corrode easily, have been replaced with ductile iron pipes of high corrosion resistance. Since 1984, the annually planned replacement schedule

Figure 11.12

Length of replaced pipes and cost per year.

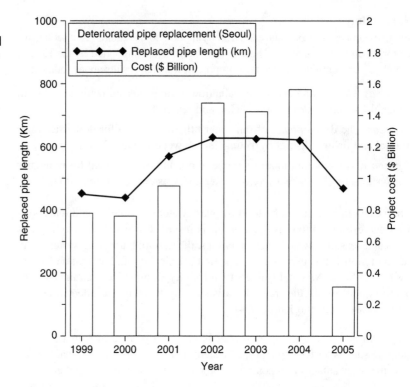

has been performed. Figure 11.12 shows the length of replaced pipes and cost per year. From 1984 to 2005, 15,696 km pipes out of 15,797 km have been replaced and the remaining 101 km will be replaced by 2007. The total project cost is about $14 billion. Due to this project, the water loss rate which was 31.8 percent in 1999 decreased to 11.9 percent in 2005. In addition, the red water problem and pipe failures have also reduced. Other local governments like Jeju, Andong, and others are performing similar pipe replacement projects for the same purpose.

The most significant improvement as a result of the deteriorated pipe replacement project is reduction of water loss. After most of the pipes were replaced, the water loss rate significantly reduced in a short period from 31.8 to 17.5 percent. It resulted from not only placing new pipes, but from many defects in the system that were fixed as the new pipes were laid. As the amount of water loss can be quantified as monetary value, it can be used to perform the cost analysis for a deteriorated pipe replacement project.

Summary

Major water security issues in Korea are focused on how to obtain reliable water sources and how to provide it efficiently to water consumers. In the last decades, Korea has faced many social and environmental changes and those changes have increased the amount of water consumption. Because the water production, including raw water, has not caught up with the water consumption, Korea is one of the water deficient countries. The climate characteristics of Korea make this situation worse.

Water security which consists of obtaining sufficient water and delivering it to the consumer reliably is one of the most critical tasks for governments and water utilities.

The only way to achieve water security is to simultaneously develop new water sources and to improve the delivering systems.

ACKNOWLEDGMENTS

The authors are grateful to the city of Seoul for the information on both the cheonggye-cheon restoration project and the pipe replacement project

REFERENCES

Gee, H. K. (chief researcher), "The Development of Rainwater Utilizing System for Domestic Water at Agricultural and Fishing District," Ministry of Agriculture and Fishery's Project Report, P. 287, 2003.

Han, M. Y., and Y. W. Kim, "Outline and Economic Valuation of Rainwater Utilization Facility in Seoul National University Dormitory," Technical Report, *Journal of the Korean Society of Water and Wastewater,* Vol. 18, No. 5, pp. 547–557, 2004.

Kim, S. (editor in chief), *Water for Our Future*, Ecolivre, Seoul, Korea, P. 287, 2004.

Kim, S., S. Jung, and H. Kim, "Temporal Variation of Precipitation Trend at Seoul, Korea," *Proceedings of Specialty Conference on Hydraulic Engineering*, ASCE, San Francisco, pp. 1771–1991, 1993.

Korea National Statistical Office, available at http://www.nso.go.kr/newnso/main.html.

Korean Ministry of Construction and Transportation, *Water Vision 2020*, in Korean, 2001.

Mays, L. W. (ed.), *Water Supply Systems Security*, McGraw-Hill, NY, P. 464, 2004.

Ministry of Environment, "Statistics: Water Distribution System," *Annual Report of Water Distribution System*, Ministry of Environment, 2004.

Stockholm Environment Institute, "Water Security for the 21st Century – Innovative Approaches", *The 10th Stockholm Water Symposium*, August 14–17, Stockholm, available at http://www.earthscape.org/r1/sws01/sws01.html, 2000.

United Nations Department for Policy Coordination and Sustainable Development, "Global Change and Sustainable Development: Critical Trends," *Advanced Unedited Text*, P. 51, 1997.

United Nations Economic Social Commission for Asia and the Pacific, "Water Resources Situation in Asia and the Pacific," *Toward an Environmentally Sound and Sustainable Development of Water Resources in Asia and the Pacific*, Water Resources Series No. 71, United Nations, NY, pp. 15–21, 1992.

Urban Flood Disaster Management Research Center, Technical Report Series (FFC03-01~ FFC03-17), 17 volumes, in Korean, 2004.

Urban Flood Disaster Management Research Center, Technical Report Series (FFC04-01~ FFC03-19), 19 volumes, in Korean, 2005.

USEPA, "Large Water System Emergency Response Plan Outline: Guidance to Assist Community Water Systems in Complying with the Public Health Security and Bioterrorism Preparedness and Response Act of 2002," USEPA.

Yoon, J. H. (chief researcher), "A Study on Monitoring, Operation and Management for Rainwater Utilization System," *Korean Society of Water and Wastewater Project Report*, P. 144, 2005.

12

WATER-BASED SUSTAINABLE INTEGRATED REGIONAL DEVELOPMENT

Olcay Unver
Kent State University, Ohio
Formerly President
GAP Regional Development Administration
Ankara, Turkey

INTRODUCTION

World experience and conjuncture at the beginning of the twenty-first century have resulted in a convergence of approaches to social and economic development owing to developments that occurred in the previous century, especially during its final decades.

The post-World War conditions, the Great Depression, continent-wide reconstruction efforts, and the emergence of new states and political systems had direct impacts on social and economic development with somewhat less direct implications on how water resources are developed and managed. The urgency to address the needs of this era, coupled with a competition between the two rival systems, brought about experimentation with different combinations of social, economic, and political tools and models. To restore or establish the infrastructure necessary to meet the very basic amenities of life justified quick action and led to the dominance of the interventionist paradigm of development, which was further supported either by the nature of the political system or by a widening poverty and the deep need for jobs or both. This era was marked by massive development projects, unprecedented investments in infrastructure, and by the direct involvement of "engineers" in the decision making processes. It was not until after the benefits of these penetrated into the different levels of the society and economy that concepts like participation, governance, accountability, bottom-up approaches etc. would become a part of the paradigm. Another similar process was needed for some of the adverse impacts of these projects to be observed or experienced before the development literature started to record issues such as environmental protection, development disparities, equity/fairness, and gender considerations. The practice, or rather the consequences of the practice were now helping produce a new theory that would later be named *sustainability,* having incorporated an intergenerational dimension.

Sustainable development is a concept still in the making. Its components have been on the global agenda since the United Nations Conference on the Human Environment in Stockholm in 1972. The concept has evolved in a decade between two United Nations Summits—the Earth Summit, Rio de Janeiro in 1992 and the World Summit on Sustainable Development, Johannesburg in 2002, supported by a number of international conferences and decisions (World Water Assessment Programme, 2003). The underlying principles of sustainable development are now largely uncontested whereas its modalities, applicability, financing, and relevance in real world situations remain to develop more fully.

This chapter advocates an integrated development approach based on the sustainable development of water resources on a regional scale. This is the area where sustainable socioeconomic development and integrated water resources management intersect and yield to a holistic formulation involving multiple sectors and multiple stakeholders. The water-based sustainable integrated regional development is covered in this chapter with its theoretical and practical aspects and through a contemporary example, the Southeastern Anatolia Project (GAP) of Turkey.

The chapter deals with the prospect of using water resources for and within the broader framework of social and economic development on a regional basis. The chapter explains the difficulties for the application of this approach in developed as well as developing countries, commenting on its value and applicability in different settings. The difficulties for a developed country include the inherent budgetary, jurisdictional, political, and paradigm-related problems that regionalization and integration typically face. Developing countries offer significantly more prospects with varying but broad scopes. A regional, integrated approach using water as an engine for development does not have lesser barriers for its application, though, in developing countries, although these are fundamentally different. They include low human and capital capacities, institutional rivalry, political short-sightedness, lack of grass-roots/democratic traditions and civil society, lack of appreciation for efficiency, corruption, and lack of incentive, among others.

The chapter addresses the role that international assistance, development banks, multi-laterals, U.N. system organizations, and the NGO community can play in overcoming these barriers. Specific reference is made to the prospects of this approach in enhancing the efficiency and effectiveness of the postconflict, postwar reconstruction efforts led or aided by the international community.

The following excerpt from a developed country, for example, would provide insight into this approach.

> President Franklin Roosevelt needed innovative solutions if the New Deal was to lift the nation out of the depths of the Great Depression. And TVA was one of his most innovative ideas. Roosevelt envisioned TVA as a totally different kind of agency. He asked Congress to create "a corporation clothed with the power of government but possessed of the flexibility and initiative of a private enterprise." On May 18, 1933, Congress passed the TVA Act.
>
> Right from the start, TVA established a unique problem-solving approach to fulfilling its mission-integrated resource management. Each issue TVA faced—whether it was power production, navigation, flood control, malaria prevention, reforestation, or erosion control—was studied in its broadest context. TVA weighed each issue in relation to the others.
>
> From this beginning, TVA has held fast to its strategy of integrated solutions, even as the issues changed over the years.
>
> (Tennessee Valley Authority Internet Web Site)

DEFINITIONS

The fundamental link between growth and water has been long recognized. Water projects have always been among the priority investment components in infrastructure development and have played an instrumental role in the social and economic development of communities, nations, and regions. It is not a coincidence that the earliest civilizations were based on or related to water resources.

Development Paradigms: Sustainable Development and Water

Since the United Nations Conference on Water in Mar del Plata, Argentina in 1977, the global community has held more than 40 international summits and conferences either on water or on issues that had implications on water. The list includes meetings on freshwater, environment, biodiversity, financing, poverty, food, desertification, climate change, and sustainable development among others.

The United Nations Conference on Environment and Development, also known as Earth Summit, in Rio de Janeiro in 1992, extensively established the interdependence between development and environment. A major outcome of this summit was Agenda 21 (United Nations, 1993). Agenda 21 represented a strong support from the governments to sustainable development, and included policy goals, priority programs, and recommendations for action. The largest chapter of Agenda 21 is Chap. 18, devoted to the sustainability of freshwater resources.

The *Millennium Development Goals* (MDGs), adopted in September 2000 during the Millennium Summit of United Nations General Assembly is comprised of eight goals, all of which can be directly or indirectly translated into water-related terms (Gleick, 2004). As an example, Goal No. 1—"Eradicate extreme poverty and hunger"—and No. 7—"Ensure environmental sustainability" have direct relevance to water whereas Goal No. 2—"Achieve universal primary education" and No. 3—"Promote gender equality and empower women" are water-related as millions of women and young girls spend many hours every day to fetch water. The health related Goals 4, 5, and 6 also have strong relevance to water, or the lack of it.

Water Resources Development and Management

The paradigm change in water resources development and management is covered with its different aspects elsewhere in this book. The general trend in time has been a shift from single-purpose to multipurpose, from infrastructure development to resource management, from supply to demand, from "hard" aspects to "soft" issues, from single disciplinary to integrated, and from tangible to less tangible. A multitude of professions are involved in water resources decisions and a three-pillared approach is standard with the social, economic, and environmental components. The general consensus on the space dimension involves the river basin as the basic unit for planning.

Integrated water resources management (IWRM) seems to be the global norm although suggestions of "parallel" or "coexistent" paradigms also exist, mainly stemming from the implementation difficulties of the IWRM process.

It is now almost standard that management and governance imply delegation, decentralization, stakeholder participation with the central government playing a regulatory role. A general agreement exists for protecting the poor, the underprivileged, and the underrepresented. Gender has become a main criterion in water resources management. On the other end of the spectrum, the availability and the nature of the financing for water projects remain highly controversial. Suggestions to augment insufficient public funds via private sector involvement are contested. Many developing nations remain in disagreement with the conditions which the loans are tied to, on the grounds that their implementation would make the projects excessively costly.

The implementation dimension exhibits much less of a standard. IWRM is still an exception rather than the rule in practice. The MDG target of halving, by the year 2015, the proportion of people who are unable to reach, or to afford, safe drinking water seems unlikely to be met. The international community has made little progress to meet a similar goal—to halve, also by 2015, the proportion of people without access to basic sanitation—adopted at the *World Summit on Sustainable Development* (WSSD), in Johannesburg in 2002 (United Nations, 2002). Very few countries have met WSSD's target for the nations to develop their IWRM and water efficiency plans by the year 2005. Developed countries are no exception to this.

Another area where much room exists for improvement is transboundary cooperation. Although the general principles to the development of transboundary water are uncontested, global-scale transborder management arrangements are few, especially in basins where there is on-going investment in infrastructure such as the Mekong, Euphrates-Tigris, the Nile, as well as those that have run out of water such as the Jordan River Basin.

Social and Economic Development

Social and economic development, like water resources development and management, has also gone through an evolutionary process, sometimes overlapping with or covering it. The economic growth focus of the post-World-War era has gradually changed to cover other aspects of development including social, cultural, and environmental issues. Emphasis has moved from massive infrastructure development to more streamlined projects, from pure public investment ventures to public-private partnerships and regulation, from direct government intervention to provision of enabling environments, and to mobilize market tools and private sector, whenever possible.

The international community, met in Johannesburg in 2002 at WSSD, 10 years after the Earth Summit, revisited the definition and the implementation of sustainable development following an assessment of the developments that occurred within the decade. The Johannesburg Declaration (United Nations, 2002) emphasized that "poverty eradication, changing consumption and production patterns, and protecting and managing the natural resource base for economic and social development" were "overarching objectives of, and essential requirements for sustainable development." In Johannesburg, the global community asserted ". . . in our collective efforts to move from commitments to action to ensure more sustainable livelihoods for all water and sanitation, energy, health, agriculture and biodiversity (WEHAB) represent five specific areas in which concrete results are both essential and achievable."

Both theory and practice of sustainable development will continue to evolve. Significant disagreement and polarization exist in terms of addressing some of the most pressing issues such as the eradication of extreme poverty and hunger, provision of basic education, supply of safe water and sanitation, fight with disease, among others. These are interwoven with more macro issues such as trade barriers, subsidies, and the Third-World debt.

The literature and the real-world application of both water resources management and sustainable development continue to grow in their virtually connected but largely separate ways.

Although there are many parallels between the two and in spite of a general recognition that water resources development/management has cross-cutting links to multiple sectors, and that sustainable development has a strong water component, the potential for bringing the two together in an integrated formulation with a regional scope remains largely untapped.

Water-based sustainable integrated regional development implies the holistic development of a region by mobilizing its water resources along with harmonious and complementary development and management in other relevant sectors and areas in a sustainable manner. Although the concept is not new, its real-world implementation has been limited (Biswas et al., 2004).

SCOPE AND SCALE ISSUES IN DEVELOPMENT: FROM CHAOS TO INTEGRATION

The term "development" covers a large spectrum and potentially applies to all economies, developed, developing, and in transition, to all settings, rural and urban, and to all sizes from local to national to supranational. The variety of actual situations for development and the possible combinations of economy, setting, and size make it very challenging to generalize or even to classify that would help policy makers and the naïve alike to any useful or practical extent. This chapter does not attempt that. However, it will be useful to look at the possible situations that may benefit from a development activity of regional nature.

Domain of Water-Based Regional Development

The domain of water-based regional development is defined by three sets of variables: scope, scale, and governance. Scope determines the breadth of the domain, scale the size, and governance the "soft" aspects of development.

Within the scope of regional development, the interest of this chapter is in water-based social and economic development. It potentially includes, from core to periphery, a number of sectors and areas. The core is comprised of the traditional sectors and areas of water resources engineering, for example, water supply, irrigation, flood and drought mitigation, and hydropower among others. These are complemented by the related sectors such as environment, biodiversity, agriculture, energy, and sanitation, and cross-cutting issues such as gender empowerment, poverty reduction, and equity and fairness among others. These usually delineate the realm of IWRM. The outer peripheral contains sectors that are related to social and economic development but not to water. These include, among others, transportation, land-use planning, trade and industry, culture, tourism/archaeology, and education.

The scale variable ranges from local to national, with intermediate zones being more relevant to water-based regional development. Unlike IWRM, water-based regional development does not have to use the river basin as its territorial jurisdiction, although this is a possibility. The scale is determined by a set of factors that may include administrative divisions, social-cultural boundaries, and those related to economy, ecology, or geography.

The "soft," governance dimension is what enhances the sustainability base of the regional development. It fundamentally comprises three "pillars": the public sector, the business community, and the people. A "healthy" regional development project has to have a mix of these. The public sector is involved in macro planning and making the investments into infrastructure and other services necessary for a resource mobilization toward development in social and economic terms. It also provides the regulatory frameworks for the players and the processes of development as well as it provides case-specific programs such as preferential credits, incentives, and direct and indirect subsidies. The business dimension, when and where available, brings multifaceted benefits to the sustainability of development. Fundamentally, government investments and interventions are planned to catalyze or initiate a process of development in which the private sector can establish, flourish, and move toward becoming the dominant sector of the regional economy. The business dimension creates employment, expands the tax base of the region, catalyzes financing, and brings innovation and technology. It may also bring rationality, efficiency, leadership, best practices, and social responsibility. The business community is now an integral part of the sustainable development worldwide (World Business Council for Sustainable Development, 1999). The third "pillar" of sustainable regional development is the people, who make the development effort meaningful as not only the beneficiaries of a project or a program but as the active players of the spectrum, from planning to management, through formal and informal governance structures. The freer and the more participative the people are, the more deeply rooted the sustainability becomes. The presence of the ways and means with which the different strata of the population, including the poor, the under-privileged, and the underrepresented can participate in the process and the willingness of the other players to incorporate their input are among the key issues in contemporary development practice.

Relevance to the Real World

The scope for developed countries in regional development exists when there are development disparities between regions and a basis for the removal of these disparities. The basis could be the availability of natural local resources, sociocultural differences, economic means, or simply a political will. Regional projects have emerged, historically, in the development process of nations, where, in the process of development, a general level has been reached, minimal infrastructure has been provided, and certain regions have performed better than others. European regional development projects and the Tennessee Valley Project are examples of this.

Regional policies within a supranational setting find their most elaborate example in European Union, where special regulation and tools exist for helping less-developed regions catch up with the average standards of the Union.

The focus in removing the regional differences in developed countries is primarily on special regulatory privileges and business incentives rather than on infrastructure. The European practice has provided a wealth of useful literature, and it is very likely to enrich and diversify substantially as Europe will have to deal with another fundamental disparity issue: that between the new and the old members, following the expansion of the European Union from 15 to 25 countries in 2004.

The potential for developing countries to embark upon regional development efforts is more varied. They may prefer to deal with lack of regional homogeneity in economic growth, or disparities in income, land ownership, and social indicators along ethnic, religious, or cultural lines via a regional development project. The latter is recognized by many governments as a fundamental reason to implement regional policies to maintain social cohesion and stability. The presence of well-defined natural resources in a region can also catalyze a regional disparity reduction program. Regional projects in Asia, Africa, and the Americas show great diversity in scope, ranging from state-supported industrial development to large-scale infrastructure mobilization. A smaller number of regional projects are water-related and involve a large spectrum from irrigation to water supply at scales from village level to large territories.

Another potential situation where regional development can be desirable is the reconstruction efforts in countries coming out of war, conflict, or disaster. The challenges typical to this situation include nation building, overcoming lack of security and central authority, and having to deal with low level or absence of democratic traditions, corruption, and low technical capability, and limited funds. In the typical case, these challenges are addressed by the international community in the form of financial and technical assistance, and by the local governments in the form of flexibility to accommodate creative solutions, organizational arrangements, and willingness to cooperate with international organizations, including *nongovernmental organizations* (NGOs), especially for projects and activities implemented by donor-supplied funds. A regional development program with a sound governance structure can help maximize financial efficiency by integrating multiple sectors and coordinating multiple donors and their executing agencies, NGOs in many cases.

As the level of national economic development goes from developing to less developed, the scope for development changes from multiple economic and social sectors to more basic amenities of life. The scale for development, likewise, moves from regional toward national, with possibility of local implementations based on the specific country conditions.

Table 12.1 helps describe some of the basic components of a development activity according to the level of organization in its management, and may aid in relating the applicability of regionalization and integration to a given set of conditions in a particular situation.

Regional development projects with a water resources core are more likely to be found in a developing country setting or a postwar/postconflict/postdisaster situation. One of the challenges these countries face in their response to the development needs and the available resources lies in the selection of the project size, the sectors to be integrated within a given project, and the organizational arrangement.

Coordination versus Integration

For the implementation of a water-based regional development project, the scale-scope relationship is to be evaluated within the possible organizational arrangement

	Standalone	Coordinated	Integrated
Area	Local	Local, regional, national	Local, regional
Project scale	Small	Medium to large	All scales
Applicable situation	Emergency, temporary	Local, regional, national	Local, regional
Sector	Single, concrete	Few sectors at a time	Multiple sectors
Stakeholder participation	Possible	Challenging	Possible
Need for government involvement	Security: Yes Implementation: low	Local: low Regional: medium National: high	Medium
Feasibility for NGO ownership	Very high	Low	Medium
Autonomous management entity	Medium	Low	High
Monitoring and evaluation	Seldom	Possible	Possible
Need for capacity building	Limited	Essential	Useful

Table 12.1

Organization versus Development Parameters

options. From an implementation point of view and in general terms, there are three options:

1. Standalone, individual projects
2. Coordination of otherwise separate projects
3. Integration of projects into holistic packages

These options may stem from, or evolve, from a common point, such as a government, or from the actions of individual entities, such as development banks, multilaterals, or aid agencies with their own agenda and procedures. It is entirely possible for an able central government planning authority to operate with a combination of these in an efficient manner. The practice has shown, though, as the need gets bigger, the capacity to deal with the multitude of arrangements deteriorates. The rationale for the use of regional development becomes more established as capacity issue is accompanied by the presence of proper project scales that involve multiple sectors. Water-based sustainable integrated regional development sits at the intersection of the above and in the presence of a water component that can be used to mobilize or catalyze economic and social development.

Table 12.1 quantitatively lists the components, or the decision parameters, of a development project in relation to the three organizational arrangements for implementation.

Standalone projects are usually justified in responding to emergency situations, in providing a basic amenity, or delivering a specific service. The provision of drinking water, shelter, and food in the aftermath of an emergency lies in this category. The same category also covers nonemergency interventions such as kitchens, health, and aid services for the needy and the underprivileged. These are usually "chaotic" programs implemented by NGOs, aid organizations, and government entities. They are typically local, relatively modest size, and in many cases undocumented and do not yield to effective monitoring or cooperation with others. Although it is possible to identify possibilities to link these programs to others in their respective sectors and/or localities, integrating them into more comprehensive and potentially more efficient broader projects is often difficult and not necessarily cost effective. The implementers of these are action oriented, time stressed, and typically reluctant to operate outside their particular domain.

Development efforts for subnational scales (regional, provincial, or local) can be formulated as *coordinated packages*. Coordination, when done properly, reduces the risk to alienate the sectoral players who would be unwilling to give up authority or to change their established mode of operation. It may improve resources efficiency, help enhance financial sustainability, and can play a positive role in promoting sustainable development. While coordination in a multisector, multiple-stakeholder setting is attractive and possible, its success depends on a number of parameters of scale and scope. A development project with well-defined and well-managed components in a manageable number of sectors, will have better chances, especially if the players accept the role of a coordinator. IWRM can be reached in a coordination arrangement, for example, when line ministries and financiers are well linked and the stakeholder environment is free and enabling. In a more typical case, the national government would lack the adequate capacity, experience, and manpower to accomplish effective coordination and the implementing-end public and nongovernmental entities would not possess the authority or, in most cases, the desire to coordinate. A major partner may be more eligible to play the role of the coordinator and may have occasional success. In a more typical case, there are multiple entities funded by or reporting to different donors/organizations and the coordination effort can easily turn into a logistical nightmare. Where accomplished, coordination may yield the data and information that, in turn, may help reduce repetition, identify bottlenecks, modify operational procedures, and fight corruption. In any case, coordination is generally an improvement over the "chaotic," standalone mode of operation.

When the type of development activity and the scale allow, an *integrated package* may be the best choice. The package usually incorporates a core, or main, sector or area with direct and indirect links to the others. Rural development is a potential area of integrated development, which, for example, may have a main irrigation component with training and extension, gender programs, basic infrastructure provision, marketing links, and micro credits. Water sector, with its various uses, is almost always suitable for integration, both with its subsectors and cross-cutting issues and areas. A multidisciplinary approach, concerted and complementary activities, a unified management, and its ability to operate in a delegated manner are among the main advantages that it provides. It is much better able to deal with multistakeholder environments, to incorporate and adapt to dynamic interrelations among subsectors/components and to monitor and evaluate development in a holistic, macro as well as micro level. IWRM can be better achieved with this arrangement as the incorporation of cross-cutting issues and sectors into water resources management is provided by the nature of integration. In water-based sustainable integrated regional development, it is almost imperative to establish a separate entity for management and implementation and to keep it free from political, irrational interventions.

Figure 12.1 depicts the relationship between scope and scale in reference to water-based sustainable regional development. Its feasible scope is in close proximity to, but larger than, IWRM, as it may sectors that are traditionally outside the realm of IWRM, such as transportation, industry, business promotion, and education. The feasible zone includes scales that range from submunicipal to river basin or larger territorial jurisdictions.

The size and the scope of a project are also usually interdependent. By changing the scale of a project, for example, the area coverage, it may be possible to delineate its sectoral composition, stakeholder base, or the implementers, or vice versa.

As a general rule of thumb, the size and the scope are interrelated; going from local to national also (usually) increases the number of sectors and stakeholders involved. The level of complexity increases in the same direction.

This is analogous to a trade-off curve that governs, on project basis, the scale decisions. Important decision variables such as manageability, efficiency and effectiveness, governance structures and levels, stakeholder participation dynamics are not linear, ever increasing or ever-decreasing functions in themselves, and in many cases they render the project scale somewhere between local and national; and the project scope between single purpose and multi sectoral.

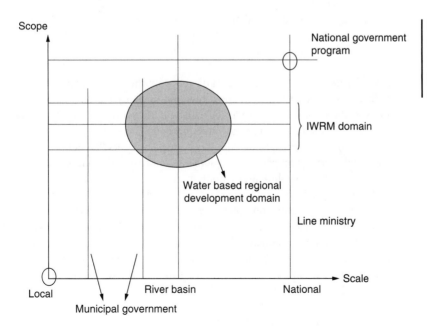

Figure 12.1
Water-based sustainable regional development domain.

CASE FOR WATER-BASED SUSTAINABLE INTEGRATED REGIONAL DEVELOPMENT

Development Frameworks

The development of water resources in a locality, a region, or a country is rarely a "standalone" project. In many cases, water resources development is not even the project itself or the objective; it is, in many cases, a component, usually a major one, in the social and economic development of an area. The foregoing statement holds whether or not a broader framework is defined or recognized.

The contemporary approach to water-based development is the development of the water resources in an organized and planned manner, and taking into account the cross-sectoral implications, either implicitly or through partnerships (Unver and Gupta, 2002). IWRM is gradually becoming the norm in this area. IWRM can be defined as ". . . a process which promotes the coordinated development and management of water, land and related resources, in order to maximize the resultant economic and social welfare in an equitable manner without compromising the sustainability of vital ecosystems" (Global Water Partnership, 2000).

A national water plan and/or policy documents will constitute the broadest sector framework. During the World Summit on Sustainable Development (Johannesburg, 2002), governments have committed to preparing national integrated water management and water efficiency plans within 5 years. This political commitment has not so far catalyzed very many plans but indicated a consensus on the need.

Legislation, in the form of a preferably unified water law, or a combination of different acts in regard to different aspects of water, is another broad framework for water resources development, local and national alike. Case-specific legislation, especially for integrated regional water-based projects, is another way of setting the framework. Many projects of this nature, including the Tennessee Valley Project of the United States and the GAP of Turkey are examples of this path.

Complexity and difficulty increase in direct proportion to the size and the scope of the project. The size of a project can be a function of many parameters, such as the area it will cover, the population that will be affected, the social and political challenges that will have to be overcome, and the funds needed. The scope of a project, on the other hand, depends on its nature, the way it is conceived by its "owners," and the stakeholders involved. Very often, projects implemented individually belong to a larger, "virtual" project due to the cross-linkages among them. An irrigation project and a water supply project belong to a broader water project if they tap the same water resource, which, in turn, can be an integral component of a sustainable resource management scheme involving environmental, agricultural and health-related issues, and governance considerations. Due to the implementation area, and with the inclusion of the other relevant components, the sustainable resource management scheme could belong to a regional development program. Conversely, it is not impossible to find large projects in which there are subprojects of modular nature. For example, certain infrastructure projects, once planned properly, can be efficiently divided into sectoral and/or local components with a relatively loose monitoring link and logistical supervision.

The practical implications of the above are not straight forward and vary according to the nature of the decision maker, that is, government, NGO, international entity, foreign-aid provider, as well as on the process itself, that is, centralized, delegated, ear-marked funding, and top-down versus grass roots.

As the project size grows, the tendency of narrowing it, or limiting its scope, also increases. This can be seen in large infrastructure projects that miss the links with related sectors or lack the "soft" aspects (such as stakeholder participation, gender balance, accountability) that could substantially enhance the sustainability and the effectiveness of the project at a fraction of the investment. For example, a large-scale water project involving dams, hydro power plants, and irrigation structures, when conceived as provision of infrastructure, may miss the opportunity to address irrigation management by water users or capacity building.

When water resources can be mobilized for development at a "meaningful" scale, the foregoing statements may render themselves to sustainable, holistic, regional projects that can be planned and managed to create a positive change in the economic and social indicators of the implementation areas and in the living conditions of the communities.

Governmental Level

Many countries, in their approach to local and regional issues, follow the path the national governments are organized. Typically, there is a planning agency/ministry, a public finance system through a finance ministry/treasury, and line ministries. Water-based development usually corresponds to further fragmentation. Water supply and sanitation component is typically distributed among the ministry in charge of infrastructure, its branches, and municipal governments. There is usually a water or natural resources ministry looking after general water resources development, river/reservoir management, sometimes irrigation, and flood control. Hydropower projects usually belong to the energy ministry or the power agency. The situation is further complicated by the presence of other related ministries, that is, ministries of environment, agriculture, health, social services, and resettlement. Implementing agencies usually operate on sector basis and according to their own program priorities. Some coordination may be possible among the agencies reporting to the same ministry and very little, if any, across ministerial boundaries. Ministries and implementing agencies compete with one another for the funds. There is usually little incentive for coordination, let alone integration. Favoritism, political irrationality, corruption, and inadequate capacity, when exist at different levels within the government, may skew the decisions and reduce efficiency and effectiveness. The system, in parts and as a whole, may exhibit resistance to or difficulty accepting stakeholder participation, accountability, and holistic approaches, vertically and horizontally.

U.N. System Organizations

The U.N. system plays a significant role in countries where the national capacity is low, the development need is high, and external funding is available. It is also highly fragmented with UNDP, FAO, WHO, WFP, UNICEF, UNEP etc covering different parts of the water-based development spectrum. Development banks, such as World Bank and Asian Development Bank, are often major players, catalyzing management arrangements in mobilizing loans. They, however, like the U.N. bodies, rely on national entities and firms in actual project implementation.

Institutional Setting and Major Players

Nongovernmental Organizations

This sector has been a major player as a result of its flexibility, action orientation, and ability to operate in hard-to-reach areas; it possesses good track record in accountability and transparency. Many donors, including governments, opt for NGOs for implementing projects that they finance. NGOs are well connected and have access to the specialized expertise and experience needed for their activities. They can more easily adapt to local conditions and work with local population. On the other hand, they are usually not partnership oriented for broader scopes and they may cause a "brain drain" from the public sector where pay scales are much lower. They are not as good in long-term planning as they are in project execution, may lack a holistic vision, and may be unwilling for coordination or integration with others.

Involved Governments

Sometimes foreign governments are also major players in development efforts. A number of western governments have assumed sector-specific responsibilities in Afghanistan, Iraq, and postdisaster aid in Asia and elsewhere. A coordination mechanism that could set an example is yet to emerge due to the top-down style involvement and the lack of adequate local capacities. On the other hand, same reasons may catalyze rational and otherwise politically unattractive solutions to development-related challenges. It may also set the scene for water-based sustainable integrated regional development if the conditions for it are present.

Governance Issues

It has been stated over and over again that the current global water crisis is a crisis of governance (Willem-Alexander, Prince of Orange, 2003; World Water Assessment Programme, 2006; International Institute for Sustainable Development, 2003) yet there is little agreement as to what water governance means and how it can be achieved.

Water governance implies, in broad terms, the social, administrative, legal, and economic systems that play a role in the management and development of water resources at different levels. It includes issues such as equity, inclusivity, gender mainstreaming, accountability, delegation, and participation among others. It involves all actors, governments, civil society, private sector, and all institutions—formal and informal, which play a role in the process.

Water governance and its reform are closely linked to the system as a whole and is a very complex issue.

Fragmentation of structures and institutions, overlapping or conflicting jurisdictions, sectarianism are some of the major signs of a lack of governance.

Water-based sustainable regional development can offer an opportunity to address this situation in a creative manner. Some of the challenges that prevent countries from reforming their overall, national structures may be more easily overcome for a subnational regional project with its own mandate and specific conditions. Solutions and structures associated with regional development can be more easily put in action and partnerships can be comparatively easier to establish.

It is a well-known and documented fact that the poor and the underprivileged suffer more from lack of safe water, sanitation, and energy. Estimates for water-related mortalities range from 2.1 to 5 million a year, depending on the criteria used (Gleick, 2004). Some 2 billion people have no access to electricity at all; 1 billion people lack access to improved water supply, and 2.4 billion people lack access to improved sanitation (World Water Assessment Programme, 2003). Women and young girls spend hours everyday to fetch water and firewood in the less-developed countries of the world, thus being deprived of educational and other opportunities and suffering from the health and livelihood impacts of environmental degradation.

Propoor Policies: Why is Special Attention Needed for the Poor?

It then follows almost automatically that development efforts will naturally help these groups the most. After all, if the starting point is sufficiently low, any meaningful improvement would imply a higher value added for them than for those who are better off.

Empirical evidence suggests that this is not a rule. Growth alone does not necessarily mean that everybody benefits, or that all beneficiaries benefit the same. On the contrary, it has been documented that economic development may actually cause a worsening of income disparities, especially in the early phases of development. There are cases where poverty—not the income disparities—has remained unchanged in spite of sustained economic growth (Lustig and Deutsch, 1998).

The hypothesis by the 1971 Nobel Prize Winner Economist, Simon Kuznets, on the relationship between economic growth and income inequality suggested that economic growth in poor countries increased the inequalities between the rich and the poor, thus making the poor relatively worse off (Kuznets, 1955). The trend was in the opposite direction in wealthier countries. It was also suggested that higher inequalities retard growth in poorer countries and encourage growth in richer countries.

A rich, diverse literature exists as to whether this hypothesis is supported by empirical evidence worldwide and for different definitions of inequality, how it would perform if measured in income distribution terms or as applied to land tenure, and how they relate to poverty reduction, as the poor can still benefit from development even when they are worse off than the rich (Anand and Kanbur, 1993).

A conclusion from the above is the need for propoor policies in water-based development that would ensure that the poor would benefit from the public investments, basic infrastructure, increased opportunities, and the growth environment created as a result of development efforts. The levels, policy elements, and implementation modalities are to be determined on case-specific basis and by ensuring that the economic viability is still maintained at large.

It is evident that the integration of the poor and the underprivileged is an integral part of a regional development effort. When regional development is based on water resources, this becomes even more important due to the direct link between water, health, and livelihood (Unver et al., 2003).

From a broader angle, the nature of growth-poverty relationship provides another rationale for the integrated approach to development as opposed to sectoral perspective.

Although propoor policies are needed in all respective sectors, a holistic approach to development would have higher chances to produce the optimal benefit to the poor on the overall investment than the summation of those resulting from individual, uncoordinated investments on sector basis.

IMPLEMENTING WATER-BASED SIRD

Development on River Basin Scale

If and when a river basin or a subbasin with development prospects can be delineated, this geographic area can become the unit for development. Depending on the conditions of the local economy, other natural resources, topography, physical infrastructure, interactions with the broader areas, and other relevant parameters, a water-based regional project can be formulated. Irrigated and dryland farming, drinking water supply, hydropower generation, fisheries, flood and drought mitigation, and recreation are among the possible components in the water dimension while the social and economic development spectrum is covered through their relevant aspects. "Hard" or infrastructure-related investments are complemented by "soft" issues that may include water users' organizations, training and retraining programs, stakeholder participation, gender-related programs, community development projects, resettlement and special projects for the poor, the disadvantaged, and the landless. Sectors and projects from the outer peripherals can be included in a holistic package that can facilitate cross-sectoral partnerships and speed up financing decisions.

Integrated Rural Development

Based on the local conditions and opportunities, rural development packages with a water component can be formulated with broader scopes. The scale here is often substantially smaller than a river basin, and therefore the project is of more modest dimensions, using multiple village areas for the typical development scale. Nevertheless, the general approach of integrated development still applies, if with more farming overtones. Local income generation, fisheries, rural-urban interactions, and market links are probably emphasized and a gender component with governance structures and capacity building and micro-to-small credit programs are among other possible components.

A Tool for Transboundary Cooperation

In case an international river basin exists with multiple riparians that can benefit from a development activity, be it a common power plant, a jointly used water treatment–water supply system, an irrigation network, or environmental protection, a transboundary project may be formulated. Although harder to plan and manage, these projects enjoy broader donor support and contribute to regional peace and harmony.

An exemplary transboundary project would be a multisector regional development program with a water core where tools such as cooperative planning, collective financing, and joint management can be implemented as a whole and/or for components of the package.

Management of Regional Development

Because water-based sustainable integrated regional development is an exception rather than the rule in development efforts and due to the complexity of the scale and scope components, it is seldom possible to find an existing entity for the management of the development project. Capacity constraints and other challenges in the government sector may make it necessary to establish a new, mandated entity for this purpose with autonomous status to some degree, to exercise public powers. It is useful, if not essential, to have the explicit political support at the highest level possible, in addition to the proper delegation of authority. The management entity has to be accountable in its operations, financially and otherwise, through a proper arrangement taking into account the local conditions. It may have a board composed of members from the government, the stakeholders, the donors of the projects, or appropriate U.N., and/or other international partners involved in the process, if any. It is imperative to provide accountability at the regional and local applicable levels to prevent potential insensitivity to local needs. Multidisciplinary staff may include, depending on the nature of the project, engineers, planners, economists, ecologists, social workers, sociologists. There may be sections or sector leaders but the hierarchy should be minimized and teamwork ensured. The staff must be hired on merit basis, should be paid from the project budget, with sufficient compensation. They should be supported by adequate local staff to facilitate proper interaction with the local groups, stakeholders, and project beneficiaries. Operational flexibility, timely decision making and fast action, freedom from political interventions are some of the key features that the regional entity should possess.

A logical frame approach can be a powerful tool for identifying, planning, implementing, and monitoring a water-based regional development project. This approach has two major advantages: it provides an analytical assessment of the process and many donor organizations are familiar with it.

Planning a Regional Development Project

A potential sequence of steps in the process include a *situational analysis*, which includes a compilation of the natural resource base (including water and land resources); existing spatial, economic, and social structures; sectoral landscape (water supply-sanitation, farming, irrigation, energy, environment, education, health); and physical infrastructure and demographics, followed by an analysis of problems, opportunities, and bottlenecks regarding development. The situational analysis may use either a "problems-oriented," approach such as water supply to a population that has no water to drink or an "objectives-oriented" approach, for example, hydropower generation, based on the conditions.

A *stakeholder analysis* is performed to identify the groups that are affected by the problems and would potentially be impacted (positively or adversely) by the solutions. When poverty, gender issues, and social development are important considerations, which they are in our case, the stakeholder analysis provides very useful information and input into the planning process. It can help minimize conflicts, reduce disparities, empower the disadvantaged, and prepare mitigation strategies.

An *analysis of objectives* is performed to allow the project address the problems identified in the earlier stages. The overall objective of "mobilizing water resources and sustainable development for positive economic and social change" is operationally described by concrete objectives and goals in this step. Means-ends relationships are developed.

Other principles, conditions, and criteria are identified and integrated into the formulation. These may include preferences set by the donors, gender-based considerations, environmental policies, and the like.

The next step is usually the *development of strategies* along with their merits and other implications. This is a critical step in decision making since it is likely to have multiple project formulations with different scope and dimensions. A decision by the "board" or the higher authority is typically required at this point.

Based on the decision regarding the strategy for regional development, the next step involves implementation elements including infrastructure design, cost schedules, and resource requirements.

TURKEY'S SOUTHEASTERN ANATOLIA PROJECT (GAP): A REAL-WORLD EXAMPLE

Turkey provides a contemporary example of water-based regional development in a multisectoral setting, through its Southeastern Anatolia Project, or GAP with its Turkish acronym.

The Turkish experience contributes valuable input to regional development due to a combination of factors that this country represents (Fig. 12.2).

A look at the geographical location of Turkey would also give a hint to the processes it has been going through. Located between the developed Europe and the developing Asia, Turkey also is a passage from single-party regimes in the Middle East to democracy,

Figure 12.2
Turkey at the crossroads of three continents.

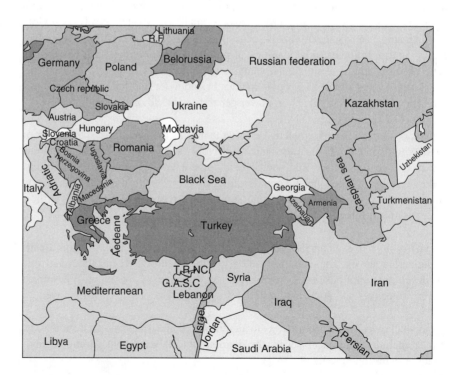

an example of economic transformation from state domination to market-driven economy, and a major emerging market with a vivid and growing civil society and major political changes, all of which is supported by a process of harmonization with the European Union.

The Turkish water-based regional development project, GAP, like the Tennessee Valley Project of the United States, stemmed from an initial desire by a leader to mobilize water resources for a broader objective. Mustafa Kemal Ataturk, the founder of the Republic of Turkey, envisioned in 1930s, transforming the waters of these rivers into "a lake of humanity and civilization" when another leader across Atlantic, was creating the Tennessee Valley Authority.

The young republic preoccupied by repairing the wounds of the recent wars and Ataturk's vision of "humanity and civilization" would have to wait a few decades before it was put on the drawing boards in 1950s.

This "wait" is typical of the national processes in both developing and developed countries in regard to the emergence of regional policies. Unless influenced by externalities, of which a good example could be the regional policies of the European Union, development on a national scale is usually the priority until some social, economic, political, or other threshold is reached that shifts the emphasis from national issues to interregional development disparities.

GAP's own process was induced by the convergence of two separate processes in Turkey; the development of water resources; and the impact of national policies on regional development levels.

Water Resources Development Component of GAP

Ancient civilizations in Anatolia and the Ottoman Empire constructed many water structures of which some of them are still operational in reasonably good condition. Remnants or ruins of water-related structures are spread over Anatolia and stand as impressive engineering techniques of their era. The first modern irrigation and drainage project, the Cumra Project, was designed and constructed between 1908 and 1914 by the Ottoman Empire. After the establishment of modern Turkey in 1923, there have been great endeavors by the state to develop the country through the development of natural resources. In this respect, many water-related projects were initiated including dams, irrigation schemes, domestic water supply, and flood control structures throughout the country. As a part of the national program, water resources projects were planned for the Euphrates and Tigris rivers, somewhat independently from one another. Early efforts to "regionalize" and integrate these date back to 1960s when the Lower Euphrates System was taken as a planning unit and a separate engineering department was assigned to the task. This process would lead, in mid-1970s, to the integration of the Euphrates and Tigris projects into a package consisting of 22 dams, 19 power plants, and the irrigation of 1.7 million hectares of land (Fig. 12.3 and Table 12.2) entitled the Southeastern Anatolia Project, or GAP. The total installed capacity of power plants is 7476 MW and the annual energy production is estimated to be 27 billion kWh at completion. This package, like other major water resources development schemes in the country, was planned and implemented by the General Directorate of State Hydraulics Works, which operates under the Ministry of Energy and Natural Resources.

Figure 12.3

Water resources development in GAP.

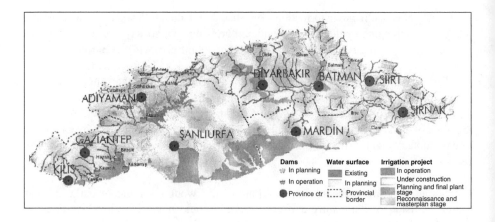

Table 12.2a

GAP Hydropower Projects

Name	Installed Capacity (MW)	Energy Production (Gwh)	Status
Euphrates	**5 313**	**20 098**	
Karakaya Dam	1 800	7 354	In operation
Atat rk Dam and HPP	2 400	8 900	In operation
Birecik Dam and HPP	672	2 516	In operation
Karkamıs Dam and HPP	189	652	In operation
Sanlıurfa HPP	50	124	Under construction
B y k ay Dam and HPP	30	84	Master plan
Ko ali Dam and HPP	40	120	Master plan
Sırımtas Dam and HPP	28	87	Master plan
Kahta Dam and HPP	75	171	Master plan
Fatopasa HPP	22	47	Master plan
Erkenek	7	43	Preliminary survey
Tigris	**2 172**	**7 247**	
Dicle Dam and HPP	110	298	In operation
Kralkızı Dam and HPP	94	146	In operation
Batman dam and HPP	198	483	Construction + operation
Ilısu Dam and HPP	1 200	3 833	Programmed
Cizre Dam and HPP	240	1 208	Programmed
Silvan Dam and HPP	150	623	Master plan
Kayser Dam and HPP	90	341	Master plan
Garzan Dam and HPP	90	315	Preliminary survey
Total	**7 485**	**27 345**	

Completed Projects	Irrigation Area (ha)
1. Hancagız Dam and irrigation	7 330
2. Sanlıurfa-Harran Plains	120 000
3. Hacıhıdır Dam and irrigation	2 080
4. Derik-Dumluca	1 860
5. Silvan	8 790
6. Nusaybin	7 500
7. Silopi Nerd s	2 470
8. Akcakale groundwater	15 000
9. Ceylanpınar groundwater	27 000
10. Devegecidi Dam and irrigation	7 500
11. Suru groundwater	7 000
12. ınar-Goksu Dam and irrigation	3 580
13. Garzan-Kozluk	3 700
14. Adıyaman- amgazi	1 000

Table 12.2b
GAP Irrigation
Projects

Partially Completed Projects	Irrigation Area (ha)
1. Mardin-C.pınar/Sanlıurfa-Harran Plains	47 795
2. Kralkızı-Dicle	24 421
3. Batman Left Bank	18 758
4. Batman Right Bank	18 593
5. Belkıs-Nizip	11 925
6. Adıyaman- amgazi	6 430
7. Kayacık Plain	19 993
8. Bozova center-pumped irrigation	1 369
9. Samsat	2 800
10. Yaylak irrigation	18 322
11. Dicle-Kralkızı gravity+ canal network	18 431
12. Dicle-Kralkızı pumped irrigation	7 845
13. Mardin Main Canal, conveyance	160 km
14. Sanlıurfa-Bozova, pumped irrigation	8 660
15. Suru Plain pumped irrigation	37 km

The regional policy in Turkey underwent a different evolution and was institutionally addressed starting in 1960s, when the country entered "the planned development era" and the government initiated the practice of making 5-year development plans. "Lesser-developed regions" were given priority status, and incentives were introduced to accelerate economic growth in these areas. The focus then was on a provincial basis and the policy decisions and follow up were carried out by the *State Planning Organization* (SPO), an agency attached to the prime minister's office.

**Regional
Development
Policies and
GAP's
Institutional
Evolution**

A regional project linking water resources development to other related sectors came in early 1980s when SPO decided to unify the respective public investments in the GAP region in a single program. A separate unit was put in charge of the project by SPO until 1989. The GAP Master Plan (Nippon Koei-Yuksel Proje Joint Venture, 1989) set the foundation for the integrated regional development in a multiple sector setting and suggested that a new entity be formed to manage it. GAP Administration was founded the same year as a body attached to the prime minister's office. This arrangement was somewhat similar to the Tennessee Valley Authority (TVA) model in that it followed the integrated multisectoral regional approach with a separate entity in charge. The major institutional difference between the TVA and GAP models was in the implementing authority and infrastructure financing. While TVA was in charge of the whole project cycle with its own budget, GAP continued to rely on the existing setting for project execution for infrastructure investments. GAP planned and implemented the "nonstructural" or "soft" aspects of development and established partnerships while it planned, coordinated, and monitored the overall execution by a variety of agencies. It produced the regional development plan and sectoral implementation programs as well as the objectives, goals, and targets for the sectors, and action plans for the public sector. On the execution side, it implemented smaller scale projects on the social dimension (projects for the poor and the landless, youth, street children, resettled population, gender empowerment), business promotion (training programs, feasibility and match-making services, micro credit), environment (wildlife and biodiversity projects, identification of hot spots, impact analyses, public and formal education, awareness raising), rural development (multivillage development programs, rural water supply and infrastructure, access to markets, alternative income generation such as bee keeping, high-value crop production, handicrafts), archaeology (research, documentation, rescue projects including the excavation and rescue of the ancient Roman city of Zeugma) among others (GAP Administration, 2003).

Following its initial organizational structuring, GAP Administration went through a partnership building period in which local organizations, farmers, municipalities, civil society, and academia were some of the partners. Joint projects and programs were established with international agencies in the same period. GAP Administration transformed the project in 1995, as a result of a participatory multistakeholder process jointly managed by the *United Nations Development Programme* (UNDP), into a sustainable human development program (United Nations Development Programme and GAP Administration, 1997), incorporating the principles set forth by the Earth Summit, Rio de Janeiro, 1992; International Conference on Population and Development, Cairo, 1994; World Summit for Social Development, Copenhagen, 1995; and World Conference on Women, Beijing, 1995.

GAP as a Water-Based Sustainable Integrated Regional Development Program

GAP is implemented as a regional, social, and economic sustainable development program with a major water development core and includes projects and programs in agriculture, education, transportation, environmental protection, manufacturing, tourism, urban and rural infrastructure, archaeology, business promotion, gender equality, and poverty reduction.

GAP-RDA's focus on sustainable human development in the region builds upon the concept of integrated regional development of the GAP Master Plan of 1989. GAP Administration, following its founding in November 1989, embarked upon a number of projects in partnership with local and national NGOs and academia looking at the economic, environmental, spatial, social, and human aspects of development.

On the social dimension of the project, socioeconomic studies and population movements (Middle East Technical University and GAP Administration, 1994), trends of social change (Chamber of Agricultural Engineers and GAP Administration, 1993), problems of employment and resettlement due to reservoir construction (Sociology Association and GAP Administration, 1991), women's status and integration into development process (Development Foundation of Turkey and GAP Administration, 1994) provided the groundwork for this dimension to be added to the project. GAP Administration then asked a group of experts and professionals from academia, civil society, and the public sector to form a working group entitled GAP Social Action Group and tasked it with a synthesis of these studies, policy advice, and implementable projects and programs. The group produced the GAP Social Action Plan (GAP Administration, 1995). The plan compiled and assessed the present situation in terms of social and cultural structures, family structure, demography, health, education and infrastructure services, settlement patterns, human-land interaction, and employment-income distribution. It then established targets, strategies, and action plans for seven major areas and issues—governance and participation, population movements and settlement, education, health, agricultural extension and capacity building, employment and income, and land tenure and land use.

On the economic dimension, a number of studies were conducted, including the feasibility for an economic development agency (Priva Consultancy and Turkish Industrial Development Bank, 1991), agricultural marketing and crop pattern studies (Tipas and AFC Joint Venture, 1992), economic analysis and credit needs for agricultural enterprises (Scientific and Technical Research Council of Turkey, 1995), and agricultural postharvest management (Compagnie Agricole et Maraichere de Sherington, 1994).

Studies for water management and environment were also conducted in early 1990s, which included irrigation technology and management of canals (Unver and Voron, 1993; Unver et al, 1993), groundwater feasibility study (TUMAS, 1990), and a comprehensive study on the management, operation, and maintenance of irrigation systems (Dolsar, Halcrow, Rural Water Corporation Joint Venture, 1993). The latter study covered a multitude of aspects of irrigation including environmental impact assessment, pricing, farmer support systems, impact monitoring, agronomy, environmental health, farm budgets, capacity building, participatory irrigation management, soil conservation, and institutional reform among others.

Physical and infrastructure studies and cross-cutting issues were studied through a number of projects. These included a comprehensive health sector study (Ministry of Health and TUSTAS, 1991), a regional education, employment, and workforce plan (Ministry of National Education and TUSTAS, 1991), regional tourism assets and planning (Ministry of Tourism and Engin Consultancy, 1991), a comprehensive regional transportation and infrastructure planning study (TDST, 1991), and a mineral resources inventory and development planning study (MTA, 1992).

In addition to these studies, GAP had created, by 1995, an extensive network with local, regional, and national stakeholders, decision makers, and opinion influencers. The network included United Nations organizations and international entities. A stakeholder consultation process was then conducted, in partnership with the United Nations Development Programme, and both the philosophy and the implementation aspects were revisited. The result was a stakeholder-led transformation of the project from an integrated socioeconomic investment program by the government to a water-based sustainable regional development project (United Nations Development Programme and GAP Administration, 1995).

This process was a major step toward a greater integration of sustainable development with socioeconomic and infrastructure projects, addressing not only the more traditional economic-social-environmental components of infrastructure development but also broader issues such as jobs, land and income distribution, population movements in a multistakeholder setting with a "new" concept, governance, introduced.

The era following the transformation is marked by efforts, activities, projects, and policy advice to governments in materializing this paradigm with its financial, coordination, networking, and implementation aspects. GAP Administration helped other organizations reformulate their projects and programs accordingly, conducted capacity building programs, worked with municipalities and local groups, and implemented a large number of pilot projects and programs, some of which became national models replicated by others.

GAP benefited from the observation that worldwide efforts over the last five decades had resulted, on one hand, in new methods, techniques, pioneering technologies, rational use of resources, and hence output growth. On the other hand, these efforts had failed to prevent greater problems in social equality, environmental destruction and the general disruption of ecological equilibrium. These conditions made it necessary to seek alternative approaches to development in general, and to the execution of development in particular. The ultimate aim of GAP is to ensure sustainable human development in the region. It seeks to expand choices for all people—women, men and children, the current and future generations—while protecting the natural system, which sustains life in all forms. Differentiating from a lopsided, economy-centered paradigm of development, this approach places people at the core and views people both as a means and an end of development. It focuses on protection against exclusion and marginalization of weak, disadvantaged members of the society. In brief, GAP aims to eliminate poverty and to promote equity and fairness in development to all through good governance. As such, it contributes to the fulfillment of human rights—economic, social, cultural, civil, and political for the downtrodden in this region.

A comprehensive overview of GAP is given in Biswas (1997).

Integrating Individual Sectors in a Holistic Program/Stock of Overall Capital Assets

While assessing the sustainability of any development process, it is important explicitly to address what is expected to be sustained, why and how long, and the mode of measurement of sustainability, among others. As a rule of thumb, a development path may be sustainable only if it ensures that *the stock of overall capital assets remain constant or increase overtime* (Dasgupta and Serageldin, 2000). These include manufactured capital, human capital, social capital, and environmental capital. The sustainable development agenda must also be concerned with both intergenerational and intragenerational equity.

The following is a cross-section of projects and activities of GAP Administration addressing the capital assets, taken from their annual status and activity report (GAP Administration, 2001).

Social Capital

- A grassroots model of irrigation management
- Land consolidation and extension services
- Participatory regional development plan
- Participatory resettlement, retraining, employment generation, and urban planning in towns inundated by Birecik Dam Reservoir
- Strengthening capacity to deal with water-related health issues
- GAP Region Public Health Project
- Project to strengthen the National Capabilities of Malaria Units in Turkey
- Projects for Community Volunteers in Malaria Control (Roll Back Malaria)
- Projects to combat leishmaniasis
- Water supply and sanitation projects
- Water supply wells for small rural settlements in the GAP region.
- Reuse of municipal wastewater in small and medium-sized communities
- Solid waste management for large municipalities
- Construction of sewerage systems and wastewater treatment plants in selected cities
- Design of sewerage and wastewater treatment plants for towns discharging into major water bodies
- GAP Urban Sanitation and Urban Planning Project

Human Capital

- CATOMs: Multipurpose community centers for women
- Youth and children's programs
- Youth to youth social development program
- Youth cultural centers
- Young entrepreneurs project
- Rehabilitation program for street children
- Environmental education in elementary schools
- Income generating activities in drylands

Natural Capital and Cultural Heritage

- Environmental impact analysis of irrigation
- Environmental impact analysis of major infrastructure projects
- Afforestation and erosion control
- Microcatchment rehabilitation and range land rehabilitation

- GAP biodiversity research project
- Wildlife project for the GAP region
- Studies on the present and prospective climatic features of the GAP Region

Interim Results of Development

GAP is a long-term program. Its financial completion rate was 52 percent or $17 billion out of the total projected investment requirement of $32 billion as of the end of 2005.

Although it is too early to expect the permanent benefits of a project this size and scope, some interim results are impressive. The completion rate in the energy sector is 73 percent and the hydroelectric benefits of the project have already exceeded $13 billion and are alone able to pay back the total investment. The physical rate of completion of the total irrigation target is 13 percent—the lowest sectoral rate—but the impacts of irrigated farming are visible and substantial. Table 12.3 shows the change in the farm incomes before and after irrigation in Harran Plain, where the beneficiaries are the 18,405 farm families, living in 107 villages over a 111,600 ha irrigation area. The per capita income in this area has tripled after irrigation.

Another impact of development can be observed in the growth of the industrial sector. The number of private enterprises has increased as a combined result of increased agricultural production, increased needs for agriculture inputs and machinery, incentives for private investors, and the presence of readily available, inexpensive labor. As seen in Table 12.4, the number of new business start-ups exceeded the number of the then existing businesses by far, following the start of irrigation in 1995.

Table 12.3

Impact of Irrigation on Harran Plain

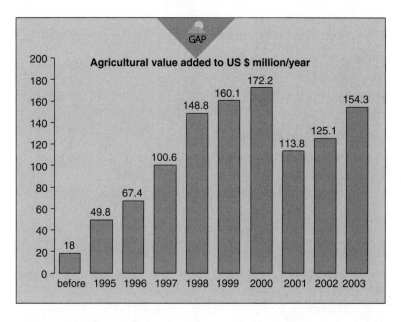

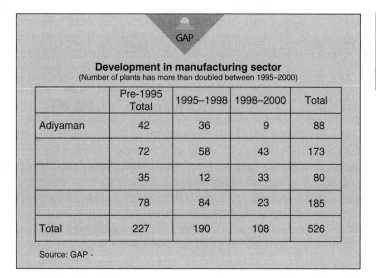

Development in manufacturing sector
(Number of plants has more than doubled between 1995–2000)

	Pre-1995 Total	1995–1998	1998–2000	Total
Adiyaman	42	36	9	88
	72	58	43	173
	35	12	33	80
	78	84	23	185
Total	227	190	108	526

Source: GAP -

Table 12.4
Impact of Development on Manufacturing Enterprises

The comprehensive investment campaign in the project resulted in major changes in other indicators as listed in Table 12.5. The rural electrification and accessibility reached almost 100 percent, literacy rates increased significantly, infant mortality figures dropped, and the land tenure system changed toward a more equitable distribution in an area, where land ownership is quite skewed due to the traditional, feudal, and semifeudal social structures. Under a special agrarian reform program implemented in the largest farming province, Sanliurfa, upper limits were imposed on the size of land that can be owned by single individuals. The government would buy, at the market price, the acreages over the set limits. This program did not result in much acquisition by the government as large land owners divided their lands and transferred the ownership of these

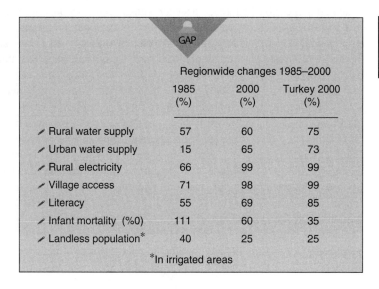

Regionwide changes 1985–2000

	1985 (%)	2000 (%)	Turkey 2000 (%)
Rural water supply	57	60	75
Urban water supply	15	65	73
Rural electricity	66	99	99
Village access	71	98	99
Literacy	55	69	85
Infant mortality (%0)	111	60	35
Landless population[*]	40	25	25

*In irrigated areas

Table 12.5
Impact of GAP on Region-Wide Indicators

to their children. A direct beneficiary emerged from this practice: women. Many women became land owners in an area where the established tradition did not grant them ownership. This was reinforced by another agrarian reform program under which over 3000 landless families were given farm land in a redistribution of government-owned land.

FUTURE DIRECTIONS AND ROOM FOR IMPROVEMENT

GAP is a contemporary example of water-based sustainable regional development. It has reached its current paradigm in a dynamic manner and benefiting the experiences of others, positive and negative, in the areas of socioeconomic development and water resources development. As in all large-scale projects, its projected benefits will take some time to materialize. The current period is one of another transition in which the infrastructure development has reached a 50 percent completion rate. Basic rural infrastructure has been put in place and substantial power production capacity has been created. Irrigation infrastructure, however modest compared to the eventual targets, has increased the farm output, which, in turn, has resulted in substantial capital accumulation. Private businesses have flourished for an interim period, a trend which was then reversed from 2001, by the worst economic and financial crisis of the country.

GAP's long-term success will depend on a number of conditions. The first and the foremost condition is the continuation of the government investments into the provision of the remaining infrastructure, which has a price tag of over $16 billion. The investment figures have dropped in recent years as a result of the austerity program of the government. A renewed commitment to investment will have to be complemented by an aggressive business promotion program and creation of new jobs. The increased per capita income and farm output are yet to trickle down to the lower segments of the income distribution. Likewise, the intraregional disparities seem to be unaffected at the current level of development. Education, healthcare, and many other social indicators have improved but remain below national averages. The "soft" programs initiated by GAP Administration need to be scaled up and established governance structures need to be institutionalized. It is imperative that a stronger cooperation between the regional programs, prepared by GAP Administration based on the Regional Development Plan (GAP Administration, 2002) and the National Investment Budget prepared by SPO, is established and maintained.

The water sector offers more optimism than despair. Some of the much needed reforms in this sector are imminent as a result of Turkey's commitment to harmonization with the European Union. As a result, decision-making processes will be more transparent, participatory, and bottom-up. Management will be decentralized and delegated. Environmental and social aspects will be more fully integrated into the water resources management and compliance will be more strictly monitored. Harmonization with Europe and a selective use of structural and other funds will help improve a number of specific areas, such as resettlement, gender empowerment, poverty reduction, and environmental protection, among others.

Transboundary aspects of GAP will be another test for GAP's success. The great experience gained and the expertise developed in the evolution of GAP from a simple water project into a water-based sustainable regional development program may provide a basis for riparian cooperation. The water resources component of GAP has not contributed positively to the relations between Turkey and its downstream neighbors. Syria and Iraq have also had serious differences and problems in regard to Euphrates (Kibaroglu, 2002). It is possible and may become politically feasible for the riparian nations to cooperate around sharing the benefits from water-based development of regional scale, region being the broader basin of the Euphrates-Tigris system, as opposed to disagreeing on water quotas or allocations of a water-sharing formula that has proved to be divisive in the past. Nonofficial, neutral Track 2 efforts may facilitate this process (CSIS Global Strategy Institute, 2005). GAP Administration and Syrian Irrigation Ministry's Euphrates Development Agency were able to agree on cooperation (Joint Communiqué, 2001; Implementation Document of Joint Communiqué, 2002) and a similar modality may be feasible for the interim time, until a functional Track 1 process can be initiated.

Last, but not the least, the project's success will be tested by the integration of the people into the development process.

REFERENCES

Anand, S. and R. Kanbur, "Inequality and Development: A Critique," *Journal of Development Economics,* Vol. 41, No. 1, pp. 19–43, 1993.

Biswas, A. K. (ed.), "Special Issue: Water Resources Development in a Holistic Socioeconomic Context—the Turkish Experience," *International Journal of Water Resources Development,* Vol. 13, No. 4, 1997.

Biswas, A. K., O. Unver, and C. Tortajada (eds.), *Water as a Focus for Regional Development,* Oxford University Press, New Delhi, 2004.

Chamber of Agricultural Engineers and GAP Administration, *Trends of Social Change in the GAP Region,* Southeastern Anatolia Project Regional Development Administration, 1993.

Compagnie Agricole et Maraichere de Sherington, *Project on Developing Post-Harvest Technologies for Fresh Fruits and Vegetables in the GAP Region,* Southeastern Anatolia Project Regional Development Administration, 1994.

CSIS Global Strategy Institute, "The Tigris and Euphrates River Basins: A New Look at Development," *Future Watch,* Center for Strategic and International Studies, Washington, D.C., June 2, 2005.

Dasgupta, P., and I. Serageldin, *Social Capital: A Multi-faceted Perspective,* Washington, D.C., The World Bank, 2000.

Development Foundation of Turkey and GAP Administration, *Project on the Improvement of Women's Status in the GAP Regions and their Integration into Development Process,* Southeastern Anatolia Project Regional Development Administration, 1994.

Dolsar, Halcrow, Rural Water Corporation Joint Venture, *Project on the Operation, Maintenance and Management of Irrigation Systems,* Southeastern Anatolia Project Regional Development Administration, 1993.

GAP Administration, *Social Action Plan,* Southeastern Anatolia Project Regional Development Administration, 1995.

GAP Administration, *Southeastern Anatolia Project: Latest State 2001,* Southeastern Anatolia Project Regional Development Administration, 2001.

GAP Administration, *Southeastern Anatolia Project Regional Development Plan,* in Turkish, Southeastern Anatolia Project Regional Development Administration, 2002.

GAP Administration, *Zeugma: A Bridge from Past to Present,* Southeastern Anatolia Project Regional Development Administration, 2003.

Gleick, P. H. et al., *The World's Water 2004-2005, The Biennial Report on Freshwater Resources,* Island Press, 2004.

Global Water Partnership, *Integrated Water Resources Management,* TAC Paper No. 4, Stockholm, 2000.

Implementation Document of Joint Communiqué, between Southeastern Anatolia Project Regional Development Administration (GAP), Turkey and General Organization for Land Development (GOLD), Syria, available at http://www.gap.gov.tr/English/Uiliski/dkp2.html, Aleppo, June 19, 2002.

International Institute for Sustainable Development, "Summary of the Pan-African Implementation and Partnership Conference on Water: 8–12 December 2003," *Earth Negotiations Bulletin,* Vol. 5, No. 20, pp. 2–8, December, 2003.

Joint Communiqué, between Southeastern Anatolia Project Regional Development Administration (GAP), Turkey and General Organization for Land Development (GOLD), Syria, available at http://www.gap.gov.tr/English/Uiliski/dkp2.html, Ankara, August 23, 2001.

Kibaroglu, A., *Building a Regime for the Waters of the Euphrates-Tigris River Basin,* International and National Water Law and Policy Series, Vol. 7, Kluwer Law, 2002.

Kuznets, S., "Economic growth and income inequality," American Economic Review, Vol. 45, No. 1, pp. 1–28, 1955.

Lustig N., and R. Deutsch, *The Inter-American Development Bank and Poverty Reduction: An Overview (Revised Version),* No. POV-101, Washington, D.C., 1998.

Middle East Technical University and GAP Administration, *GAP Region Population Movements Study,* Southeastern Anatolia Project Regional Development Administration, 1994.

Ministry of Health and TUSTAS, *GAP Health Sector Implementation Project,* in Turkish, Southeastern Anatolia Project Regional Development Administration, 1991.

Ministry of National Education and TUSTAS, *GAP Regional Education, Employment and Manpower Planning Study,* in Turkish, 1991.

Ministry of Tourism and Engin Consultancy, *GAP Regional Tourism Inventory and Development Planning Study,* in Turkish, 1991.

MTA, *Research on Feasible Mineral Resources Potential in the GAP Region,* in Turkish, Southeastern Anatolia Project Regional Development Administration and MTA (General Directorate of Mineral Research and Exploration), 1992.

Nippon Koei–Yuksel Proje Joint Venture, *GAP Master Plan Study*, Republic of Turkey Prime Ministry State Planning Organization, April, 1989.

Priva Consultancy and Turkish Industrial Development Bank, Model for Economic Development Agency, Southeastern Anatolia Project Regional Development Administration, 1991.

Scientific and Technical Research Council of Turkey, *Economic Analysis of the Agricultural Enterprises in the GAP Region, Determination of the Short, Medium and Long Term Credit Needs,* Southeastern Anatolia Project Regional Development Administration, 1995.

Sociology Association and GAP Administration, *Survey on the Problems of Employment and Resettlement in Areas to be Flooded by Dam Lakes in the GAP Region,* Southeastern Anatolia Project Regional Development Administration, 1991.

TDST, *GAP Regional Transportation and Infrastructure Development Project,* Southeastern Anatolia Project Regional Development Administration and TDST (Temel Eng. Inc., Dapta Eng. Inc., Su-Yapi Eng. & Consulting Inc., Temelsu International Eng.), 1993.

Tipas and AFC Joint Venture, *GAP Agricultural Marketing and Crop Pattern Study,* Southeastern Anatolia Project Regional Development Administration, 1992.

TUMAS, *Feasibility Study on Groundwater Reserves in Ceylanpinar Area,* in Turkish, Southeastern Anatolia Project Regional Development Administration, 1990.

United Nations, *Agenda 21: Earth Summit— The United Nations Programme of Action from Rio,* United Nations Publications, 1993.

United Nations, *Report of the World Summit on Sustainable Development, Johannesburg, South Africa, 26 August–4 September 2002,* United Nations Publications, 2002.

United Nations Development Programme and GAP Administration, *Sustainable Development and the Southeastern Anatolia Project Seminar Report,* Vol. 1, pp. 27–29 March, Sanliurfa, 1995.

United Nations Development Programme and GAP Administration, *Project Package on Strengthening Integrated Regional Development and Reducing Socio-economic Disparities in the GAP Region,* Southeastern Anatolia Project Regional Development Administration, 1997.

Unver, O., and R. Gupta (eds.), *Water Resources Management: Cross-Cutting Issues,* Middle East Technical University Press, Ankara, 2002.

Unver, O., R. Gupta, and A. Kibaroglu (eds.), *Water Development and Poverty Reduction,* Kluwer Academic Publishers, 2003.

Unver, O., and B. Voron, "Improvement of Canal Regulation Techniques: The Southeastern Anatolia Project-GAP," *Water International,* Vol. 18, pp. 157–165, 1993.

Unver, O., B. Voron, and T. Akuzum, "Improvement of Field Water Distribution and Irrigation Techniques: The Southeastern Anatolia Project-GAP," *Water International,* Vol. 18, pp. 166–174, 1993.

Willem-Alexander, Prince of Orange, "Water: A Crisis of Governance," *ADB Review,* Vol. 35, No. 1, pp. 4–5, Asian Development Bank, Manila, January-February, 2003.

World Business Council for Sustainable Development, "Sustainability through the Market," *Sustain, The Quarterly Newsletter of WBCSD,* Issue 8, pp. 8–9, February, 1999.

World Water Assessment Programme, *Water for People, Water for Life, The United Nations World Water Development Report,* UNESCO Publishing/Berghahn Books, 2003.

World Water Assessment Programme, *Water: A Shared Responsibility, the United Nations World Water Development Report 2,* UNESCO Publishing/Berghahn Books, 2006.

13

COMMUNITY MANAGEMENT OF RURAL WATER SYSTEMS IN GHANA: POSTCONSTRUCTION SUPPORT AND WATER AND SANITATION COMMITTEES IN BRONG AHAFO AND VOLTA REGIONS

Kristin Komives
Institute of Social Studies, The Hague, Netherlands

Bernard Akanbang
TREND, Kumasi, Ghana

Wendy Wakeman
World Bank, Washington DC

Rich Thorsten
University of North Carolina, Chapel Hill

Benedict Tuffuor
TREND, Kumasi, Ghana

Alex Bakalian
World Bank, Washington DC

Eugene Larbi
TREND, Kumasi, Ghana

Dale Whittington
University of North Carolina, Chapel Hill

INTRODUCTION

Over the last two decades it has become widely accepted that communities in rural areas of developing countries should assume most management responsibilities for their water projects. Community management means different things to different people, but at a minimum it involves the delegation of operation and maintenance of rural water supply systems to the community of users (Harvey, 2005). More comprehensive community management begins with active involvement of rural communities in decisions about choice of service level, siting of facilities, tariffs, and management structures. After project completion, communities assume control and responsibility for the operation, maintenance, and repair of the system (Schouten and Moriarty, 2003). Although the community management model shifts much of the responsibility for system sustainability from the facility provider to the end users (Harvey, 2005), government and donors still play an important role. Higher levels of government typically pay for the capital costs of system installation, and after construction is completed communities may receive assistance through government-organized or financed postconstruction support programs, such as training, spare parts delivery systems, and technical and administrative support visits.

One of the driving forces behind the shift to community management was mounting evidence of the poor performance of centrally managed rural water supply programs (Therkildsen, 1988, Briscoe and Garn, 1995). Some empirical studies showed that community-managed water systems fare better (Narayan, 1995, Gross et al., 2001). The advantage of community management is usually attributed to the sense of ownership and responsibility that this approach instills in local water users. Proponents of community management believe that villages will take better care of their water supply system if households feel that they have chosen it, must pay for it, and will be responsible for maintaining and repairing the system when it breaks. In addition to these advantages, the community management model appeals to donors and governments who are trying to do as much as possible with limited resources. Asking communities to contribute their time, labor, and money helps stretch limited program budgets and personnel resources (Lockwood, 2004).

Community management and participatory approaches to development, more generally, have received much scholarly attention in recent years. This work suggests that community management is not an unqualified success. Community-level factors (such as local power dynamics, social capital, the extent and quality of participation, and management capacity) and external factors (such as system design, construction quality, and spare parts access) affect the performance of community-managed water supply programs (Lockwood, 2004, McCommon et al., 1990, Isham and Kahkonen, 2002, Manikutty, 1998, Prokopy, 2005).

One important actor receives surprisingly little attention in this literature: the village water committee. In most rural water supply projects, a higher level of government charges the village water committee with putting community management into practice. In addition to the challenging tasks of managing operations, collecting tariffs, and organizing repairs, the village water committee also provides the critical link between the community and external support structures. Most postconstruction support programs require proactive, effective village water committees to secure external assistance (Komives et al., 2006).

There is little evidence in the literature as to how village water committees (rather than the services they manage) fare over time or about the factors that influence their performance and sustainability. The objective of this chapter is to begin to address these gaps in the literature by examining data from the 200 communities in the Brong Ahafo and Volta regions of Ghana that we collected in 2005. We seek to answer two questions: How are village water committees faring 5 years after the installation of new rural water supply systems; and does postconstruction support help improve the functioning of these committees? In the next, second section of the paper we present background information on the rural water sector in Ghana. The third section briefly describes the fieldwork.

In the fourth and fifth sections, we present our results. We describe the status of village water committees in communities where boreholes with hand pumps were installed at least 5 years ago. We then investigate to what extent postconstruction training and two models of postconstruction support visits help improve watsan-community relations and payment rates. The sixth and final section discusses the significance of these findings.

BACKGROUND AND STUDY CONTEXT

In the early 1990s, the Government of Ghana established a *National Community Water and Sanitation Program* (NCWSP) with the objective of providing improved and sustainable water and sanitation services to rural communities. Today the Ghanaian *Community Water and Sanitation Agency* (CWSA) is the body responsible for facilitating and coordinating rural water supply investments in accordance with the NCWSP. The financing of rural water supply investments in each region of Ghana is undertaken mostly by different donor agencies. The donor projects are expected to follow the project CWSA's guidelines for project planning and implementation.

CWSA estimates that 52 percent of Ghana's rural population currently has access to improved water services. Its national strategy for expanding service coverage promotes a "demand-driven" planning approach that emphasizes participatory project design and implementation. Donor-financed rural water supply projects are expected to include consultation with communities about relevant technology and management choices, and to ensure the participation of women in these decisions. Communities may also be involved in overseeing construction and participate in the construction process. Once the borehole projects are built, communities establish village water and sanitation ("watsan") committees to manage the systems. It is strongly encouraged that women serve as members. Watsan committees are provided management training during project implementation, and two village-based "caretakers" per borehole are trained to handle routine repairs and maintenance.

The rural water supply projects in Volta and Brong Ahafo adopted this general approach to project development, though there were some differences between the projects. Villagers in Volta, for example, were less likely to have been involved in decisions about technology choice or borehole location than villagers in Brong Ahafo.

The DANIDA-funded VR-CWSP in the Volta Region began in 1993. The first phase of the project concluded in 2003. Between 1995 and 2000, the years of interest for our study, VR–CWSP installed 664 new boreholes and rehabilitated 238 boreholes in the Volta Region. The International Development Association (IDA)-funded CWSP 1 project in Brong Ahafo was a 5-year project initiated in 1994. Two hundred and fifty four boreholes with pumps, 82 hand-dug wells with pumps, and 8 facilities without pumps were installed through the CWSP between 1996 and 2000.

Once boreholes and handpumps are installed, the watsan committees in both Volta and Brong Ahafo have access to a well-developed, multifaceted (although underfunded) system of postconstruction support. A central actor in the postconstruction support system is the *District Water and Sanitation Team* (DWST), a roving multidisciplinary team employed by the district assembly. The DWSTs are not supposed to do either minor or major handpump repairs themselves. Their primary responsibility is rather to help the village watsan committees obtain the support and training they need to run and repair the systems, and to resolve any management and water use conflicts that arise. The DWSTs visit watsan committees on request, and may assist communities in finding spare parts. They may also visit some communities on their own initiative to check on conditions and organize training sessions on topics they consider to be relevant. However, the financial resources available to the DWSTs to carry out these functions are limited and the activities of the DWSTs vary across districts.

An important resource for watsan committees are the "area mechanics" living in the district. These individuals were originally trained during the project implementation process to do routine maintenance or repair work on boreholes at the request of communities. Area mechanics are frequently called upon to obtain the spare parts needed by the community and then to install these parts. Communities must pay for the services of the area mechanics from revenues collected from village households or money obtained in some other way. The DWSTs may help watsan committees link up with an area mechanic when major repairs are needed.

Area mechanics, caretakers, and watsan committee members obtain spare parts from a well-developed system that includes a central spare parts warehouse in Tema, Ghana, and three subnational warehouse outlets in the northern, central, and southern regions of the country. The warehouses and outlets are needed to ensure the availability of pump parts for the four standard handpumps used in Ghana: Nira, Afridev, Ghana-modified Indian Mark II, and French Vergnet.

Watsan committees in Volta also receive assistance from another *postconstruction support* (PCS) program initiated and supported by DANIDA: *monitoring of operations and maintenance* (MOM). This is a PCS program of quarterly visits to communities by the district *Environmental and Health Assistants* (EHAs). During their visits, the EHAs do a technical assessment to determine how well the boreholes are functioning, review financial records, and check on payment practices. The records of these quarterly audits are compiled at the district level, in theory giving district-level officers a systematic picture of what is happening in the district. The contact between the EHAs and watsan committees also provides an opportunity for the EHAs to recommend improvements to the systems in place for managing and collecting revenue from water users, to provide information on improved health and hygiene practices as well as to potentially assist communities to obtain other forms of PCS (e.g., DWST, spare parts, and area mechanics). In 2002 and 2003, DANIDA was solely responsible for funding MOM in the Volta region. Since 2004, the responsibility for the program has fallen to the district governments in Volta. Many districts have reduced the frequency of these EHA visits due to resource constraints.

FIELDWORK

The data were collected from villages in four districts in Volta (Ho, Jasikan, Kadjebi, and Nkwanta) and five districts from Brong Ahafo (Asunafo, Dormaa, Kintampo, Tano, and Wenchi). The four Volta districts were selected because their district governments had sustained the quarterly MOM visits after the end of DANIDA funding. We chose the five districts in Brong Ahafo because district-level socioeconomic data from these five districts matched well with the Volta districts in our study.

We restricted our sample frame to villages in these nine districts that had received no more than two project boreholes with handpumps at least 5 years ago. This effectively also limited the size of the villages (from approximately 200 to 5000 people). These sample selection criteria yielded a potential sample frame of 97 villages in the Volta region and 120 villages in the Brong Ahafo region. All 97 villages for the Volta region were selected, and 103 out of the 120 villages in the Brong Ahafo region were randomly selected.

Five different data collection instruments were used in each village: (1) a household questionnaire (administered to 25 randomly selected households in each community), (2) a watsan committee questionnaire, (3) a questionnaire for the borehole caretaker in the village, (4) a guide for a focus group discussion with village leaders, and (5) a guide for focus group discussion with women. In addition, DWST engineers conducted a technical assessment of the villages' handpumps and observed the amount of water collected from the most heavily used handpump in the village (for a single day). Interviews were conducted in two languages—Twi and Ewe. The survey of the watsan committees is the source of most of the information included in this chapter.

Fieldwork began in late March 2005 and concluded in early May 2005. A team of enumerators, the DWST engineer, and field supervisor/focus group facilitator typically spent one entire day in a village collecting data, and then the following day moved on to the next village in the sample. A total of four teams were employed in the Volta region and five teams in Brong Ahafo.

CURRENT STATUS OF WATER SYSTEMS AND WATSAN COMMITTEES

Profile of Villages and Water Projects in the Two Regions

Table 13.1 presents a comparison of the study villages in Brong Ahafo and Volta along a number of different dimensions. As shown, villages in Brong Ahafo are quite similar to those in Volta in several respects. The average village has a population of close to 1000. The mean distance to a post office or public telephone in each region is between

Table 13.1

Profile of Villages and Water Supply Projects in Brong Ahafo and Volta

	Source	Brong Ahafo	Volta
Average number of people in village (mean)	Leaders	1154	1076
Mean distance to nearest post office (km)	Leaders	14	14
Mean distance to nearest paved road (km)	Leaders	17	5
Mean distance to nearest public telephone (km)	Leaders	15	12
Mean distance to river or stream (km)	Leaders	1.5	1.2
Percent of villages without electricity	HH	82%	54%
Mean percent of households in village with electricity, in villages that have electricity	Hh	33%	49%
Mean percent of households in village who own land	HH	0.62%	0.41%
Average size of landholdings in village in hectares (only including landholders)	HH	13.6	18.9
Median household monthly cash expenditure in village (in US$)	HH	73	54
Percent of villages that always have an alternative to boreholes in dry season	Women	20%	11%
Percent of villages where village leaders say they definitely or probably could obtain 3 million cedis from outside the village for repair or improvement of the boreholes	Leaders	25%	48%
Mean percent of households in village who say they trust their neighbors	HH	75%	75%
Mean percent of households in village who say they trust their leaders	HH	74%	80%
Percent of villages with only one borehole installed in Danida or CWSP project	HH	62%	50%
Average number of years since the pumps were installed on the new boreholes	District records	6.2	5.8
Average number of liters of water withdrawn from community's most-used borehole during day of observation at source	Source observation	7900	5524
Average number of users who collected water during day of observation at source	Source observation	328	163
Average number of liters collected per user during day of observation at source	Source observation	24	34
Percent of villages in which at least two out of three data sources say the community was involved in making the decision about what technology to install in the project	Combo	58%	25%
Percent of villages in which at least two out of three data sources say the community was involved in making the decision about where to site the boreholes	Combo	65%	18%

	Source	Brong Ahafo	Volta
Percent of villages in which at least two out of three data sources say the community was involved in making the decision about what tariff to charge	Combo	95%	99%
Percent of villages in which at least two out of three data sources say the community was involved in making the decision about who would manage the system	Combo	91%	94%
Percent of villages in which the women's focus group agrees women were encouraged to participate in preproject meetings	Women	72%	73%
Percent of villages in which women's focus group says women participated less than men in preproject meetings	Women	35%	56%

Table 13.1

Profile of Villages and Water Supply Projects in Brong Ahafo and Volta (*Continued*)

12 and 14 km. On average three quarters of the households in a village say they trust their neighbors and their leaders. There are, however, some differences across regions. For example, the mean distance of villages in Brong Ahafo to a paved road was 17.2 km v. 5 km in Volta. More villages in Brong Ahafo lacked electricity (82 percent) than in Volta (54 percent). Average monthly cash expenditure by households in Brong Ahafo is higher than in Volta, and a larger percentage of households in the Brong Ahafo villages are land owners.

Table 13.1 also compares the water supply projects and project methodologies in the two regions. Half of the Volta villages in the sample, and just below 40 percent of those in Brong Ahafo, had only one borehole installed in the village through the rural water supply project. The boreholes in Brong Ahafo are slightly older than those in Volta: the average age of the pump installations on the project boreholes in Volta is 5.8 years v. 6.2 years in Brong Ahafo.

Households in almost all of the villages in both regions felt that they controlled decisions about the composition of the watsan committee and about the tariff adopted after construction. In a third of cases in both regions, women who participated in the women's focus group stated that women had been actively encouraged to participate in preconstruction meetings. The largest differences in preconstruction project implementation occur in community involvement in the decision about the technology to use (58 percent were involved in Brong Ahafo v. 25 percent in Volta) and where to site the borehole (65 percent of villages in Brong Ahafo were involved versus 18 percent in Volta).

Our fieldwork revealed that most boreholes were "working" in the study villages in both Volta and Brong Ahafo. The exact numbers of boreholes classified as "working" depends, however, on the definition adopted. We considered boreholes not to be working if they were "completely" broken down, if the handpump was broken, if the handpump produced no water after 30 strokes, or if the handpump did not have water year-around. Most of the boreholes we classify as failures using these criteria seemed to have been broken for a long time and effectively abandoned. Using this definition, all project boreholes were "working" in 92 percent of the study villages in Volta, and 88 percent in Brong Ahafo at the time of our field visit. A borehole that we classify as "not working" may well be repaired in due course by an active watsan committee.

Not only are most boreholes in the study villages working, they are also being heavily used by households in the village. In the average study village in Volta, 96 percent of the households interviewed reported collecting at least some of their water from the project borehole. In the average village in Brong Ahafo, 98 percent of households were using water from the borehole. Mean reported per capita water use during the dry season for those households collecting water from the boreholes was about 28 L in Brong Ahafo and 33 L in Volta, which is very close to the average withdrawals recorded during the source observation activity in each region.

The results of the source observation of the most heavily used borehole in the village suggest that more water is withdrawn on average from the boreholes in Brong Ahafo (7900 L per day) than from those in Volta (5524 L a day on average). There were also twice as many individuals pumping water from the boreholes in Brong Ahafo (327 on average versus 163 in Volta) on the day of observation. Consistent with the self-reported water use from the household interviews, the amount withdrawn by each individual, however, was higher in Volta than in Brong Ahafo (34 L on average versus 24 L on average in Brong Ahafo). These levels of per capita water use are quite high for rural areas of Africa when people carry water from a source outside their home (White et al., 1972, Katui-Kafui, 2002). Thus, in both Volta and Brong Ahafo the borehole projects have succeeded in terms of supplying relatively large quantities of water for household use.

Most households in both regions use their village boreholes as a primary water source in the rainy and dry seasons. However, the boreholes are not the only source of water for drinking and cooking in the villages. About half of the sample households in both regions report using a river, stream, or other surface water source during the rainy season, and a quarter report using that water for drinking. Rainwater collection is also very common (roughly 75 percent of households).

Given that most boreholes are working and households are using them, it should not be surprising that households are satisfied with their improved water services. For example, in Volta 83 percent of the households interviewed in a typical village reported they were satisfied with the pressure at the borehole during the dry season (v. 82 percent in Brong Ahafo), and 90 percent were satisfied with the repair and maintenance service at the boreholes (v. 86 percent in Brong Ahafo). In 80 percent of the women's focus groups in Volta, participants agreed that they were satisfied or very satisfied with their water system (v. 84 percent in Brong Ahafo).

Profile of Watsan Committees in the Two Regions

Table 13.2 summarizes key information about the watsan committees and caretakers in Volta and Brong Ahafo. Most villages in the sample have functioning watsan committees: only three of our study villages in Volta (3 percent) and seven in Brong Ahafo (7 percent) had been disbanded or relieved of their duties. Another four villages in Volta (4 percent) and seven villages in Brong Afaho (7 percent) are currently inactive or dormant. There are a variety of different explanations for these instances of watsan failure or inactivity. In some cases, the committee had stopped work due to conflicts with the community or village leaders (usually over revenue collection,

	Volta	Brong Ahafo
Percent of watsan committees that have been disbanded or been relieved of their duties	3% (3)	6.7% (7)
Percent of watsan committess that are domant/inactive	4% (4)	7% (7)
Percent of watsan committees holding regular meetings with the community	75% (74)	70% (69)
Percent of these watsans who say they hold meetings "as needed" as opposed to on regular intervals	38% (28)	32% (22)
Percent of these watsans who say they hold meetings at least once a month	39% (29)	19% (13)
Number of people who usually attend meetings	114 (range 4 to 2000) (Median 50)	132 (range 4 to 800) (Median 100)
Percent of villages where the women's focus group says they dissatisfied with the work of the watsan committee	12% (12)	20% (20)
Percent of villages where the village leaders focus group says they dissatisfied with the work of the watsan committee	19% (18)	22% (23)
Percent of villages where women's focus group says the watsan committee is trustworthy	74% (72)	83% (81)
Percent of watsans in which present members were selected through election[*]	46% (45)	38% (36)
Percent of watsans in which present members were selected because they volunteered	0% (0)	18% (19)
Percent of watsans in which present members were selected through appointment	36% (35)	44% (36)
Percent of functioning watsan committees in which the members have undefined terms	72% (67)	89% (90)
Number of committee members	8.5 (St. dev 2.0)	7.6 (St. dev 2.9)
Percent of committee members who are women	41% (St. dev 0.1)	41% (St. dev 0.2)
Range of percent of women in committee	14% to 100%	0% to 100%
Percent of watsans with female chair	6% (6)	6% (6)
Average age of committee chair	53 (Range 26—107)	54 (Range 27—90)
Percent of villages in which some watsan members are paid for their work	1% (1)	0% (0)
Percent of watsans that have never received training to help fulfill their administrative, technical, or managerial functions (excluding training for caretakers)	7% (7)	26% (27)
Percent of villages with no caretakers	10% (8)	26% (27)
Percent of villages with no caretakers in 2001	7% (6)	6% (5)
Percent of villages with caretakers that have at least one female caretaker	29% (24)	35% (22)

Table 13.2

Profile of Watsan Committees in Volta and Brong Ahafo

*With few exceptions, the method used to select committee members does not differ between 2005 and the year of construction. In five cases in which the watsan committee members remember how the selection was done, the first committee was elected and the current committee appointed. In three cases, the villages moved from a system of volunteers or appointments to elections.

[†] Five cases.

the use of collected revenues, or unsuccessful repairs). In others, the committee was dormant because there was "no work to do" (the borehole had either not broken down or had not functioned in a long time). In some villages, another village-level institution had assumed the responsibility for the water system.

Most villages still have at least one caretaker. Nonetheless nearly a quarter of villages in Brong Ahafo and 10 percent of those in Volta were without any designated caretakers at the time of our survey. The most common reason for not having a caretaker is that those who were trained at the time of the project have since left the village. Among villages with caretakers, one-third has at least one caretaker who is a woman.

Watsan committees typically have about eight members (7.6 in Brong Ahafo v. 8.5 in Volta). The typical committee is headed by a 53-year-old male. On average in both the Brong Ahafo and Volta study villages, 41 percent of the watsan committee members were women. Only rarely is the head of the watsan committee female (in 6 percent of the villages in both Brong Ahafo and Volta). In a substantial minority of villages, the members of the watsan committee were selected by direct election. In other villages watsan committee members were appointed. In 18 percent of the villages in Brong Ahafo, committee members volunteered but this never happened in the study villages in Volta. In most villages members serve undefined terms. They are almost never paid for their work on the committee. Most watsan committees report having received training to help them fulfill their functions, but a quarter of the current watsan committees in Brong Ahafo have had no training at all.

Nearly three-quarters of the watsan committees hold meetings with the community. A third of those watsan committees say they hold meetings "as needed," while 39 percent in Volta and 19 percent in Brong Ahafo hold regularly scheduled meetings at least once a month. These meetings are large and attended by many people who are not members of the watsan committee. In villages in Brong Ahafo the watsan committees reported that an average of 132 people usually attended its meetings (median = 100 people). In Volta the watsan committees reported an average attendance of 114 people (median = 50 people). In about half of the villages, in both regions, committee members said that women participated actively at the meetings.

As shown in Table 13.3, in the majority of villages in both Brong Ahafo and Volta, watsan committee members have a clear sense of their responsibilities. Most watsan committees understand that they are responsible for hiring and supervising area mechanics. They know that they must manage the finances, keep financial records, report on finances to the community, and pay for spare parts and repairs. There is more disagreement as to whether the watsan committee is responsible for new projects and system expansion, and the majority does not feel that facilitating training of caretakers or technicians is their responsibility. Only three quarters of watsan committees see it as their responsibility to collect fees from households, and even fewer say it is their duty to set tariffs.

Table 13.3 suggests that watsan committees in Volta have a broader vision of their responsibilities than those in Brong Ahafo. For example, 97 percent of watsan committees in Volta see supervising area mechanics as part of their duties v. 81 percent in Brong Ahafo. Eighty-two percent of committees in Volta, versus 56 percent in Brong Ahafo feel responsible for facilitating training or education for households. Committees

	Percent of Existing Watsans Who Say That This Is the Committee s Responsibility		Table 13.3
	Volta	Brong Ahafo	Watsan Committee Members Perceptions of their Responsibilities
Supervising area mechanics who maintain and/or repair the system	97% (91)	81% (82)	
Maintaining a bank account, financial records, etc.	96% (90)	90% (91)	
Paying for repairs and spare parts	95% (89)	90% (91)	
Financial reporting to the community	94% (88)	81% (82)	
Resolving water use conflicts in the community and with other communities	88% (83)	77% (78)	
Soliciting financial or technical help for repairs	85% (81)	70% (71)	
Hiring and/or supervising the caretakers of the system	82% (71)	73% (74)	
Maintaining log books, and maintenance records	82% (77)	74% (75)	
Facilitate training or education for households	82% (77)	56% (57)	
Collecting fees from households	76% (72)	78% (79)	
Planning for system expansion or new wells and boreholes	70% (66)	54% (55)	
Deciding what fees and tariffs to charge	69% (65)	57% (58)	
Hiring and/or supervising the attendants of the system	65% (62)	56% (55)	
Supervising new construction or expansion projects	63% (59)	45% (45)	
Facilitate training caretaker or technicians	41% (39)	27% (28)	

in Volta are similarly more likely than those in Brong Ahafo to identify system expansion and the supervision of new construction as part of their duties and to recognize their responsibility to provide financial reporting to the communities they serve.

Most communities trust their watsan committees and are aware of how the watsan spends its funds. Nonetheless, the fieldwork reveals some evidence of poor communication or poor relationships between watsan committees and community members. In about 20 percent of the villages in each region, village leaders expressed overall dissatisfaction with the work of the watsan committees. Women in Brong Ahafo seemed less satisfied than those in Volta with their watsan committees (20 percent of women focus groups were unhappy with the performance of the watsan committee in Brong Ahafo v. only 12 percent in Volta).

Although more watsan committees in Volta listed financial reporting to the community as one of their responsibilities, the results of the women's and village leaders' focus groups suggests that transparency is higher in Brong Ahafo. Forty percent of the women's focus groups in Volta said they do not know how the watsan committee spends its money. The same was true in only one quarter of the villages in Brong Ahafo in which watsan committees actually collect revenue. The results of village leaders' focus groups show the same pattern: 24 percent of the village leaders' focus groups in Volta said they did not know how the watsan committee spent its money, versus only 5 percent of groups in Brong Ahafo.

Technical Breakdowns and Postconstruction Support

Almost all of the study villages in both Volta (96 percent) and Brong Ahafo (96 percent) have experienced borehole breakdowns since their system was constructed. This in itself is not an indication of project failure or neglect; it is simply a feature of the technology. On average it took villages in both Volta and Brong Ahafo 18 days to restore service after a breakdown. Part of the reason that villages have been successful in both Volta and Brong Ahafo in responding to borehole breakdowns appears to be that they have functioning watsan committees, and most of these watsan committees generally have the financial resources on hand to pay for services they feel they need.

The watsan committees in both regions have taken advantage of available postconstruction support services to keep their systems running. However, only 40 percent of watsan committees in Volta versus 64 percent of villages in Brong Ahafo feel they have sufficient access to spare parts to fix any problem they may encounter. Sixteen percent of watsan committees in Volta (as compared to 9 percent in Brong Ahafo) say either that spare parts for their pump model are not available or that they do not know where to obtain parts. Ten percent of villages have received substantial outside assistance in the form of system rehabilitation or grants for major repairs in the last five years. The Church of Latter Day Saints is the most common source of this assistance, but NGOs, cacao buyers, and other nongovernmental entities are also responsible for this external assistance to some villages.

When asked what external support has been most helpful to them, many watsan committees mentioned the work of the area mechanics, the assistance finding spare parts, and the technical training provided to caretakers (Table 13.4). In Volta, however, watsan committees were also very appreciative of the help they had received on administrative and financial matters through the MOM program and of the health and hygiene information provided by the EHAs. Watsan committees were also asked what external assistance or services they are not currently receiving would be most useful to them. The most common response was a request for financial assistance or for the provision (presumably free of charge) of repairs, spare parts, or system upgrades. The second most common request (by an eighth of all watsan committees in the sample) was for additional training for caretakers, either to replace caretakers who had left or to upgrade the skills of those who had stayed. Some watsan committees wanted to replace the area mechanic by training the caretaker to do major repairs. A handful of watsan committees asked for easier access to spare parts or support enforcing payment systems or regulations in the community.

Watsan Committees' Collections and Expenditures in the Two Regions

Revenue collection in Volta and Brong Ahafo is very uneven, but most watsan committees generally feel they have the financial resources on hand to pay for services. Seventy-one percent of watsan committees in Brong Ahafo and 60 percent in Volta feel that the revenue they collect from households is sufficient to cover operation and maintenance and small repairs. However, more than one quarter of the watsan committees surveyed reported not collecting any revenue from households in the last year. Among those watsan committees that did collect revenue from households, the committees in Volta reported collecting an average of US$169 annually from households (v. US$173 in Brong Ahafo). The range of reported collections is very wide, however, with some committees saying they collected less than US$1 from households during the entire year and others reporting household contributions above US$400 (Fig. 13.1).

	Number of watsans that mentioned each answer	
	Volta	Brong Ahafo
Regular visits....		
....from DWST		1
....from EHA	7	
Financial support	3	1
Preventative maintenance, repairs to borehole, work of area mechanic	11	16
Training for watsan or caretaker	11	5
Help getting spare parts/location parts	3	11
Instructions on use and cleanliness	4	
Hygiene and health information	9	1
Help with financial management, record keeping, billing, and payment	10	1
Help with system management	7	
Refresher info via news bulletin	1	
Provision of tools	3	
Audit done by DWST	1	

Table 13.4

Responses by Watsan Committees to the Question: "What Technical or Administrative Help from the Outside that you Receive do you Consider the Most Valuable/Useful?"

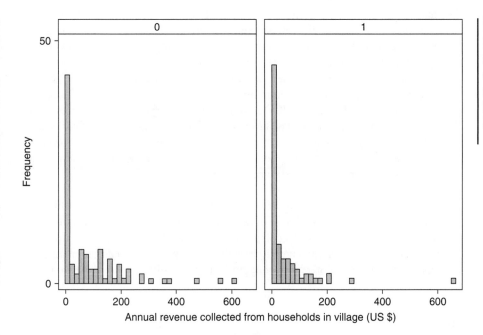

Figure 13.1

Frequency distribution of watsan committee's annual revenue collection from households in US$—by region.

Annual revenue collected from households in village (US $)

In Volta, almost all watsan committees reported having a regular payment system in place. Fifty-four percent of villages in Volta used a "pay-as-you-fetch" system in which households pay for each bucket of water collected at the borehole. Another 29 percent charged some form of fixed fee on a regular basis. In Brong Ahafo, many more watsan committees (22 percent) chose to collect money from households only when needed, or simply stated that they did not collect revenue from households (20 percent). Another 23 percent used a pay-as-you-fetch payment system, and the remainder said they charged fixed fees. Virtually all of the watsan committees using a pay-as-you-fetch system reported that the charge for an 18 to 20 L container of water was 50 or 100 cedis (100 cedis is approximately US$0.01[1]). The most common fixed fees reported were $0.11 and $0.22 per household per month. The highest reported fixed fee was $3.30 per year or $0.28 per month.

Having a payment system in place does not, however, necessarily guarantee that households are paying for water service. Roughly a third of the watsan committees in Volta and Brong Ahafo that purportedly have a regular payment system in place did not report collecting any revenues from households over the last year. The results of the household interviews also raise questions about the efficacy of revenue collection systems in the villages. In 44 percent of villages in Brong Ahafo and 2 percent of villages in Volta, none of the households interviewed who used the boreholes reported paying for water service. In most of these villages, the watsan committee either admitted to not collecting revenue from households or said they made a collection only when needed. Twenty percent of the villages where no households were paying for water, however, supposedly had regular payment systems in place. In villages where some households did report paying for water service, average payment rates were the same in Volta and Brong Ahafo: about 85 percent of households who collected water from the boreholes reported paying for it.

The majority of watsan committees in both Volta and Brong Ahafo reported expenditures during the last 12 months (90 percent in Volta v. 81 percent in Brong Ahafo). On average 60 percent of watsan expenditures in Brong Ahafo and 69 percent in Volta were for repair fees and spare parts. Other significant expenditures include payments to borehole attendants, chemicals, and, in a small number of cases, new facilities. The mean annual expenditure by a watsan committee in a study village in Volta in which repairs were made was about US$73 (v. US$124 in Brong Ahafo). When expenditures and collections from households over the last 12 months are compared, roughly half of the watsan committees in Brong Ahafo and Volta have a negative balance. The mean and median difference between expenditures and collections in Volta over the last 12 months were +$29 and −$2.22 respectively (−$10 and $0 in Brong Ahafo).

Making sense of the relationship between watsan collections from households and expenditures is difficult for a number of reasons. The first is data problems: many committees had trouble with recordkeeping, and information on savings is not available. Second, many watsan committees rely heavily on sources other than community contributions. Twelve percent of watsan committees in both Volta and Brong Ahafo reported

---

[1] US$1 = 9000 cedics

supplementing revenue from households with grants received for new boreholes, repairs, and preventative maintenance, but qualitative answers provided during the interview process suggest that there are many other ways that watsan committees "make ends meet." These include delaying payments to area mechanics and other suppliers, taking out loans from community members, and using resources from the community's central development fund. Third, a number of committees that have been successful collecting revenue from households mentioned that they have difficulty holding on to those funds. There can be much pressure from village leaders to release accumulated funds for other village projects. In one village the charges for water are explicitly used as a source of general village tax revenue.

POSTCONSTRUCTION SUPPORT AND THE FUNCTIONING OF WATSAN COMMITTEES

This description of watsan committees suggests that in general the watsans in Volta are more active, have a broader vision of their responsibilities, and do a better job of revenue collection than those in Brong Ahafo. In this section of the chapter, we investigate whether the differences in watsan committee activity can be attributed to differences in the external support that the committees have received. The analysis considers the support visits from DWSTs, the regular quarterly monitoring visits from EHAs (through the MOM program in Volta), and training that watsan committees have received in the postconstruction period.

We look at two indicators of watsan activity and acceptance in the community. The first is the perceived trustworthiness of the committee. We classify watsan committees into two groups—those that both the village leaders' focus groups and the women's focus groups reported to be "trustworthy" or "very trustworthy" and all other committees. By this definition, 61 percent of the watsans enjoy broad support in their communities. The second indicator is the percentage of households in a community that are paying for water service. Payment rates reflect differences in the ability of committees to set up and manage a collection system as well as to enforce payment.

We use logit and OLS regression models to attempt to explain these two dimensions of watsan performance in communities with active watson committees. There are two sets of models. The first includes a restricted set of explanatory variables. These models explain performance as a function of three groups of variables:

- Village characteristics (whether the boreholes are working; village population; level of infrastructure development measured by the percent of households with electricity; village income level measured by median household expenditure; and social capital measured by the percent of households who say they trust their neighbors)

- Watsan committee characteristics (percent of committee members who are women; whether the committee members joined through election, appointment, or another approach)

- Postconstruction support (training in the postconstruction period, number of visits from the DWST in a 1-year period, participation in the MOM support program, and assistance from the Mormons or other nongovernmental entity)

The second set of models retains these variables and adds a fourth group of explanatory variables:

- Watsan committee management (regular meetings with the community; system of regular charges or fees; committee takes responsibility for setting tariffs; whether the village still has at least one caretaker)

Watsan committees in the Volta region are more likely to have instituted all of these management practices, possibly because of the influence of the MOM program. This second set of models, thus, separates out the influence on performance of management practices that watsans have adopted from any additional effect of the MOM support visits. Summary statistics for all variables are listed in Table 13.5.

Table 13.6 presents the regression results. The explanatory variables have the expected sign in most cases, though relatively few variables emerge as significant. In terms of trust, the reduced model suggests that trust is higher when watsan committee members are appointed, possibly because they are appointed by political or tribal leaders who also enjoy community support. Trust of the watsan is lower in communities that have received significant outside

Table 13.5

Definitions, Means, and Standard Deviations of the Variables to be Used in the Regression Models

	Mean (stand dev)
Trusted	0.61
Equals 1 if both the women's focus group and village leaders' focus group say that they find the watsan committee to be trustworthy or very trustworthy	(0.49)
Percent paying	0.64
Percentage of households in the village that have nonzero expenditures on water (estimated based on household survey results)	(0.44)
Population	1.82
Population of village in 1000s, as reported by village leader's focus group	(274)
Percelec	0.14
Percent of hhs in village with electricity	(0.23)
Medexpend (in 1000s of cedis)	579
Median hh expenditures (from hh survey)	(198)
Perc trust neighbors	0.75
Percentage of hhs in village that report trusting their leaders	(0.14)
Elected	0.42
Equals1 if current watsan committee members were elected	(0.50)
Appointed	0.40
Equals 1 if current watsan committee members were elected	(0.50)
Training_post	0.56
Watsan committee received training in postconstruction period	(0.50)
Financial training	0.45
Watsan received training on financial management matters in postconstruction period	(0.49)

	Mean (stand dev)
DWST visits	4.09
Number of visits by DWST officials per year	(3.16)
Volta/MOM	0.46
Village is in Volta and therefore received support from MOM program	(0.50)
NGO/Mormons	0.10
Equals 1 if village has received major financial or technical support (e.g., system rehabilitation) from the Church of Latter Day Saints or another nongovernmental donor	(0.31)
Committeewomen	0.41
Percent of watsan committee members who are women	(0.15)
Boreholes working	0.90
Equals 1 if all project boreholes were working at the time of the fieldwork	(0.30)
Regular payment system	0.65
Equals 1 if watsan committee has implemented a pay-as-you fetch system or requires households to pay a fee each month to cover water service	(0.48)
Regular meetings	0.71
Equals 1 if the watsan committee reports holding regular meetings with the community	(0.45)
No caretaker	0.17
Equals 1 if the village no longer has a caretaker	(0.38)
Decide fees	0.63
Equals 1 if the watsan committee sees it as its responsibility to set the fees that are charged	(0.48)
Flat fee	0.31
Equals 1 if the watsan committee charges a flat fee per household or a fee based on hh size	(0.46)
Pay as you fetch	0.38
Equals 1 if the watsan committee has instituted a pay as you fetch system	(0.49)

Table 13.5

Definitions, Means, and Standard Deviations of the Variables to be Used in the Regression Models (*Continued*)

assistance from the Mormons or other nongovernmental agencies. This could be because the water supply systems in these villages had deteriorated (under watsan committee management) to the point where major system repairs or rehabilitations were necessary. The poor quality of the systems was, in most cases, the reason they were targeted for repair grants.

When watsan management activities are included in the model, level of trust enjoyed by the committee is associated with two factors. Having regular meetings with the community increases transparency, and is associated with higher levels of trust. Not having a caretaker for the village is associated with lower levels of trust, perhaps because it reflects the inability of the watsan committee to quickly handle technical problems or to engage in regular routine maintenance. Training, DWST visits, and participation in MOM all have insignificant coefficients in this full model. If postconstruction support helps improve the

Table 13.6

Factors Associated with Trust and Paying for Water—Regression Results

	Trust (logit)		Percent households paying for water (OLS)		
	Reduced model	Full model	Reduced model	Full model	Full model: only villages where at least some hh's pay for water
Population (in 1000s)	.047 (.073) 0.521	-.003 (.078) 0.974	-.005 (.012) 0.673	-.007 (.001) -0.356	-.008 (.007) -0.231
Percelec	-.791 (.784) 0.313	-.861 (.848) 0.310	.337 (.143) 0.020**	.166 (.097) 0.089*	.168 (.083) 0.046**
Medexpend (in 1000 cedis)	.001 (.001) 0.332	.001 (.001) 0.328	-.000 .000 0.359	-.000 (.000) 0.004	-.000 (.000) 0.002***
Perc trust neighbors	1.349 (1.297) 0.298	735 (1.421) 0.605	-.302 (.235) 0.200	.095 (.160) 0.552	-.170 (.145) 0.242
Elected	.471 (.518) 0.362	.011 (.577) 0.984	.184 (.097) 0.59	.113 (.067) 0.092	-.0181 (.064) 0.780
Appointed	1.290 (.555) 0.020**	.875 (.615) 0.155	.051 (.102) 0.615	.063 (.070) 0.364	-.069 (.0671) 0.305
Committeewomen	1.495 (1.116) 0.180	1.531 (1.200) 0.202	.138 (.193) 0.474	.051 (.130) 0.698	.106 (.119) 0.374
Training_post	.183 (.362) 0.613	-.117 (.397) 0.767			
Financial training			.098 (.067) 0.142	.083 (.045) 0.065*	.113 (.040) 0.006***
DWST visits	-.062 (.054) 0.250	-.0930 (.0613) 0.130	-.009 (.010) -0.329	-.008 (.007) 0.233	-.001 (.006) -0.928
Volta/MOM	.123 (.443) 0.782	-.206 (.502) 0.681	.162 (.082) 0.050**	-.029 (.058) -0.612	-.174 .052 0.001***
NGO/Mormons	-1.013 (.554) 0.067*	-.643 (.596) 0.280	-.043 .099 0.666	.007 (.067) 0.920	.018 (.059) 0.759
Boreholes working	.562 (.575) 0.328	.129 (.634) 0.838	.238 (.114) 0.039**	.147 (.077) 0.057	.144 (.077) 0.066*

Table 13.6

Factors Associated with Trust and Paying for Water—Regression Results (continued)

	Trust (logit)		Percent households paying for water (OLS)		
	Reduced model	Full model	Reduced model	Full model	Full model: only villages where at least some hh's pay for water
Regular payment system		.410 (.417) 0.326			
Flat fee or fee based on household size				.510 (.052) 0.00***	.344 .058 0.000***
Pay as you fetch				.780 (.056) 0.000***	.5252 (.058) 0.000***
Regular meetings		1.295 (.392) 0.001***		.062 (.046) 0.178	.094 (.043) 0.031**
No caretaker		-1.161 (.507) 0.022**		.053 (.057) 0.354	.039 (.053) 0.465
Decide fees		.5602 (.408) 0.170		-.005 (.046) -0.913	-.013 (.044) -0.772
Constant	-2.602 (1.434) 0.070	-2.242 (1.643) 0.172	.473 (.252) 0.062*	.030 (.178) 0.866	.620 (.171) 0.000***
Number of observations	179	179	179	179	137
Pseudo R^2 [R^2]	0.082	0.1728	0.26	0.69	0.59

relationship between the committee and the community, it may be because it encourages committees to increase the transparency of their operations and helps committees retain caretakers or train new ones.

The models explaining the percentage of households in a village that are paying for service indicate that payment levels are higher in villages with other infrastructure services, after controlling for other factors. Villages with higher levels of electricity coverage have higher payment rates for water, possibly because households in these villages are more accustomed to service payments. The correlation between electricity coverage and water payment rates does not indicate that households in richer villages are more likely to pay for water. To the contrary, after controlling for other factors, payment rates are actually lower in villages with higher median household expenditures. Whether or not all project boreholes were working at the time of the field work is another important explanatory factor: if boreholes were broken, households were not fetching water and therefore not paying for it.

In the reduced model, participation in the MOM program also appears to have a significant and positive effect on payment rates. However, this effect disappears when watsan committee management variables are added to the full model. MOM villages are more likely to have adopted regular payment systems, and this is associated with the higher payment rates in these communities. Payment rates are highest in communities with pay-as-you-fetch systems, but communities with flat fee systems also do significantly better than those that only collect money from households when needed for expenditures. The financial management training that watsan committees have obtained is also a significant determinant of payment rates: payment rates are higher in villages where the committee has received this training, even after controlling for payment system.

The third and final model explaining payment rates excludes villages where no households are paying for water. It examines the factors that are associated with higher payment rates in villages where households do pay for water. In addition to the factors mentioned above, this final model shows that holding regular meetings with the community is associated with higher payment rates. The dummy variable for Volta (participation in the MOM program) is negative and significant.

DISCUSSION

Our field investigation found that the vast majority of both the boreholes with handpumps and the watsan committees in two regions of rural Ghana are still operational 5 years after the initiation of the project. It is not possible to conclusively attribute the success of the water projects to the operations of the watsan committees, but clearly the community management model seems in large part to be working as designed. Given the sad history of many handpump projects in rural Africa, these results are encouraging news that well-designed rural water projects can be sustainable.

There is, however, an important caveat. Our results suggest that the watsan committees in these two regions of Ghana are having considerable difficulty operating their financial systems to recover the costs of operation and maintenance. Many villages in Brong Afaho have abandoned any attempt to collect money from households using the handpump(s), and instead rely on *ad hoc* revenue collections when breakdowns occur. Watsan committees in villages in both Brong Afaho and Volta that have adopted pay-by-the-bucket or fixed monthly fees are having difficulty enforcing their cost-recovery rules. Many households simply do not ever pay; others do not pay on a regular basis. These findings suggest that free rider problems remain a serious threat to the long-term performance of these rural water projects, and that postconstruction support services from higher-levels of government need to focus their efforts on such collective action problems.

The results of the regression models suggest that convincing watsans to move to a regular payment system could increase payment rates, although our fieldwork also suggests that some watsan committees have faced difficulties managing and holding on to cash reserves. Financial management training and instituting the practice of regular community meetings may help watsan committees achieve higher payment rates regardless of the payment method they chose to apply. The analysis provides less evidence for a direct link between postconstruction support and the quality of the community-watsan relationship, but postconstruction support that helps committees retain trained caretakers and encourages them to hold regular meetings would appear to contribute to improving relationships.

ACKNOWLEDGMENTS

Research for this article was funded by the Bank-Netherlands Water Partnership, a facility that enhances World Bank operation to increase delivery of water supply and sanitation services to the poor. The conclusions reported are those of the authors and donot necessarily reflect the views of the funding organization. The authors would like to express their appreciation to Jennifer Davis, Linda Prokopy, Rob Chase, Robert Roche, Arthur Watson, Jennifer Sara, and CWSA officials for their assistance and guidance on this study.

REFERENCES

Briscoe, J. and H. Garn, "Financing Water Supply and Sanitation Under Agenda 21," *Natural Resources Forum*, Vol. 19, No. 1, pp. 59–70, 1995.

Gross, B., C. van Wijk, and N. Mukherjee, "Linking Sustainability with Demand, Gender and Poverty: A Study in Community-Managed Water Supply Projects in 15 Countries," World Bank/UNDP Water and Sanitation Program, Washington, DC, 2001.

Harvey, P., "Operation and Maintenance for Rural Water Services: Sustainable Solutions," Background Report for WELL Briefing Note 15, WELL Resource Center, WEDC/LSHTM/IRC, 2005.

Isham, J., and S. Kähkönen, "How do Participation and Social Capital Affect Community-Based Water Projects? Evidence from Central Java, Indonesia", in C. Grootaert and T. van Bastelaer (eds.), *The Role of Social Capital in Development: An Empirical Assessment*, Cambridge University Press, Cambridge, UK, 2002.

Katui-Kafui, M., *Drawers of Water II: Kenya Country Study,* International Institute for Environment and Development, London, 95 pages, 2002.

Komives, K., B. Akanbang, R. Thorsten, B. Tuffuor, M. Jeuland, W. Wakeman, E. Larbi, A. Bakalian, and D. Whittington, "Postconstruction Support Activities and the Sustainability of Rural Water Projects in Ghana," *Draft Report to the World Bank*, February 2006.

Lockwood, H., *Scaling up Community Management of Rural Water Supply Delft*, International Water and Sanitation Centre, The Netherlands, 2004.

Manikutty, S., "Community Participation: Lessons from Experience in Five Water and Sanitation Projects in India," *Development Policy Review*, Vol. 16, No. 4, pp. 373–404, 1998.

McCommon, C., D. Warner, and D. Yohalem, "Community Management of Rural Water Supply and Sanitation Services," UNDP-World Bank Water and Sanitation Services, 1990.

Narayan, D. "The Contribution of People's Participation: Evidence from 121 Rural Water Supply Projects," *Environmentally Sustainable Development Occasional Papers Series No. 1*, World Bank, Washington, DC, 1995.

Prokopy, L. S., "The Relationship between Participation and Project Outcomes: Evidence from Rural Water Supply Projects in India," *World Development*, Vol. 33, No. 11, pp. 1801–1819, 2005.

Schouten, T. and P. Moriarty. *Community Water, Community Management; From System to Service in Rural Areas*. Intermediate Technology Development Group (ITDG) Publishing, London, UK, 2003

Therkildsen, O., *Watering White Elephants: Lessons from Donor-Funded Planning and Implementation of Rural Water Supplies in Tanzania*, Scandinavian Institute of African Studies, Uppsala, 1988.

White, G., D. Bradley, and A. White, *Drawers of Water: Domestic Water Use in East Africa*, University of Chicago Press, 1972.

14

IS PRIVATIZATION OF WATER UTILITIES SUSTAINABLE? LESSONS FROM THE EUROPEAN EXPERIENCE

Antonio Massarutto
Dipartimento di scienze economiche, Università di Udine - IEFE, Università Bocconi, Milano
Via Tomadini 30/a, 33100 Udine, Italy
antonio.massarutto@uniud.it

INTRODUCTION

Water supply and sanitation services (WSS) have received a growing attention from economists in the last few years, for the greatest part because of the potential of opening to competition and *private sector participation* (PSP) (Rees, 1998; Cowan, 1993 and 1997; Finger et al., 2006; Massarutto, 2006b).

Expectations in the water policy community on this respect have been rather high. PSP and liberalization is regarded as a way to improve efficiency and economic viability of the WSS industry in developed countries, but even more as a necessary strategy for improving accessibility to water resources in less-developed countries.

International institutions such as the World Bank, the OECD, the WTO, and the IMF have strongly pushed in favour of an increased *private sector involvement* (PSI), sometimes through moral persuasion, sometimes imposing it as a precondition for supporting financial aid to developing countries. The strategy for the achievement of Millennium Development Goals, with its compelling targets concerning water, is for a large part relying on private sector initiative (Winpenny et al., 2003; OECD, 2000; The World Bank, 1999 and 2005).

In more recent times, however, this attitude has been strongly criticized. Empirical evidence is definitely not supporting the claim of superior performances of the private sector, while it also demonstrates many failures in private provision of WSS (Lobina and Hall, 2003; Gleick et al., 2002; Hall, 2001). In fact all experiences show that introducing competition in the WSS is quite difficult, and even when this occurs in some market segments, state regulation and direct management of other phases continues to be fundamental; it also shows that WSS is a very risky business and it is difficult—and above all increasingly costly—to ask the private sector to carry on this risk (Bayliss and Hall, 2002). The international literature assessing WSS performance is not conclusive with respect to the preferred strategy for PSI, showing that public sector–based solutions may still have an important role to play (Massarutto, 2005).

In this perspective, WSS as other network industries and public utility sectors have been thoroughly analyzed in order to clarify liberalization models that are applicable, regulatory needs and patterns of private sector involvement (Renzetti and Dupont, 2003; Finger et al., 2006; Rees, 1998). On this basis, a huge literature in the past 15 years has examined the outcomes of liberalization and private sector involvement both from a theoretical perspective and on empirical grounds (Massarutto, 2005).

On the other hand, little attention has been dedicated to WSS as a key element for achieving sustainable water management. This may appear strange, since WSS represents in fact one of the most important intermediaries between natural water resources and human consumption. Outcomes of privatization are often discussed only in terms of short-run efficiency and financial performance.

We have argued elsewhere (de Carli et al., 2003; Massarutto, 2006) that an understanding of sustainability in the water sector requires that different components of "water capital" are considered simultaneously: not only the natural capital, but also man-made infrastructure, water-related knowledge base and "social" water capital, intended as the set of beliefs, common values and governance rules that allow society, in any given context, to allocate water and the related costs among competing uses. Natural resources can be substituted (to some extent) by other water resources or by artificial infrastructure; the technical and economic feasibility of this substitution, as well as social acceptability of its implications, however, cannot be taken for granted, since it is constrained both by economic efficiency, political legitimacy, and social acceptability. In this perspective, assessing sustainability of WSS becomes a problem in itself, and poses specific issues.

First of all, artificial water assets have a very long economic life and their correct depreciation is a key requisite of sustainability. If the present generation does not set aside adequate economic resources, the next generations will have to pay twice, because they will also inherit the public debt that has financed first-time infrastructure. *Asset management and development* (AMD) is therefore a key dimension of WSS to be monitored, and that should be performed efficiently.

Second, water industry viability is not only a matter of reproducing physical assets. What should be guaranteed, in turn, is the overall capacity of WSS systems to attract adequate resources in the future in terms of professional expertise, capacity to invest, capacity to maintain and improve the knowledge base of water systems.

Third, the output of WSS cannot be measured only in terms of quantity of water supplied or pollution abated; its sustainability depends on the fact that the whole artificial water cycle does not impact on the natural environment in terms of permanent and continuous depletion; on the contrary, it should be able to contribute from its own part to the recovery of a good ecological status of water bodies.

A fourth issue regards pricing and, more generally, the way the water industry obtains its revenues and covers its costs. While the capacity to recover costs in the long run is an obvious prerequisite of sustainability, accessibility and affordability issues are also important, at least for the components of WSS that deals with uses that are recognized as basic social rights.

In this perspective, we argue here that an involvement of the private sector in WSS management is, to some extent, a consequence of the structural change that the water industry has to face in order to achieve sustainability goals. On the other hand, once we consider WSS not simply as a utility delivering services to consumers but incorporate sustainability issues, constraints that limit PSP potential also assume importantance and should be considered with great attention.

First the drivers that link sustainable WSS management to private sector involvement are discussed. (Then the main options for private sector involvement are analyzed by referring to the main management models and the regulatory issues that characterize each of them. Finally, propose an analytical grid is proposed to be applied for the assessment of WSS sustainability. This analytical grid is used in order to discuss the main implications of WSS management models.

The WSS Industry: Developments across Centuries and Drivers of Private Sector Involvement

For a long time, WSS have been basically local issues, managed quite simply by local communities with relatively small efforts and correspondingly low input of technology and investment. Human settlements used to be conditioned and constrained by the availability of water resources as well as by the proximity of rivers or other "natural" ways to dispose of wastewater. Availability of water represented one of the most important "limits to growth" for human settlements.

Of course, this statement is true only in general terms. During history, powerful examples of water infrastructure still capture our admiration for their skilful technological solutions and the massive effort provided for realizing them. Urban development always called for the creation of waterworks aimed at making available enough water for an increasing and increasingly demanding population. The diffusion of sanitary standards during the nineteenth century in parallel with massive urbanization has been identified as the main driving force that led to the institution and overall diffusion of WSS services as we know them today, and the starting point of a specialized water industry.

Until recent times, however, the matching between water demand and water availability used to be searched through an "extensive" management model based on physical infrastructure, long-distance transfers, dams, reservoirs, aqueducts, and other similar devices whose main philosophy was that of bringing in "from elsewhere" water that was not available locally, because of quantitative limits or inadequate quality. On the other hand, the problem with wastewater management was mostly that of safely getting rid of it in order to avoid consequences on public health; once wastewater could be brought far from urban settlements, it could be given back to the environment without too many concerns for the final quality of watercourses.

Technology for this kind of "extensive" water management system is quite simple and developed on the basis of scientific knowledge that was already in place at the time of ancient Romans or during the Middle Ages; the main problem lies in the availability of economic resources for realizing and later operating and replacing water infrastructure. As Barraqué, 1995, puts it, what is "scarce" is not water, but money. In many countries, this model has lasted until very recently; many developing countries are now engaged in a similar strategy.

This feature helps us understand why the typical organizational pattern for this kind of water management system relies highly on centralized institutions and on the role of the state as the principal provider of economic means.

The most important sources of inputs to water management were therefore construction works and engineering of long-distance supply systems, dams, and reservoirs. Very often, the necessary investment for these water resources management facilities was carried on by the State and financed from the public budget, eventually after a private initiative starting centralized WSS systems in the wealthiest parts of the cities. Massive positive externalities arising from connecting all urbanites to the system (e.g., for health reasons) are at the origin of this trend (Barraqué, 2006).

The water policy network is dominated by "iron triangles" between water users, construction industry, and public administration spending money in order to achieve consensus

(Bressers et al., 1997). The role of the private sector in this model is basically that of supplying public works and construction.

At a certain point, the "extensive" model enters a crisis: available water resources are too distant and too costly to gather; construction of new water systems meets increasing social opposition; environmental quality of local water resources becomes a problem in itself; and more generally, urban development and social demands become so widespread and conflicting that the extensive model is not practicable anymore without severe external effects that turn out into increasing social costs.

This turning point starts a new phase of water management that can be labeled as "intensive"; the focus is concentrated on integrated management of locally available resources; technology becomes more complex and specialized in qualitative rather than quantitative issues; stronger and stronger emphasis is put on information, quality control, pollution prevention. This transformation of the object of water policy has obvious effects on the structure of the water industry, determining strong pressures toward PSI and thus generating as a feedback effect the need for new governance mechanisms. See Fig. 14.1.

New water policies impact on the equilibria of the water industry, basically, through five channels, each on its way leading to some PSI, unless the public system is able to provide the required functions directly and effectively.

First, *urban water systems* (UWS) require fixed costs that are for the most part sunk; these costs grow as far as UWS provision becomes more difficult, and demand (arising both from service users and from environmental policy) is more exigent. This represents a formidable pressure to growth of the territorial size of water undertakings (in order to share these costs on a larger base of customers) and outsourcing of activities implying economies of scale or high labour costs (e.g., public works, customer services, billing, emergency management, project and design, research and technological development (RID),).

Second, the increase of technical complexity of water management forces local water management systems to delegate decisions to professionals having the necessary expertise. This leads to the creation of specialized water management organizations, whose belonging to the private or public sector follows a good deal national paths and traditions. On the other hand, specialization and increased complexity also means that less and less critical inputs can be produced "on site" and more and more are incorporated into "technological objects" and managerial skills that local operators will select and purchase on the market. WSS operators are required to be in the best position to dialogue with external suppliers of technology, and this in the end might represent a strong incentive toward corporatization and an increasing recourse to the market for the provision of inputs. The value chain of water services includes more and more value added that is produced by specialized firms on the market rather then directly by WSS operators.

In some countries, like France, the need for professional skills has led quite early to delegation of operation to private companies, instead of developing the same skills

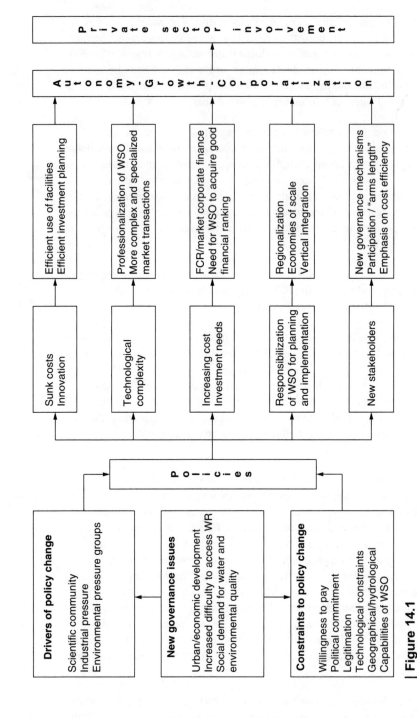

Figure 14.1

Drivers of private sector involvement in the WSS industry.

in-house.[2] This is not an obligate route, however, Italian and German municipally owned water companies, as well as the British Water Authorities before privatisation are well-known examples of WSS undertakings developed within the public sector, though with some sort of entrepreneurial autonomy. On the other hand, it means for sure that the public management system has to incur fixed costs whose relevance in the service economy is growing higher. The more sunk costs are required in order to acquire and maintain appropriate managerial skills, the more these undertakings are forced to search for strategies aiming at keeping the average cost low. This is achieved for example by an increase of territorial size, in order to acquire economies of scale, or by diversification (multiutilities), or by an increasing recourse to outsourcing of activities that require economies of scale and specialization. Of course, this also means that in order to achieve greater efficiency, WSS undertakings should also become increasingly autonomous from political power and oriented toward a businesslike behaviour.

At a certain point, the trade-off might be unsustainable: maintaining acceptable levels of efficiency is possible only by increasing degrees of freedom and entrepreneurial autonomy, until public control on activities becomes only nominal. The sale of shares to the market, often forced by budget constraints, completes the process, until companies finally behave as private companies.

The third channel is the financial one. The increasing cost of water services in the present phase is paralleled by the crisis of public finance and the imposition of upper limits to taxation and government budgets. Self-finance through full-cost recovery becomes necessary. Tariffs and prices paid by consumers will serve as a cash flow for sustaining market-based finance, for which the WSS operator becomes the intermediary.

Once again, this does not necessarily lead to private operation of services, since the public sector might have the capacity to access financial markets at even more favourable conditions than private companies; on the other hand, what matters for financial markets is the reputation of water companies and the guarantee of their faithfulness to a business-oriented philosophy.

The decision to privatise therefore can be "political," in the sense that governments and local authorities might wish to divest in order to give a signal that eventual losses in the future will not be transferred again on the public budget; privately owned companies, from this point of view, will need to achieve a good balance between revenues and costs, and therefore will not engage in new expenditure unless they will be able to finance it through prices paid by consumers.[3]

---

[2] Peculiar institutional mechanisms might influence this development; for example, in the French case, delegation to private companies was often made necessary by the difficulties that the French law opposes to the creation of local publicly owned companies and by the difficulty for municipalities to have access to credit (Barraqué, 1995).

[3] The history of British Water Authorities is enlightening. When they were created in 1973, they were required to operate on a cost recovery base; however, charges used to be kept low for political reasons, while significant new investment had to be carried on; so the WA experienced a substantial rise of their debt, what was possible only because financial markets treated this debt as if was warranted by the central government. At the moment of privatising the Water Authorities in 1989, the British government had to write off this debt and absorb it into the public debt, otherwise the whole operation would be a failure (Barraqué, 1995; Andrews and Zabel, 1999).

The fourth channel regards the division of competences between WSS operators and water resources planning. In the traditional model, WSS management is mostly concerned with operation, while infrastructure is planned, financed, and realized in the frame work of central government policies or municipal public works departments.

In the new paradigm the water industry is required to play a more important and responsible role, with substantial degrees of freedom; one important consequence is that WSS operation should become responsible for making investment decisions and therefore acquire both suitable geographical scale and vertical integration.

In the new division of competences, public functions are more concerned with regulation of water uses, setting quality objectives and enforcement, while WSS companies become responsible for achieving these objectives through an integrated management of infrastructure together with service operation. This is, once again, a mighty force that changes the adequate territorial size (from municipal to intermunicipal or regional dimension) and, above all, multiplies the tasks that operators have to perform and the level of economic risk that it is required to bear.

Finally, the last channel regards the increased conflictuality of the water policy arena and the emerging need to establish a *super partes* regulatory system. WSS operation, while taking care of a service of general interest, cannot claim to represent the whole of the public interests concerned with water. In order to guarantee a credible enforcement of environmental and quality regulation, autonomy of WSS undertakings from water policy is an important prerequisite. Once again, this can be achieved within the public sector; yet many examples tend to demonstrate that some sort of privatization of the WSS significantly increases the degree of antagonism between regulators and water companies and thus of regulatory effectiveness.[4] This is also likely to take place because citizens and public administrations tend to be more exigent and attentive when services are managed by private companies.

LIBERALIZATION AND PSI OPPORTUNITIES IN THE WATER SECTOR: A FRAMEWORK FOR THE ANALYSIS

WSS services are now provided under a high variety of organizational systems throughout the world, with many different and alternative arrangements for involving the private sector. While being commonly recognized as an activity that requires state intervention due to a multiplicity of market failures, they are also increasingly being considered as an elective terrain for private sector initiative. This occurs either because of the evidence of public sector flaws in the management of WSS (and, more broadly, of public services or more simply, because the public sector faces constraints (financial, technological, and so forth) that limit their capacity to invest and manage, particularly, but not only, in developing countries.

---

[4] The British case is again enlightening. Before privatization, the Water Authorities were responsible both for WSS operation and for administering water policy. The level of compliance with the standards that they themselves had established was estimated below 15 percent, and no prosecution did ever occur. After privatization, compliance levels have rapidly increased to close to 100 percent (Summerton, 1998).

Axis	Description	Regulatory Issues / Market Failures
I	Transactions between the WSS operator and public entities holding the responsibility for service provision	Incomplete contracts Transactions costs Sunk costs Information aymmetries
II	Transactions between the WSS operator and suppliers of inputs along the value chain	Vertical integration Cost of capital for long-run undertakings Principal-agent relations in procurement
III	Transactions between WSS operator and entities holding the property rights on natural resources	Externalities Long-run sustainability of water management systems Transactions costs in the trade of water rights
IV	Transactions between WSS operators and final consumers	Natural monopoly Public good dimensions (e.g., health issues) Accessibility and affordability issues Resilience and flexibility

Table 14.1

Transactions in the WSS Value Chain and Related Market Failures

Different types of transactions can be identified among relevant actors, each of whom raises specific governance issues (Finger et al., 2006). See Table 14.1.

In order to understand which PSI models are suitable for the WSS sector, refer to an "economic policy roadmap" (Massarutto, 2005a; see Fig. 14.2) that suggests to explore, first, the potential for competition in the market and its compatibility with the dimensions of general interest that characterize the industry; second, the potential to separate different phases in order to have competition in those that are not natural monopolies; third, to consider competition for the market via competitive tendering; and finally, to explore the potential for developing incentive regulation through benchmarking.

Applied to the water industry, this analytical grid shows that the main features of WSS with respect to this check can be summarized as follows (Massarutto, 2005a):

- *Dimensions of general interest* are widespread and fundamental, including public good dimensions, externalities and equity issues.
- *Risk dimensions* are correlated in particular to the very long life of assets; being investment sunk for a long time, guarantee that costs are actually recovered depends on the credibility of commitment of public entities to ensure water company viability as well as from the actual willingness and ability of water users to pay (Table 14.2). Market risks are present in the segments open to competition.
- Limited potential for *competition in the IV market* due to physical indivisibilities in the WSS network, economies of scale in service provision, and negligible importance of variable costs; some competition is made possible by innovative technical

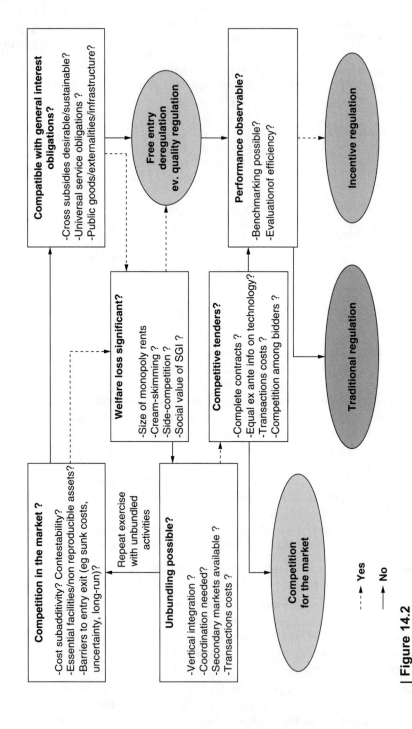

Figure 14.2

The roadmap for regulating services of general interest.

Typology of Risk	Description
Operational	Quality of actual management system
	Status of knowledge on infrastructure and investment needs
	Performance standards to be guaranteed and allowed flexibility margins
	Patterns of water demand in relation with tariff structure (e.g., water consumption if tariffs are based on volumetric charges)
Tariff	Perspective tariff dynamics v. operators' commitments
	Cost-based (RoR) v. price caps
	Cost pass-through and guarantees of minimum total revenue
Profitability	Expectations on capital remuneration; measurability of risk exposure
Competition	Contract length
	Barriers to entry and exit
	Post-contract arrangements (e.g., how are sunk costs compensated after contract termination)
Regulatory	Is regulation clear and coherent?
	Is the legislative framework stable/predictable?
	Is there a discretional margin for regulators?
	Is new environmental/quality regulation envisaged, and does the contract take this into consideration? How will be the corresponding cost be transferred into tariffs?
Municipal	Financial standing of municipality and capacity to respect financial obligations
	Patterns of urban development
Political commitment	Are regulators committed to maintain prices at a level that covers quasi-rents and not only short-run marginal costs?
	Political attitude toward private companies
	Political commitment to ensure financial viability of operating company
Instability	General economic situation, currency issues, financial rating of the Country etc
Civil society	Social attitude toward private companies and pricing policies
Country interest	Expectations on GDP growth, water market development, long-term perspectives

Table 14.2

Principal Typologies of Risk in the Water Industry and Their Main Determinants

Source: Adapted from Pricewaterhouse and Coopers et al., 2004.

solutions enabling self-provision of at least some components of the service (e.g., drinking water treatment; septic tanks; rainwater infiltration; in-house recycling), on an individual or community base (Massarutto and Paccagnan, 2006).

- Limited interest for *third-party access* to infrastructural networks, since essential facilities represent the most part of the total cost.

- *Unbundling* between phases (raw water extraction, drinking water production, distribution, wastewater collection, and sewage treatment) possible, though most segments remain natural monopolies even if isolated.

- *Competition for the market* is very practicable and in fact very commonly practiced; yet the sector fails to satisfy requirements for perfect contestability. In particular, information asymmetries and transactions cost protect incumbents against new entrants and are the cause of incomplete contracts along the I market. For this reason, tenders cannot be based on price only, unless they are effected only for very simple and short-termed tasks, what would ultimately require that the public be responsible for long-term asset management and planning. These conditions are better satisfied, in turn, once operation and asset management are treated as separate businesses and/or if private commitment in AMD and its assumption of risk is limited and corresponded by adequate compensation; on the other hand, it tends to create a dualism between investment decisions and management that can contrast with long-term efficiency of the system. In the case of delegation of both asset management and operation, contracts need to be long-term and frequent renegotiation of conditions is needed, which reduces competitive incentives toward cost reduction.

- *Incentive regulation* is an adequate solution provided regulators are strong enough so as to reduce information asymmetries; and quality regulation is detailed and is credible. Efficient capital markets (for takeover), systematic benchmarking, and yardstick competition are fundamental and require a skilled, independent, and well-informed regulator. Concerns on long-term sustainability are raised nonetheless, since the regulatory system might be concerned with short term and depreciation of existing assets not guaranteed.

Opportunities for PSI are well present, yet severely constrained; they require in any case that the state be able to fulfil complex regulatory tasks and acquire an adequate contractual power. On the other hand, however, a higher regulatory capacity also increases regulatory risks and reduces interest for the private sector to participate.

Solutions to this trade-off are not simple. An important reason for this difficulty is the interdependence of the four markets: solving for market failures in one of them is well possible, but it tends to create the case for more fundamental market failures in one or the other markets.[5]

The consequence of this trade-off between the requirements for competition in the four markets can be described more clearly with the aid of Fig. 14.3, representing the value chain of WSS services. In the figure, is distinguished a *service core sphere* (responsibility, asset management, and operation), a *market sphere* (activities to be supplied in order to provide WSS services), and a *regulatory sphere* (activities that are performed by the public sector). Different layouts are used for representing transactions in the four markets.

---

[5] For example, the economic theory suggests that in order for competitive tendering to be efficient, tenders should be based on single performance indicators (e.g., price), contracts should be complete (they should foresee all potentially relevant events), and all relevant information should be shared by participants to the tender (also meaning that this information should be handled by the public authority and communicated to all bidders). It also suggests that short-term contracts are more suitable for this purpose. This model can be applied more easily for single activities (e.g., managing a wastewater treatment plant) than for the integrated service as a whole: but this means that in order to maximise the scope for competitive tendering, all the more risky and uncertain activities (such as AMD) should be provided by the public sector.

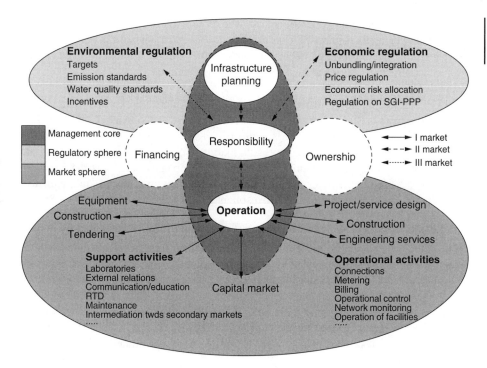

Figure 14.3
The value chain of WSS.

Theoretical analysis and empirical evidence underline two main trade-offs. The first one concerns primary and secondary markets. The more the primary market is based on market transactions (tenders, contracts, and the like), the more the industry seems to tend toward vertical integration in the secondary market; vice-versa, when the primary market is based on the incumbent operator's monopoly, this shows a greater tendency to purchase inputs from the market and outsource, eventually via the recourse to partnerships with the private sector for realizing the main facilities (Kraemer, 1998).

This can be explained by considering that local monopolies are not necessarily reaching a scale that is optimal for all phases. Especially if historical, institutional, and political reasons force local monopolies to small scales, their capacity to perform activities in-house will be reduced, and the likelihood of a recourse to the market higher. This will be more probable, if price regulation does not allow operators to transfer costs to consumers.

On the other hand, competitive tendering in the primary market increases risks; vertical integration helps private operators to reduce the risk by increasing the economic margin of the transaction and by raising information barriers against potential new entrants or even against regulators.[6]

The second trade-off concerns risk assumption, with particular reference to AMD. Here a reversed mechanism seems to apply: the more the private sector is involved in operation

---

[6] Industrial organization theory normally considers transactions costs as the main explanatory factor for vertical integration. We suggest in turn that market risk reduction is a more powerful driving force in this case; transactions costs in the II market do not appear to be significant, even if it should be emphasized that efficiency in procurement is directly linked with knowledge and technical capabilities of operator.

and faces some market and operational risk, the more it tends to refuse responsibility of AMD; vice-versa this responsibility is more easily accepted when these operational and market risks are reduced (e.g., because contract terms are much longer or even eternal, because contractual clauses allow to pass extra costs onto tariffs).

This can be understood by recognizing that AMD contracts are exposed to significant transaction costs arising from information asymmetries and opportunistic behaviour. On one side, once the investment is sunk, responsible entities might be tempted to impede price increase over the short-run marginal cost, thereby appropriating the "quasi-rents" and making it impossible for private companies to recover the fixed cost (Noll et al., 2000); on the other side, the water company might easily capture regulatory decision and force technological and investment choices (Lobina and Hall, 2003).

In fact asset ownership, management, and development are almost never definitively allocated in the service core or in the external spheres; rather, they are posed somewhere in-between, with a continuum of solutions based on sharing competences, responsibilities, and risk. The more they are allocated on the private sector, the higher is normally the time horizon and the correlated risk premium; on the other hand, when these competences are separated from operation and remain in the public sector, contracts are released more easily for operation. The public sector can later on adopt other arrangements for PSI in that particular phase (such as project finance agreements), while pursuing autonomous strategies for operation.

ALTERNATIVE MODELS FOR ORGANIZING WSS SERVICES

The existence of a multiple dimension of market failures helps understand why so many alternative solutions have been experienced throughout the world. Fundamentally, alternative solutions differ on the basis of the answers to the following questions:

- Who is the owner of water resources? On what basis is a WSS system entitled to use them and what degrees of freedom are available for eventually trading these rights in favour of other subjects?
- Is there a legal responsibility for providing services (i.e., is there a subject who is responsible to guarantee that service is provided at certain conditions)? Is this subject a public entity or a water company?
- Who is the owner of assets and infrastructure used for providing the service?
- On what basis is the operator chosen? Is it a public entity or a private firm?
- What are the main sources of economic risk, and how is this risk shared among company shareholders, public entities (taxpayers), and consumers?

We can distinguish three main alternative models, whose main features are resumed in Table 14.3.[7]

---

[7] Kraemer, 1998, refer to them as French, British, and German model, because these countries represent their most typical example; even though, an evolution is taking place in each of them, while other countries exhibit similar or alternative combinations. We prefer to consider the three models as archetypes.

Table 14.3

WSS Management Models and Their Most Important Features

	Delegated	Regulated monopoly	Direct public management
Main example	France	England and Wales	Germany
Other examples (involving variants)	ITA (few), ESP, East European countries		NL, SWE, ITA AUT, GRE
Ownership of water resources	Public		
	Variant I (US): water rights can be privately appropriated and transferred		
	Variant II (some limited cases in Europe): licensed users can trade use rights under public supervision		
Responsible entity	Municipality, eventually voluntarily or compulsorily associated	Water company	Muncipality, ev. associated
Ownership of water companies	Private	Private	Public (responsible entity) Often multiutilities Minority shares for private partners or other public bodies
Legal form of operator	Private company with concession contract Variant: specific purpose company jointly owned by municipalities and private companies	Private-law company contract and being owned by a parent company	Public law arrangements with different degrees of autonomy and governance structure Private law companies (with obligation for unbundling in case it is also active on other markets)
Ownership of assets and infrastructure	Responsible entity, either directly or through dedicated private law companies (poss. mixed during the contract lifetime)	Operator Variant: assets owned by public entities and/or consumers' associations	Responsible entity Variant: users' associations and individual consumers
Choice of operator	Tender (based on a mixture of economic offer and "beauty contests") Variant (ITA): tender for partner in mixed venture companies, with majority shares in public hands but operational decisions fully delegated to private partner	Direct (usually resulting from partial or total privatization of previously publicly owned firms) Variants: municipality may retain shares and even the majority of shares; governance rules protect the private	Direct

(Continued)

Table 14.3
WSS Management Models and Their Most Important Features (*continued*)

	Delegated	Regulated monopoly	Direct public management
Structure of the industry along the value chain	Vertical integration aimed at maximising the value added produced in-house and minimize contract, operation and market risks	Vertical integration on make/buy considerations "Contract companies" usually produce only activities that require sunk costs that are specific for that contract; other activities are purchased from the parent company and/or from the market	Vertical disintegration Market for procurement Joint ventures and PPP for specific initiatives (DBFO) Variant/trend: • Increased outsourcing of service activities as well as procurement • Horizontal integration of distribution with gas/electricity; sewage treatment with waste management; dedicated companies for raw water production
Patterns of competition	Competition for the I market	No competition in the I market Inset appointments (possibility for new customers to choose their preferred operator with ev. bulk supply from the main one) Competition for company ownership in the stock exchange market Yardstick competition (benchmarking)	Competition in the II market Procurement and outsourcing

Figures 14.4 a–c represent each model in terms of the value chain structure analyzed previously in Fig. 14.3. Transactions between the different components of the value chain are represented with different layouts in order to distinguish market transactions

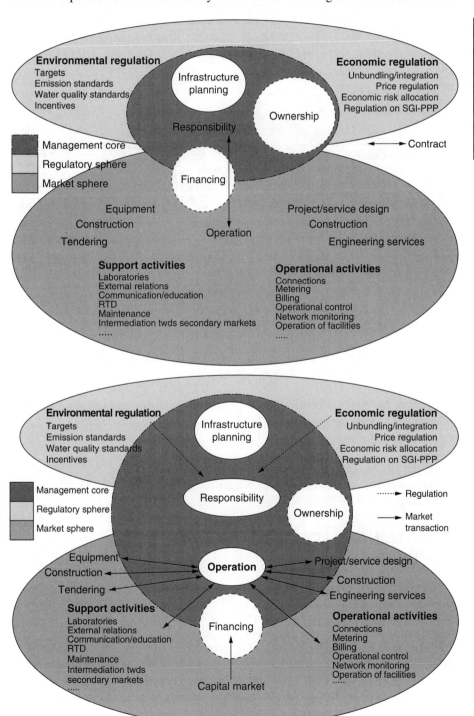

Figure 14.4

Organization of the WSS value chain (a) Delegated model (b) Regulated monopoly (c) Direct public management.

Figure 14.4
(*Continued*)

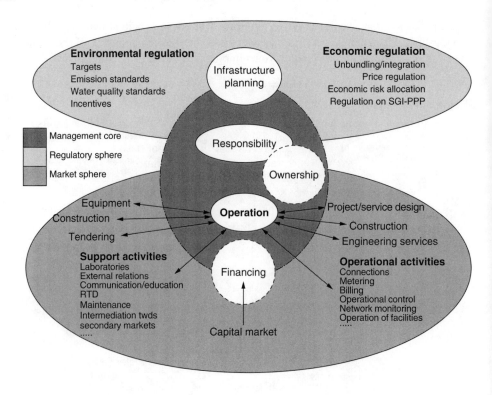

(normal lines), regulatory transactions (bold lines), and hierarchical transactions (belonging to the same sphere without linking rows).

Each model provides different solutions to the trade-off discussed in the section on "Alternative Models for Organizing WSS Services" and entails different regulatory issues that are summarized in Table 14.4.

Fundamentally, the French model relies on tenders and contracts, the British one on regulatory authorities specifying service obligations and price limits, the German one on hierarchical control by the responsible entity.

ASSESSMENT OF SUSTAINABILITY OF WATER LIBERALIZATION MODELS

One of the reasons why an assessment of privatizations (and more generally of regulatory reforms) is so difficult is the fact that performance involves a great number of critical dimensions regarding different aspects of individual and social welfare. A mistake that should be avoided is that of referring only to some of these dimensions and neglecting the others. In particular, many studies on privatization are limited to considering the outcome of managing companies in terms of profitability and cost recovery (Massarutto, 2005). In fact it is quite common (and also easy to understand) that privatized companies show better

Table 14.4

Regulatory Issues

	Delegated	Regulated Monopoly	Direct Public Management
Regulatory System	Quality standards and environmental standards exogenously imposed No price regulation (tender + direct agreements between responsible entity and private partner) Voluntary instruments (e.g., service charters)	Quality regulator Environmental regulator Economic regulator (price-caps on 5 year base) Service charters Variant: service contracts	No explicit regulation Soft-regulation and benchmarking for providing comparative information Variant: public sector contracts; service charters
Main Regulatory Instruments	Tenders Contracts Performance standards	Price-caps and performance standards enforced by independent regulatory authorities	Hierarchical control Benchmarking Voice
Main Regulatory Issues	Contract completeness Regulatory capture in the renegotiation phase Dealing with sunk costs after termination of contracts Conflict resolution Barriers to competition in the next bids due to information advantages of incumbent	Site specificities in the cost function Transfer prices and in-house transactions Easy to parameter OPEX and cost of assets, very difficult to assess AMD as a whole	Cost inflation and limited incentive to enhance productivity Gold plating and excessive investment Less credible enforcement of quality targets
Noneconomic Factors	Corruption Collusion between private companies and elected people (may be enhanced in case of PPP)	Reputation Conflict between regulators	Extra costs required for meeting political agendas of elected people Corruption

economic performance; however, this can be due to improved efficiency, but also to higher prices, reduced investment or quality levels, higher transfers from the public budget, and the like. If this occurs, a better economic performance of water companies corresponds to a shift of costs and risks on other social actors, or onto future generations.

An evaluation of WSS cannot take place without a comprehensive assessment of all these dimensions, seen in a long-term perspective. Following Barraqué (1999), three main dimensions of sustainability can be explored (the "3Es"): ecologic, economic, and ethical.

The first dimension concerns the relation between WSS and the water environment. WSS should be managed in such a way as to comply with the environmental sustainability targets. These targets are site specific, but it can be assumed that a good indicator is the capacity to meet the targets set by the WFD and to guarantee a continuous improvement of environmental quality (de Carli et al., 2003). Resilience and vulnerability of WSS systems to future events (e.g., diffused pollution of drinking water sources, urban growth) add further ecological dimensions of sustainability.

The second dimension concerns industry viability and its capacity to sustain in the long-term investment and operational expenditure needed for keeping the water system up to the required performance. This has to do especially with long-term capacity to replace aging infrastructure and expand it in order to meet newly emerging demands, either in quantitative or in qualitative terms; but it also requires that the water industry will continue to be able to attract economic resources (labour, capital, RTD, and so forth) in order to maintain and improve its management capacity in the future.

It is important to note that this requirement has both a local and a general dimension. From a general perspective, the water industry as a whole should be able to adequately develop. This means that the whole system composing the "value chain" of WSS (not only including water companies, but also all the other actors on which the final output depends), should be wealthy and economically sound.

From a local perspective, in turn, it should be guaranteed that each water management system is capable of having access to appropriate solutions developed by the water industry and adapt them to local exigencies.

Finally, the third dimension concerns affordability and accessibility. WSS development cannot exceed a threshold that is represented both by the *collective* ability to pay (measured, for example, by the ratio of WSS expenditure on the annual GDP) and the *individual* ability to pay (of poorer households, regions, and so on).

Following this line of analysis, an approach has been developed that is based on a definition of social welfare (entailing all possible stakeholders, including future generations) and adopted a procedure based on two steps:

- Assessing the "net social dividend" of regulatory reform: it is the net overall difference between service costs, before and after the reform, and regardless who actually pays for them. The net dividend could be positive (meaning an increase in overall efficiency) or negative; it can arise both from a reduction of x inefficiencies (production costs closer to theoretical optimum) or from the elimination of deadweight losses.

- Assessing how this net dividend is shared among stakeholders: this depends on regulatory arrangements, but is not completely neutral (in the sense that not all stakeholders can benefit from the same net dividend in each scenario). Net benefit for a given stakeholder may be negative even if the net social dividend is positive.[8]

In Table 14.5 the main components are summarized that have considered, the related stakeholders, and some indicators.

---

[8] For example, tariffs might increase even if service is more efficient, because of higher profits or higher investment.

Stakeholders	Main Concerned Outcomes	Indicators
Companies and shareholders	Profitability Corporate value	Financial indicators (ROI, EVA etc) Net asset value
Responsible entities	Positive externalities (e.g., public health) Synergies w/public goods provision (e.g., rainwater management) Coherence with urban development patterns Financial transfers (royalties, canons, revenues from privatization etc.)	Health parameters (e.g., water-borne diseases) Capacity to follow urban development (e.g., % of connections during time) Canons and royalties
Customers	Value for money Affordability Accessibility Sizing/Capacity to meet peak demand	Annual cost per hh % of water bill on family budget Service interruptions
Environment/ other water uses	Achievement of water quality objectives Compliance with environmental and health protection standards	Compliance with Environment Regulations Contribution of WSS to water environment degradation/improvement
Suppliers of inputs (e.g., workers)	Volume of economic transactions with the water sector Levels and quality of employment Salaries Technological development	Value added and its composition Mean labor cost
Next generations	Maintain the system financially viable in the long term Guarantee service functionality/ resilience Adopt state-of-the-art appropriate technologies	Actual investment/real depreciation of assets Degradation of water environment
Taxpayers	Reduce impact of WSS on general budget	Net contribution from/to the public budget

Table 14.5
Stakeholder of WSS and Concerned Outcomes of Policy Reform

Table 14.6 summarizes the main results of the analysis. Fundamentally, the three models perform differently with respect to the various components of costs. Some reduction of operational cost due to increased productivity is likely to be stronger if competitive incentives are stronger. This is not simply due to privatization, but rather depends on the guarantee that the operator has on revenues. In any case that these incentives are higher in the *regulated monopoly* (RM) (especially if price caps are adopted) and lower in *direct public management* (DPM). The delegated model has some potential for cost reduction, but awarding criteria and contract negotiation patterns may favor instead some cost inflation that is profited by suppliers belonging to the same group as the operator.

Table 14.6

Management Models, Economic Risk, and Main Expected Outcomes on Performance

	Delegated	Regulated Monopoly	Direct Public Management
Main risk dimensions for the private	Market risk (tender) and recovery of sunk costs Operational risk (initial information missing or wrong; emerging new issues during contract lifetime) Commitment of public authority to ensure cost recovery and viability	Regulatory risk Takeover Unforeseen investment Public reaction forces regulators to keep down unpopular price increases	Limited to DBFO and market for procurement
Main risk dimensions for the public	Information asymmetries Technological lock-in	Regulatory capture	Lower efficiency More vulnerable to pressures from workers and consumers
Main risk dimensions for consumers	Collusion leads to extraction of monopoly rent shared by municipality and private company Quality reduction if contracts are not fully specified and/or badly enforced	Higher cost of capital Cost pass-through Quality reduction corresponding to what quality dimensions are actually specified by regulations and service charter and enforced	Lower credibility of quality standard enforcement may lead to deterioration of service quality
Main risk dimensions for future generations		Underinvestment induced by unwillingness to raise tariffs in the short term	Underinvestment Slowdown of environmental and quality expenditure due to public budget pressures
Main risk dimensions for suppliers/workers	Market power of operator face to suppliers Vertical integration	Pressure for lower salaries—outsourcing and for staffing reductions	Higher competition on procurement and reduced profit margins for suppliers
Public subsidies and likelihood of self-sustaining WSS finance	Obligation for FCR Mutuality systems financed by ear-marked taxes Variant: public budget contributes to investment with specific grants	In principle no subsidies and obligation for FCR; new obligations only when tariff increase allow investment to be viable Variant: public sector can assume part of the risk for long-term infrastructure renewal in order to guarantee against risk of bankruptcy	Water tariffs and charges intended as local taxes and aimed at long-run FCR Variants: public accounting does not consider depreciation and capital costs; public budget finances investment

| Patterns of risk allocation | Investment risk separated from operational risk | All investment responsibilities on the water company (variant: creation of specific purpose companies for the ownership of assets, also responsible for fund raising and owned by public or consumers)
Responsibility for regulators to ensure industry viability
Price caps and cost pass-through in order to share risk of unexpected events with consumers | Entirely on the public
Cost-based tariffs ⇔ economic risk shifted to consumers and/or taxpayers
Some limited assumption of risk by private firms in DBFO arrangements |

On the other hand, capital costs and depreciation costs are likely to be lower in the case of DPM, since long-term schedules can be adopted and interest rates are lower.[9]

A further cost component on which regulatory models impact differently is transactions costs that are presumably higher in the delegated model (tender, conflict resolution, consultants, and so forth) and lower in the DPM.

The resulting net effect on costs is not easy to predict. Empirical observation is inconclusive for two reasons: the first one is that service costs are not the same everywhere because of local specificities in the cost function; a benchmarking should be made on individual cost components, what is actually impossible with the available data. The second and more important reason is that different models correspond also to differently to the way the costs are calculated and levels of recovery are defined (Massarutto, 2004).[10]

Table 14.6 analyzes in more detail the structure of economic risks that is likely to occur in the three models. Each model entails a specific typology of risks that are typical of that model. The way they are actually shared among different actors depends on the concrete regulatory instruments adopted and the set of incentives provided to actors.

Once again, a confrontation of Tables 14.7 and 14.8 shows the importance of the trade-off discussed in the section on "Alternative Models for Organizing WSS Services." More effective regulations (e.g., detailed contracts and quality specification, frequent tenders, and credible enforcement) are also putting higher pressures on the private and increasing its risks (and consequently the expected returns on investment, which reflect on total costs). In turn, a looser regulation and a weaker incentive system (cost-based regulation, poor specification of duties, renegotiation of contracts, and the like) lead to higher profitability and lower pressure on cost reduction.

PRIVATE SECTOR PARTICIPATION, WATER PRICES, AND AFFORDABILITY

Liberalization and more generally PSI are often blamed for causing rising water bills. For instance, the empirical evidence clearly shows that after privatization, prices almost always rise. Required returns and profit earned by private companies repeatedly are indicated as the main reason for this.

Obviously, delegating any activity to the market implies that market operators earn a profit, and this comes naturally with the very idea of competitive markets and a capitalist economy. In principle, this is definitely not a bad thing, providing cost savings and

---

[9] This statement depends of course on the financial rating of the country (and of municipal authorities). In developed countries, governments can normally obtain better conditions since they do not have to face market risks; on the other hand in developing countries the country risk might be higher than the market risk, resulting in a lower rate for private companies. However, in this case it is likely that private companies will require a supplementary risk premium for the country risk (e.g., for preventing monetary inflation) and thus a higher remuneration.

[10] In the UK the regulatory asset base is represented by what private investors have actually paid for the assets; in Germany, it is the full reconstruction value, while in France it corresponds to the financial costs paid by municipalities (loan repayment). These different ways of calculating the RAB lead to huge differences in the final cost (Massarutto, 2004).

Table 14.7

Expected Overall Net Dividend: Drivers of Cost Increase and Reduction

	Delegated	Regulated Monopoly	Direct Public Management
Operational cost	Depends on awarding criteria: *low incentive* if tenders are based on beauty contests and/or protect incumbents; *high incentive* if based on fixed price	Permanent incentive to reduce costs particularly with price-cap based regulation	Overall weak incentive to reduce costs in order to avoid conflicts with trade unions Quasi-market mechanisms (e.g., subsidy caps), price caps and appropriate management rewarding schemes can provide higher incentives, but lower than in the RM due to the reduced risk of bankruptcy Threaten to privatize can be effective if credible
Infrastructure cost	Lowering factor: willingness to reduce conflicts with local people and avoid unnecessary investment; concessions provide permanent disincentive to invest if not explicitly foreseen in the contract Increasing factor: operator is vertically integrated with construction industry, consultancies, equipment manufacturing	Lowering factor: price-caps provide permanent incentive to minimize capital expenditure Increasing factor: cost pass-through and cost-based regulation provide permanent incentives to expand investment; market often requires shorter repayment schedules	*Lowering factor*: possibility to depreciate over longer time schedules *Increasing factor*: gold plating, possibility to invest (enhanced if the WSS system is more autonomous from local administration and tariffs are based on FCR); lower incentive to make agreements with neighbouring services in order to share infrastructure and sunk costs
Cost of capital	High, proportional to risk effectively borne	Medium-high; risk that regulator underestimates it for keeping price low	Low, but constrained by public finance conditions, country rating etc.
Transactions costs	High (tender, contract, enforcement, monitoring, conflict resolution)	Medium-high (regulatory agencies, reporting)	Medium-low (higher if some of the above remedies are adopted)

Table 14.8
Strategies and Options

	Delegated	Regulated Monopoly	Direct Public Management
Main available strategies for improving performance	Soft regulation providing benchmarking and info Separate operation from AMDP Simplify awarding criteria, tender objects (e.g., single activities vs. integrated service) and contract duration; trade-off with level of PSI Increase contractual power of local authorities (associations etc)	Yardstick competition and econometric benchmarking Improve accountability through information, benchmarking, reporting and public participation Impose outsourcing through unbundling (trade-off with coordination costs) Reduce risks by providing guarantees and/or by keeping some part of the risk in public hands	Separate operation and management from enforcement Improve accountability through information, benchmarking, reporting and public participation Private-law arrangements and contracts Outsourcing and delegation of tasks
Potential for introducing more competition	Reduce duration of contracts (requires solutions to avoid sunk costs; Reduce size of contracts (requires that public authorities provide strategic planning of the WSS system) One-dimension bids based on economic performance (requires very detailed contract specification and separation between operation and asset management)	Inset appointments New customers Allowing bigger consumers to bypass the utility	Outsourcing DBFO
Potential for outsourcing	Low due to permanent incentives to vertical integration	High, but requires unbundling and transfer price regulation	High; can be further expanded by legal provisions and/or tight price regulation
Potential for community management and individual solutions	Depends on the degree of development of the system, technical availability of low-cost solutions (e.g., self-treatment of effluents) and local circumstances (e.g., population density). Suitable for less developed service areas (e.g., drinking water production where the system still relies on local water)		

increased efficiency are transferred partly to consumers and the profit margin equals the market cost of capital and does not incorporate monopoly rents. The question should not be whether it is just or fair for a private company to earn a profit for providing an essential service, but rather to evaluate whether water users and society as a whole are better or worse off after liberalization.

The answer to the last question is not straightforward. As argued previously, the WSS is a very capital intensive industry: while opportunities for productivity gain certainly exist, they could be overturned by private investors' remuneration so the final outcome is not so beneficial to consumers; in turn, public sector–based management and financial models could allow longer-term repayment schedules and lower cost of capital, and hence lower total costs.

Price increases after PSI do not necessarily depend on profit. Rather, they often are due to discovering costs that were previously unaccounted for and the need (or the political choice) to recover them from the water bill instead of taxation. Raising prices at cost-recovery levels is also a way to be fairer to the next generation, particularly concerning the correct depreciation of assets value with the aim of restoring their capital value. Very often, PSI is not the cause of price increase; for example, sometimes the need to increase prices is at the origin of privatization since local authorities might not want to be responsible for raising the water bills. Privatization is also a way to raise higher barriers against the temptation to once more pour public money into the system.

Finally, price increases might be due to the need to finance new investment. The WSS industry is undergoing a dynamic phase, given the contemporary necessity to replace aging assets and to develop them further in order to catch up with the demanding targets imposed by quality regulation and environmental policy.

In order to decide whether consumers and society are better off after privatization, water bills are not the only element to be evaluated; the net effect on taxpayers (present and long term) and on future generations as well as improvements in service quality and environmental performance should also be considered.

Affordability is just another issue. Threshold values over which WSS costs pose affordability problems are considered conventionally in the reach of 1 to 3 percent of family income (Gleick, 1998). Whatever the management model and the way private operators are involved, many indicators show that these thresholds are likely to be reached, if not for the average household, at least for the weakest categories of water users, even in the developed world (Barraqué, 2006). Because WSS demand is income-inelastic, cost-recovery prices might represent a considerable burden for poorer families.

Solutions for sharing this burden in a more equitable way are known and practiced in many countries. In many EU countries there are specific equalization systems that have been developed within the water sector (Massarutto, 2004). In France, for example, the Agences de l'Eau collect water taxes from all water users, and spend the resulting budget as grants-in-aid at zero interest for contributing to the capital expenditure of local authorities. In the Netherlands, a special purpose bank has been created, with the guaranty of the state, in order to provide long-term loans at low interest for water investment. In England and Wales, as well as in Italy (though at a smaller scale), equalization is searched through the creation of large territorial units. In Germany and in Italy, cross-subsidization and self-finance allowed by joint management of urban network services have also provided an important source of funding.

Finance of first-time investment through public budget and taxation has represented the rule in all EU countries, and this has allowed a more equitable distribution of costs. It is important to stress, however, that the affordability issue is transversal to the privatization/liberalization debate. Equalization of water bills, public subsidies, and water-dedicated financial institutions can exist under any model of PSI, and are in some cases a very useful complementary tool.

FUTURE SCENARIOS: THE WAY AHEAD

It is impossible to conclude with respect to the desirability of any of the described alternative regulatory models; quality of regulation in all cases is a more important determinant for overall efficiency gains (net dividends) than the model itself; on the other hand, regulatory instruments have a decisive impact on the set of incentives, risk sharing, and ultimately for deciding what social actors are most to benefit from reform.

One of the fundamental criteria of welfare economics is thus violated. We cannot simply concentrate on overall efficiency gains, since the way they are distributed also affects critically the likelihood of their occurrence. This consideration has some important consequences for policymakers and helps us identify some alternative scenarios for sustainable private sector involvement in the water industry.

The first one is that, instead of choosing *among* regulatory models, it would be wiser to leave all the models available for choice and create the conditions that allow municipalities to change from one model to the other. Reversibility of choice is a key issue, since the best competitor—in an industry in which competition is not easy nor frequent—is the threat to move to a different regulatory model (e.g., from DPM to RM or from D to DPM).

The second is that each model is more suitable for certain segments of the service and less suitable for others. Dividing the industry in different segments (both horizontally and vertically) could be very helpful for the sake of designing appropriate incentive schemes and to reduce market risks. On the other hand, some diseconomies (coordination costs) may arise, especially if the public sector is not sufficiently skilled for providing coordination on its own. In particular, it seems fundamental that operation and asset management responsibilities are separated in order to reduce risks; but on the other hand, it is extremely dangerous to separate decisions in these two spheres since this might result in inefficient planning of assets. Starting a virtuous circle of skill accumulation enabling the public sector to provide these tasks is therefore of utmost importance, as well as finding instruments that allow sharing of the risks for asset management while maintaining operator's role in choosing technologies and shaping investment strategies.

In a perspective of risk-sharing, *private-public partnerships* (PPPs) can play an important role since they provide a flexible solution that can be adapted to the specific risk conditions that characterize each context. As an alternative to delegation, PPPs can be organized as tenders for the choice of an industrial partner that complements professional skills and management expertise of the local system. As an alternative to divesture,

It might attract private financial investors into enterprises that are controlled and managed by the local public system, thereby adding a stronger commitment to economic efficiency and viability. As an alternative to outsourcing, it might be functional to large investment projects and help divide investment responsibilities from operation, reducing the risk of dualism.

The third lesson is that model performance surely can be improved by adopting innovative regulatory solutions. The most promising of these solutions appear to be those in which different competitive patterns are introduced "laterally" (e.g., through benchmarking or quasi-markets) and those in which consumers have a role in assessing service performance and controlling both elected people and operating companies (through public participation to decision making, self-responsibility for specifying certain service components, or even through the ownership of managing systems, for example, through consumers' cooperatives). Another development that could provide interesting results is self-regulation of water companies through officially recognized certification of water management systems to be implemented in management contracts.

Finally, if the "pure" models of liberalization (in particular, delegation with competitive tendering; full divesture with independent regulation and yardstick competition) do not provide optimal solutions, some combination could generate better outcomes. For example, if tenders cannot be based only on ex-ante definition of parameters and require renegotiation, a regulatory authority providing informative means and arbitration is required.

An important result of our analysis is that it is impossible to single out one solution that is better than another. All the different scenarios have strong and weak points and in all, the results depend on the regulatory choices made and on the patterns of allocation of the incentives and related risks.

Another important "transversal" result is to point out the many trade-offs among the different societal objectives (e.g., efficiency, equity, environmental effectiveness) and the different stakeholders (consumers, company shareholders, workers, taxpayers, future generations, and so forth). The choice of whether to favour one or the other side of the trade-off is mostly a political one. It is fundamental, in any case, that analysis of outcomes of WSS liberalization keeps long-term sustainability of water management as a key reference point.

ACKNOWLEDGMENTS

The chapter derives from the three year experience in the EU-funded Euromarket project. Full reports are available at http://www2.epfl.ch/ mir/page18246.html. Support from the FIRB project *"Competition and regulation evolution in the local utilities market in Europe"* is also acknowledged. The author wishes to thank all the colleagues involved in both projects, although none of them bears responsibility for the judgments and the errors contained herein.

REFERENCES

Andrews K., Zabel T., Sustainability of the water services industry in the UK, in Barraqué, ed., 1999.

Barraqué, B., *Politiques de l'eau en Europe*, La Découverte, Paris, 1995.

Barraqué, B. (ed.), "Water 21. Towards Sustainable Water Management in Europe," Final Report to the European Commission, Bruxelles, 1999.

Barraqué, B. (ed.), *Urban Water Conflicts in the International Experience*, UNESCO, Paris, 2006.

Bayliss, K., D. Hall, *Can Risks Really be Transferred to the Private Sector? A Review of Experiences with Utility Privatization*, Psiru, University of Greenwich, available at www.psiru.org, 2002.

Bressers H., O'Toole J., Richardson J., Networks of Water Policy, Frank Cass, London, 1997.

Correia, F. N. (ed.), *Water Resources Management in Europe*, Eurowater, Balkema, Rotterdam, 1998.

Cowan, S., "Regulating Several Market Failures; The Water Industry in England and Wales," *Oxford Review of Economic Policy*, Vol. 9, No. 4, 1993.

Cowan, S., "Competition in the Water Industry", *Oxford Review of Economic Policy*, Vol. 13, No.1, pp. 83–92, 1997.

de Carli A., Massarutto A., Paccagnan V., "La valutazione economica delle politiche idriche: dall'efficienza alla sostenibilità", Economia delle Fonti di Energia e dell'Ambiente, n.1-2, 149-187, 2003.

Dinar, A. (ed.), *The Political Economy of Water Pricing Reform*, Oxford University Press, 2000.

Finger, M., J. Allouche, P. Manso (eds.), *Water and Liberalization: European Water Scenarios*, IWA Publishing, Ashland, OH, 2006.

Gleick, P., "The Human Right to Water," *Water Policy*, n. 1, pp. 487–503, 1998.

Gleick, P., G. Wolff, E. Chalecki, R. Reyes, *The New Economy of Water. Benefits and Risks of Globalization and Privatization of Freshwater*, Pacific Institute, San Francisco, 2002.

Hall, D., *Water in Public Hands*, PSIRU, London, 2001.

Kraemer, R. A., "Public and Private Water Management in Europe," in Correia (ed.), *Selected Issues in Water Resources Management in Europe*, Balkema, Amsterdam, 1998.

Lobina, E., and D. Hall, "Problems with Private Water Concessions: A Review of Experience," available at www.psiru.org, 2003.

Massarutto A., *Water pricing: a basic tool for a sustainable water policy?* in Cabrera E., ed. *Challenges of the new water policies for the XXI century*, Balkema, Amsterdam, 2004,

Massarutto, A., "A Policy Roadmap for the Evaluation of Liberalization Opportunities and Outcomes of Regulatory Reforms," *Working Paper Series in Economics*, 02-05, Dipartimento di scienze economiche, Università di Udine, 2005a.

Massarutto, A., "Assessing Regulatory Reforms in the European Water Industry: Insights from the Economic Literature and a Framework for Evaluation," *Working Paper Series in Economics*, 05-05, Dipartimento di scienze economiche, Università di Udine, 2005b.

Massarutto, A., V. Paccagnan, "Competition in the Market, Self-Supply and Community Systems: Alternative Solutions for Liberalizing the Water Industry," *Working Paper Series in Economics*, Dipartimento di scienze economiche, Università di Udine, 2006.

Noll, R., M. Shirley M., S. Cowan, "Reforming Urban Water Systems in Developing Countries," *SIEPR Discussion Paper 99-32*, Stanford Institute for economic policy research, University of Stanford, Stanford, CA. 2000.

OECD, *Global Trends in Urban Water Supply and Wastewater Financing and Management: Changing Roles for the Public and the Private Sector*, CCNM/ENV(2000)36/Final, Oecd, Paris (www.oecd.org), 2000.

Pricewaterhouse and Coopers, OFFICE INTERNATIONALE DE L'EAU, BIPE, "MEIF – Evaluation Methods for Investing in the Water Sector: Forward-Looking Financial Strategies and Water Pricing," European Commission, 5th Framework Programme, Final Report, available at http://www.meif.org, 2004.

Rees, J. A. "Regulation and Private Participation in the Water and Sanitation Sector," *Natural Resources Forum*, Vol. 22, No. 2, pp. 95–105, 1998.

Renzetti, S., and D. Dupont, "Ownership and Performance of Water Utilities," *Greener Management International*, n. 42, pp. 9–19, 2003.

Summerton, N., "The British Way in Water," *Water Policy*, n. 1, pp. 45–65, 1998.

The World Bank, *Water. Competition and Regulation,* The World Bank Group, Washington D.C., 1999.

The World Bank (PPIAF), *Approaches to Private Participation in Water Services: A Toolkit*, 504, 2005.

Winpenny, J., "Financing Water for All," *Report of the World Panel on Financing Water Infrastructure (Camdessus Commission)*, World Water Forum, Kyoto (available at: http://www.gwpforum.org), March 2003.

Index